OLDUVAI GORGE 1951–61

VOLUME I

Olduvai Gorge

OLDUVAI GORGE
1951–61

VOLUME I

A PRELIMINARY REPORT ON THE
GEOLOGY AND FAUNA

BY

L. S. B. LEAKEY

WITH CONTRIBUTIONS BY

P. M. BUTLER, M. GREENWOOD, G. GAYLORD SIMPSON
R. LAVOCAT, R. F. EWER, G. PETTER
R. L. HAY AND M. D. LEAKEY

CAMBRIDGE
AT THE UNIVERSITY PRESS
1967

CAMBRIDGE UNIVERSITY PRESS
Cambridge, New York, Melbourne, Madrid, Cape Town, Singapore, São Paulo, Delhi

Cambridge University Press
The Edinburgh Building, Cambridge CB2 8RU, UK

Published in the United States of America by Cambridge University Press, New York

www.cambridge.org
Information on this title: www.cambridge.org/9780521105170

First published 1965
This digitally printed version (with additions) 2009

A catalogue record for this publication is available from the British Library

ISBN 978-0-521-05527-7 hardback
ISBN 978-0-521-10517-0 paperback

CONTENTS

LIST OF MAPS AND TEXT-FIGURES

LIST OF PLATES

LIST OF PLATES

Introductory Note to the 50th Anniversary of the Discovery of 'Zinjanthropus'

The Olduvai Gorge in the Republic of Tanzania came to the attention of the world shortly after my mother Mary discovered the 'Zinjanthropus boisei' skull on July 17th 1959. The field of African prehistory, and in particular the study of human evolution, has changed and developed dramatically over the past 50 years. I am particularly pleased that Cambridge University Press have decided to republish the 5 monographs that comprehensively cover the many scientific studies that have been undertaken on the Olduvai material collected by my parents, Louis and Mary, working with a number of colleagues. As the Golden Anniversary of the discovery approaches, it is timely to reflect on the importance of that find.

I was lucky to arrive at Olduvai two days after the discovery and I well recall the excitement of the occasion. My parents were operating on a very tight budget and the field season was short. Fortunately, on hand was world-renowned photographer Des Bartlett who, aided by his wife Jen, fully recorded on film the first few days of excavations and reassembly of bone fragments back in camp. As pieces were glued back together, and the shape of the skull and its morphology became clear, my parents showed uncharacteristic and unrestrained emotion! At the time, ages for fossils were wild guesses and radiometric dating had not been done anywhere in Africa. The best, guessed age for Zinj was a little more than 500,000 years. Some months later, a real Potassium/Argon date was obtained by Jack Evenden and Garniss Curtis, and the 1,750,000 age was announced. This ignited huge excitement worldwide and for the first time my father was able to raise financial support for extended field work at Olduvai. Everything changed. The unqualified enthusiasm and support of the National Geographic Society from 1960 onwards had a major impact on the later work at Olduvai, and indeed on the growing international interest of Africa as the cradle of humanity.

Since those first exciting years at Olduvai, the investigation of human origins has gone forward and extended to many other sites in Africa. The age of hominins has been taken back to beyond five million years and the collected fossils and lithic records are now numerous. International multi-disciplinary teams are working in many parts of the world and, with the exception of a few fundamentalist 'flat earth' types, the acceptance of the fossil record of our past is widely accepted. Much of this has come about because of the initial Olduvai finds.

The pioneering work at Olduvai was the launch of this fantastic 50-year period when we as a species have come to realize and appreciate our common evolutionary past. Olduvai, conserved and protected by the Republic of Tanzania, remains as a landmark in the epic story of humanity, and these monographs are a wonderful testimony to that landmark.

Richard Leakey, FRS

FOREWORD

BY PROFESSOR G. GAYLORD SIMPSON

There is no spot on earth more fascinating and more deeply significant for all of us than Olduvai Gorge. It has, to begin with, extraordinary scenic beauty. In one direction stretches the great, open Serengeti Plain, broken here and there by hills that are massifs of old, weathered rock piercing the younger sediments. In the opposite direction is the rifted Balbal depression and, beyond it, the towering slopes of the Ngorongoro Caldera. Other volcanoes, extinct and living, crenellate the long horizon. The varicoloured gorge itself displays high vertical cliffs, steep, sweeping slopes, and fantastic eroded forms. For those of us who enjoy occasional escape from our own teeming species, the gorge has an added charm: few places in the world are so free from recent works of man. It has been thousands of years since any human maintained a permanent dwelling there. The occasional visitors are all sporadic nomads: wandering tribesmen and palaeontologists.

The desert grandeur of the scene is, however, only an unexpected bonus at Olduvai. The true significance lies literally deeper, in the successive strata exposed in the walls of the gorge. The stunning dimension of the gorge is not in space but in time, not in the seemingly ageless and unchanging face of pristine Africa but in a dated sequence of major events with repercussions everywhere on earth. Here, one after another, are chapters in history ranging from sometime in the early Pleistocene to the present day. The span in years cannot yet be considered established with sufficient accuracy, but by any count it is extremely impressive. The best evidence so far available indicates at least one and three-quarter million years.

Practically every literate person in the world now knows that the Leakeys and their associates have found at Olduvai a succession of human (or to be more conservative and technical, of hominid) cultures beginning with one of the earliest and most primitive yet surely identified. Now probably even better known is the Leakeys' discovery within the most recent years of skeletal remains of several distinct kinds of humans and near-humans at various levels in time. The present status of those studies is the subject of other volumes of this work, and I am not directly concerned with it here. Suffice it to say that this sequence of cultures and of hominid remains is unique and that no other one place on earth has yielded so much information about the early history of the human family.

The present volume deals with discoveries less generally appreciated but far greater in extent and at least as important. The beds at Olduvai also contain an extremely rich succession of fossil faunas, mainly of mammals but also including 'the richest find of avian fossils known to date from the whole of Africa' (p. 71) and a more limited variety of lower vertebrates and molluscs. On present evidence, the oldest of these faunas probably belongs somewhere in the latter part of the Villafranchian in correlation with the European sequence—that is, sometime toward the end of what is now formally and somewhat arbitrarily designated as early Pleistocene. That oldest fauna at Olduvai contains ancestors or close relatives of mammals still present in the same region, but it is also notably archaic. Almost all the species and a number of the genera differ from any now living. Here also are some strange, ancient, wholly extinct groups surviving later than in most other parts of the world: clawed ungulates (chalicotheres, *Metaschizotherium*); enormous, distantly collateral relatives of the elephants (deinotheres, *Deinotherium*); three-toed horses (*Stylohipparion*).

Later Olduvai mammalian faunas show, as would be expected, increasing approach to the recent African fauna, but the story is not that simple. What is seen is not merely ancestral forms evolving into the present species but also, and extensively, more complex changes suggestive of movements of range and fluctuations of climatic and other ecological conditions as yet imperfectly understood. Examples of the unexpected and extraordinary developments include the great diversity and striking peculiarity of the pigs (Suidae) in Bed II and the prevalence of gigantism among the animals of that bed. It is evident that Olduvai will eventually give us a standard, dated sequence for most of the Pleistocene in Sub-Saharan Africa, as well as a priceless paradigm for evolutionary processes in faunas in general.

In comparison with previous publications, especially Leakey (1951), the present work marks a great step forward. In 1951 (written in 1949) the mammals of Beds I and II were treated as if they belonged to a single fauna. Later work by Leakey soon made it clear that such is not the case, but his cautionary notes have been overlooked in some quarters, and only now is fully adequate evidence available and published. The most important general point in the present volume is conclusive proof that quite different faunas occur in Beds I and II. Indeed it is strongly suggested but not yet worked out in detail that these two 'beds' contain not merely two faunas but a whole series of related and transitional but distinguishable faunas. Geological evidence, also included here in a preliminary way, agrees in indicating that these are not single beds by a strict geological definition of 'bed' but are in fact complex sedimentary sequences, stratigraphic members or formations, covering a long span of time and also with numerous hiatuses of varying extent. It should be further emphasised that although collection, preparation, and study have not yet reached a point permitting the most refined distinctions of faunal succession, the data for such analysis are being gathered. Recent collections of specimens *in situ* have meticulous records of precise levels *within* each of the 'beds'.

This is a progress or interim report in two different aspects. In the first place, study of the collections involved has not been completed. Palaeontology is necessarily a slow science, costly in labour and time in each phase from discovery through collection and laboratory preparation to comparison, identification, and analysis. Thus although some groups, such as the Suidae, are here given rather full treatment, others, such as the rodents, are discussed only in a highly preliminary and vague way. Secondly, work at Olduvai is continuing apace, but in order to issue a report without immoderate delay a terminal date for it had to be set arbitrarily. This volume includes only collections and field observations through the Leakeys' campaign of 1960–1. Since then work both geological and palaeontological has continued, and while it does not seem to contradict the essential conclusions of the present study, it has already resulted in highly significant clarification and addition of details.

For work of less complexity, of less urgent importance, and of less worldwide interest to anthropologists, archaeologists, palaeontologists, and geologists, we might be willing to await more nearly definitive monographic publication. We are not willing to wait for available information on Olduvai, and indeed cannot do so if we are to continue our related studies effectively. All will therefore applaud Leakey's decision to bring out the present strictly preliminary publication, and will admire the energy and devotion that have given us so prompt and useful a record of results up to 1962.

G. G. S.

Harvard University
Cambridge, Massachusetts, U.S.A.

February 1963

INTRODUCTION

In the first chapter of my earlier book (Leakey, 1951) on Olduvai Gorge and on the evolution of the Hand-axe culture there, I outlined the sequence of events which led to my first expedition to Olduvai in 1931. I also summarised the work carried out from 1931 to 1947. There is no need to refer here to any of that earlier part of the story, since it can easily be found in the first book and I shall discuss the sequence of the events from 1947 onwards.

In 1947 the first Pan-African Congress of Prehistory met in Nairobi and at its conclusion I arranged an excursion to sites in Tanganyika, including Olduvai Gorge. One of the first places to which I took the visiting scientists on that occasion was FLK I, for it was there, during the 1931–2 season, that we had first proved the existence of a primitive Stone Age culture *in situ* in Bed I. This is the culture which I later described under the name of Oldowan and which has sometimes been erroneously referred to as the 'Pebble culture'. Other parts of the Gorge were, of course, also visited, but both on that occasion and on many others, from 1947 to 1958, I took visiting scientists to see the site FLK I. Thus the place which eventually gave us such a wealth of new information about earliest man and the evolution of the Oldowan culture is one which many leading Prehistorians and Pleistocene geologists had seen, before these discoveries were made.

Although the earlier book on *Olduvai Gorge* was only published in 1951, it was completed for press in 1949. Until the first book was finished, all our efforts, ever since 1931, had been devoted to a preliminary survey. This consisted of:

(*a*) establishing a sequence of evolutionary stages of culture within the geological horizons exposed in the Gorge;

(*b*) locating as many different sites as possible, at as many different levels as possible, with a view to carrying out more detailed excavations when funds and time became available;

(*c*) obtaining an overall picture of the geological history of the Gorge and its possible link with the climatic fluctuations of the Pleistocene period, which we were trying to elucidate elsewhere in East Africa.

When the preliminary work was over and the first results had been sent to press, I decided that we could start upon the next phase of our study. What we needed most, was to locate and uncover in Olduvai Gorge a series of living-floors of early Hand-axe man and also of the preceding Oldowan culture. We hoped to find living-floors similar to those studied at Olorgesailie during the period 1940–6, which had revealed the more evolved stages of the Hand-axe culture.

In 1949, therefore, we did a further brief period of exploration, this time reviewing the various sites we knew, with a view to starting excavations in 1951. My wife and I decided that the most profitable place to start a detailed excavation was a site which we had found in 1935 and which is known as BK II (Leakey, 1959*a*). This site is two and a half miles up the southern fork, or 'side gorge' of Olduvai. We set up camp there for the first time in 1951, and in a few weeks it was clear that we had a rich living-floor in Bed II. At that time we believed this belonged to Chellean Stage 1 and that it was at the base of Bed II. We now know that it is very much higher in the sequence and the preliminary report published in 1959 will therefore need much modification (Leakey, 1959*a*). During the exploratory stages of our work at Olduvai we had needed relatively few workmen, but to excavate we required a larger labour force and it became a major difficulty to supply enough water for the staff, with the limited funds that were then available. At Olduvai itself it seldom rains, but in wet weather there is a little water in the Gorge from rain which falls on the neighbouring mountains. This water seldom lasts more than a few weeks and mostly not more than a few days, so that water has to be transported from many miles away. For our work at the site BK II in 1951 and subsequent years we had to bring our water a distance of a little over thirty miles. The drain upon our financial resources caused by the cost of water transport for the workmen was a major limiting factor in the amount of excavation that could be carried out in any one season.

Because of this and other difficulties the excavations which we carried out from 1951 until 1958 at site BK II, and also at a nearby site SHK II, were not conducted with the same degree of detailed recording as has been possible in recent years, now that we have ample funds at our disposal. Our principal aim during this early period was to obtain a large series of stone tools and other associated material (including fossil fauna) *in situ* on living-sites, in order to achieve a better understanding of the nature of a full assemblage of stone tools of the various stages of the Hand-axe culture.

Most of the funds for this period of our work came to us through the generosity of Mr Charles Boise; first, in the form of direct financial contributions, and later through the Boise Fund, which he set up for this purpose at Oxford University. We are very deeply indebted to Mr Charles Boise for all the help he has given us. We also received assistance from other sources during this period. Most of it was for the purpose of capital equipment such as the purchase of vehicles, tents, water trailers, but some was also used to augment the money received from Mr Boise for the recurrent costs of excavation. Our other benefactors were the Wenner-Gren Foundation, the Wilkie Brothers Foundation, the Shell Company of East Africa, the newspaper *Reveille* and Mr Malin Sorsbie, to all of whom we acknowledge our deep gratitude.

We also wish to thank the Tanganyika Government and, especially, the administrators of the Northern Province, as well as the director and staff of the Tanganyika National Parks, for all the help given to us in connection with water supplies and in many other ways too numerous to detail. I am very grateful to the Trustees of the British Museum of Natural History, as well as to the staff in the department of palaeontology, for making available to me study material in the fossil collections. I also wish to thank the staff of the department of zoology for providing me with comparative material during the course of many weeks of study spent at the Museum. We also express our warmest thanks to the Trustees of the Coryndon Memorial Museum for allowing me to go to Olduvai for several weeks at a time to take part in this research work.

During this period we had a number of voluntary helpers in the field and to them we also offer our thanks; among them were Mrs Jean Brown, Miss Jane Goodall, Miss Gillian Trace, and our son Jonathan. Throughout this time our African staff was headed by Mr Heselon Mukiri, my very able senior technical assistant. The other African staff varied from year to year.

By the end of 1958 my wife and I felt that for the time being we had done enough work on these sites in Bed II and that we should turn our attention to the Oldowan culture in Bed I and to the deposits at Laetolil, south of Olduvai. In 1959, therefore, we first of all revisited Laetolil, a site which we had discovered and worked for a few weeks in 1935 and where Dr Kohl-Larsen had subsequently collected fossils. These were described by Dr Dietrich, who referred to this site as *Vogelflüss*.

After three weeks at Laetolil it became evident that we would not find any living-floors of the Oldowan culture there and we, therefore, transferred our camp to Olduvai Gorge. We began examining sites in Bed I, with a view to choosing one of them for more detailed excavation. We revisited many of the sites which had been located by us in earlier years and we had still not decided where to start an excavation, when Mr Heselon Mukiri unexpectedly found a very well preserved hominid lower molar in a block of hard rock, at the site known as MK I, which had yielded Oldowan tools from 1931 onwards. This discovery was so important that we decided that excavations must be put in hand as quickly as possible.

Our funds for that year were then exhausted so we returned to Nairobi, where I successfully arranged for an overdraft on my research account. We then went back to Olduvai with the intention of excavating at site MK I for about three weeks. It had been agreed that Mr Des Bartlett, a partner of Armand and Michaela Denis, should come to Olduvai with my son Richard on 17 July in order to make a film of the stages of the excavation at the MK I site. Since we did not wish to start excavations until Des Bartlett had arrived, my wife and I spent 15 and 16 July doing general survey work.

On the morning of 17 July I was ill with influenza and my wife went out alone and spent the morning on the site FLK I, where we both considered there might be a living-floor of the Oldowan culture. At about 11 a.m. she noticed what seemed to be part of the petrous portion of a human temporal bone. Looking closer, and brushing away the surface scree a little, she exposed two pre-molar teeth. Satisfied that these were hominid she hurried back to camp to fetch me. We had, at length, found promising remains of a hominid in Bed I.

Since Mr and Mrs Bartlett and my son were due to arrive that afternoon, equipped with ciné cameras, we most reluctantly delayed work on this exciting new find until they arrived. The next day Bartlett began to photograph each stage of the process of uncovering the remains and, later, of fitting together the pieces of the skull. Only a small area could be excavated with the money (overdraft) available, but sufficient work was done to establish that we had discovered a living-floor of the Oldowan culture. This will be described in another volume.

I gave a preliminary report on this important discovery at the opening meeting of the third Pan-African Congress of Prehistory at Leopoldville in August (1959*b*) and also published a short note in *Nature* (1959*c*). This was dated 15 August, but did not actually appear until September, owing to a printers' strike in London. The find was also described in the *Illustrated London News* in August (1959*d*).

Renewed grants by the Wilkie and by the Wenner-Gren Foundations enabled me to pay off my overdraft and initiate new work at the site. In October I went to America to take part in the Darwin Centenary celebrations at Chicago University and I took the opportunity to try to raise funds for extensive work at Olduvai during 1960. In this I was successful, as the Research Committee of the National Geographic Society agreed to support our work. Early in February 1960, therefore, we were able to start extensive excavations at site FLK I, where the *Zinjanthropus* skull and a tibia had been found. That season continued until the end of February 1961—a little over twelve months—during which some 92,000 man-hours of work were completed, compared with only about 40,000 man-hours in all the preceding thirty years. We are, indeed, deeply grateful to the Research Committee of the National Geographic Society for thus making it possible for us to work on such a large scale. We are also very grateful for the generous help that the Wenner-Gren and Wilkie Brothers Foundations gave to us during the interim period at the end of 1959, which enabled us to start work on the site before the National Geographic grant became available.

In 1958, a year before the discovery of *Zinjanthropus*, Dr Jack Evernden of Berkeley, University of California, had become interested in the possibility of dating the Olduvai deposits by means of the potassium–argon technique. I took him to the Gorge on a special visit, so that he himself could collect samples of the rocks. At intervals, after this, I sent further samples to him and in 1961 Dr Garniss Curtis (Evernden's partner in the potassium–argon studies) came to East Africa to collect additional specimens. The results of this study are given in chapter VIII, in which the article published in *Nature* on 29 July 1961 is reprinted (Leakey, Evernden and Curtis, 1961) together with one or two other notes.

As a result of the success of the 1960–1 season, the Research Committee of the National Geographic Society made further generous funds available to continue the work in 1961–2. The Wilkie Brothers Foundation also made another contribution towards the capital costs. We are deeply grateful for this help.

The 1960–1 season resulted in uncovering a very large area of the living-floor, or camp-site, of *Zinjanthropus* and in the discovery, at a nearby site, of juvenile and adult hominid remains. The former is represented by pieces of the skull, the mandible, an upper molar and hand bones, and the latter by the greater part of the left foot, parts of a hand and a clavicle. Whilst *Zinjanthropus* is undoubtedly an Australopithecine, preliminary work suggests that the other remains represent a distinct type of hominid.

At the close of the season a hominid skull was found in Bed II at site LLK, and a site rich in cultural and fossil remains was located at the top of Bed I at FLK N I.

The new evidence for dating the various Olduvai deposits which has become available during the past few years is of very great importance. This is, in part, the result of the development of the potassium–argon dating technique. Fully as important, however, is the extensive fossil fauna found *in situ*. As will be seen in the subsequent chapters, this indicates an Upper Villafranchian (Lower Pleistocene) age for Bed I. The lower part of Bed II probably also belongs to this period, whilst the upper part of Bed II, upwards to Bed IV, is Middle Pleistocene.

PERSONNEL DURING THE 1960–1 SEASON

by M. D. Leakey

The 1960–1 season at Olduvai lasted almost exactly twelve months. Work at FLK I, the *Zinjanthropus* site, was begun on 24 February 1960, and camp was finally packed up on 26 February 1961. From the beginning of the season until the end of November 1960 I was most fortunate in having the able assistance of my eldest son, Jonathan, who undertook vehicle maintenance, supervision of water and food supplies (both for ourselves and the African staff), radio operator duties, etc. Moreover, it was he who discovered and excavated site FLK NN I. When Jonathan left at the end of November to open the Snake Park attached to the Coryndon Memorial Museum, my second son, Richard was fortunately able to take over general camp duties. Our special thanks are due to him for tackling the long-outstanding problem of ascertaining, by levelling, the difference in height between the Balbal depression and the top of the fifth fault.

I would like to express my thanks to the African staff, headed by Mr Heselon Mukiri, for the excellent work they did, under conditions which were often very trying.

We were most fortunate in having the help of Mrs S. C. Coryndon, my husband's assistant at the Coryndon Memorial Museum, and of Miss Margaret Cropper who has since left Kenya to study Prehistory in England.

To Mr M. J. Tippett, who came to us early in January 1961 and remained until the close of the season, are due my special thanks for his reliability and efficiency in carrying out the duties assigned to him.

Dr Maxine Kleindienst, who joined us the following season to carry out excavations at site JK 2 in Bed IV, was with us during the final three weeks of the season. I am most grateful to her for her help on the excavations, and for taking charge during my enforced absence in Nairobi for one week.

NOTES ON SECOND IMPRESSION

(1) Page 32: it would appear that the name *intermedius* for a species of *Potamochoerus* is preoccupied, having been used by the late Dr Lonnberg for what is in fact not a new species, nor even a race of the living bush pig. However, my name *intermedius* is thus invalid, and will have to be changed.

(2) Page 92: the account of subsequent volumes has been brought up to date.

L.S.B.L.

January 1967

CHAPTER I

THE GENERAL GEOLOGICAL EVIDENCE

In my report on Olduvai, which was published in 1951, a brief chapter was included which had been written by the late Professor Hans Reck. In it he summarised the geological evidence which was then available as the result of his own 1913 expedition and of his work with us during 1931. This preliminary chapter had, in fact, been written at a time when he had not yet worked through all his field notes. He had planned to prepare a much more detailed report, accompanied by drawings of selected sections of the geological deposits, after his return from his expedition to Portuguese East Africa. Owing to his sudden death on that expedition, this full report was never written, and after the end of the war in 1945 it proved impossible for his widow to trace his many books of field notes. Thus, all that was available for me to use in the 1951 volume was the original preliminary note prepared before the war.

Dr P. E. Kent was a member of my 1935 expedition to Olduvai and he subsequently published a brief general note on the geology of a small part of the exposures which he had examined.

In 1961 Dr R. Pickering, a member of the Tanganyika Geological Survey, spent some four and a half months with us at Olduvai, by kind permission of the Director. He concentrated, mainly, upon making a detailed map of the FLK region where *Zinjanthropus* had been found in 1959. This will be published in volume 3 of this series. He also prepared a general map of the gorge based on aerial photographs and some rapid geological reconnaissance work. This map, with some modifications, appears in the present volume. It had been intended that his report should also be presented in this volume. There was, however, much that did not seem to me to fit the visible facts, while many of the conclusions were also unsatisfactory. Dr Pickering has, of course, every right to his own views both on the nature of the deposits and upon the conclusion to be drawn from what he has observed. However, if his report had been included here, it would have been necessary for me to add so many notes that these would have been more voluminous than the report itself. I therefore invited the Director of the Geological Survey and the Chief Geologist to visit me at Olduvai to discuss the matter. As a result, I decided not to publish Dr Pickering's report in this volume.

Subsequently, I invited Dr Richard Hay of the Geological Department of Berkeley Campus of the University of California to come out to East Africa in 1962 and make an independent study. His preliminary note has been published in America and has, by permission, been included as appendix 1 in this volume. I feel that he has already achieved a greater measure of understanding of the details of the complex Olduvai deposits than any of those who studied the area before him, and I strongly commend his preliminary note as giving an excellent general picture of the Olduvai geology. If I disagree with him on a few minor points—as will become apparent in this chapter and in the chapter on climate—it certainly does not mean that I have anything but praise for his work as a whole.

Since a general outline of the geological position is necessary to an understanding of the palaeontological evidence that forms the main part of this volume, a summary of the present state of our knowledge follows. In particular, an attempt will be made to indicate in what matters our greatly increased knowledge has necessitated modifications of some of the statements which were made in the 1951 report.

In the first paragraph of his report, Professor Reck rightly stressed the differences in the morphological appearance of the gorge in its westerly reaches (where it runs, in main, across the underlying basement complex rocks) and in the eastern part, where it cuts through the full series of the lacustrine and other sediments which are exposed in the gorge as it is today.

He then went on to say 'the valley is thus a derivative of the original drainage system of the Central African plateau and similarly is, at least, of early Tertiary origin, showing that, even at that time, the drainage was towards the east'.

This statement about the direction of the pre-faulting drainage of the area seems to require some modification, for the present drainage system appears to be due to the post-Middle Pleistocene faulting, which affects the whole of the area. If the conformation of the land prior to the faulting is reconstructed, it is difficult to see how the drainage could have been 'to the east' at the time when the Olduvai sedimentary series was first being formed. The very much older Tertiary drainage may, or may not, have been in a generally eastward and seaward direction but, in Bed I and Bed II times, the greater part of the local drainage was from the east and south towards the west and north.

Since we know that a valley of mature form was cut through the sedimentary series of Beds I to IV after the faulting, the mature shallow valley in the upper reaches of the gorge may well be nothing more than the westerly extension towards El Garja lake of this same geologically young valley.

It seems impossible to see how the drainage of this part of the Serengeti Plains could have been 'to the east' prior to the faulting. Since we know that the Victoria basin contained a lake standing some 300 ft. above the present water level and that the Serengeti is part of the drainage basin it seems much more probable that the drainage was westwards, towards the lake. It was only when the faulting resulted in the creation of the Balbal depression that a drainage pattern towards the east and south developed.

THE STRATIGRAPHIC SEQUENCE

The deposits which are exposed in Olduvai Gorge in the central part consist of:

(*a*) Consolidated tuffs under the lava, about which little is known.

(*b*) A thick basaltic lava with a very irregular upper surface. Its lower surface is visible only at one place.

(*c*) Bed I.

(*d*) Bed II.

(*e*) Bed III.

(*f*) Bed IV.

(*g*) Bed Va.

(*h*) Bed V.

(*i*) The main steppe limestone or caliche.

(*j*) Bed VI.

To the east, between the second fault and the Balbal, as well as in the Balbal depression itself, the rocks underlying the sedimentary series are not visible. West of the junction of the main and side gorges the base of the sedimentary series cannot be seen for very many miles. When, at last, it is once more exposed it is found to rest upon a welded tuff which, in turn, overlies rocks of the basement complex.

Let us consider each of the deposits forming the sequence.

(*a*) *The deposit under the lava*

There is only one place where the base of the lava sheet is exposed. This is in a low cave, beneath a waterfall, some 300 yards to the west of the third fault. Professor Reck illustrated the under-surface of the lava, as it can be seen in this cave, in plate VI of the 1951 report and he wrote: 'The nature of the underlying beds is not known because the floor of the low cave excavated in the rock, to which the lava forms the roof, is covered with sand and debris.' This was true at the time of Reck's visit in 1931, but subsequently floods scoured out the cave so that the underlying deposits became visible. These were found to consist of stratified tuffs not unlike some of those within Bed I above the lava. In 1962 Dr Hay had a small pit dug near the mouth of the cave. Some 9 ft. down a hard tuff was encountered. Dr Hay believes that the deposits over this tuff consist of *in situ* sediment; but it may well be that they are, in part, material from the main Olduvai series which had been re-deposited by water action in fairly recent times. Dr Hay believes that this tuff may be the same as that which underlies Bed I (in Reck's sense) far to the west. He, therefore, suggests that these underlying tuffs, together with the lava itself, be included in a new definition of Bed I.

This seems to me to be unwise, at the present time. We know that Bed I (in Reck's sense) is of Upper Villafranchian age, on the basis of its fossil fauna. There is, however, no fossil evidence upon which we can date the lava or the tuffs which underlie it. They may well also be of Upper Villafranchian age, but they may be older. Consequently, in this chapter and also in the rest of the book (other than the appendix by Dr Hay), the lava as well as the beds beneath it are not treated as a part of Bed I.

(b) The lava

The basaltic lava which underlies Bed I can only be seen between the second fault and the junction of the main and side gorges near VEK–FLK. To the east, the faulting has buried it deep below the channel of the gorge but it can reasonably be assumed that the lava does extend towards the volcanic highlands, which lie east and south and which were presumably the origin of the flow. Westwards the lower margin of the sedimentary series is not visible for very many miles, and when it does reappear it does not rest upon the lava but upon a welded tuff.

(c) Bed I

Bed I is visible for a few hundred yards east of the second fault and it can be traced from there westwards for many miles. Its upper limit was defined in 1931 by Reck and Leakey as the flagstone bed, which had then been traced from the second fault through to the region of site VEK. This horizon was chosen as a matter of convenience because it could easily be followed, and has been retained for the same reason by Dr Hay. We now know that it can also be seen in places some miles up the side gorge and that it reappears in the main Gorge at the fifth fault, whence it can be traced westwards.

For a long time it seemed as though this arbitrary dividing line between Bed I and Bed II coincided to some extent with a major faunal change. This view is no longer tenable. It is now clear that the lower part of Bed II belongs with Bed I geologically, faunally and culturally, and that the break occurs within Bed II. When, therefore, the term Bed I is used in this chapter and in the succeeding ones, it is used in the sense in which it was originally defined, as from the lava to the flagstone or 'marker bed'.

To the east Bed I is very thick, reaching as much as 120 ft. in places. It is composed mainly of coarse volcanic material. At both the second and third faults, the lowest deposits of Bed I consist of fine-grained clays and tuffs, and these contain fossil remains of crocodiles and some fish in addition to the mammalian fauna. They are overlain in this region and also to the west, as far as JK I, by a thick volcanic tuff, which becomes finer-grained to the west. It seems to represent an eruption of considerable magnitude. Further to the west, for example, in the region of HWK, FLK and VEK, the tuffs of Bed I are very much finer-grained and they are interbedded with clays and temporary land surfaces such as those upon which the living floor of *Zinjanthropus* was discovered. Still further to the west, near the fifth fault, even the 'marker bed' at the top of Bed I becomes finer-grained, while the underlying beds are much more clayey, with far fewer intervening land surfaces.

At one time Bed I was believed to contain little fossil material, but this is now known to be incorrect. Even when there is no human occupation level, fossils occur in the clays, while the numerous living-floors have yielded an immensely rich and varied fauna of mammals, birds and some reptiles. This fauna will be briefly considered in subsequent chapters but the detailed reports cannot be ready for many years in view of the quantity of material, both in actual numbers and in variety of species.

(d) Bed II

Bed II was defined by Reck as 'the deposit lying over the marker Bed at the top of Bed I and beneath the Red Bed', which he called Bed III. This definition is accepted and used here and it is also used by Dr Hay in his appendix, but it involves certain practical difficulties. In the first place, to the west Bed III is not red; and, moreover, it is not always easily recognisable. Secondly, it is now known that Bed III was, in some places, removed by erosion channels prior to the deposition of Bed IV, so that Bed IV may be found resting directly upon Bed II. Furthermore, there are 'red beds' both in Bed II and in Bed IV and these have sometimes been mistaken for Bed III.

Reck considered that Bed II consisted of a single and relatively continuous series of deposits without any noticeable depositional break. This is now known to be untrue. The lower and upper parts of Bed II are separated by a break of considerable magnitude. This can be established by geological evidence and is also represented by a major faunal change.

Bed II can be traced from the second fault westwards over an extensive area. There are probably better exposures of this horizon than of any of the other Pleistocene deposits of Olduvai. It varies in thickness from a few feet in the region of the second fault to over 80 ft. in the west.

A little to the east of the second fault the dip of the deposits, which resulted from the fracture, has caused Beds I, II and III to disappear below the level of the present-day gorge. Bed II, however, does reappear near the first fault at the edge of the Balbal.

The view is taken here that Bed II may be divided into three parts. The lower series, which is lacustrine and fluviatile, belongs geologically and faunally with Bed I. It represents the closing stages of the Villafranchian. The middle series is represented by aeolian sands, sub-aerial soils and weathered clays. It marks a dry period which is followed by a major change of fauna. The upper series begins with river channels filled with torrential gravels, followed by deposits which are partly fluviatile and partly lacustrine, depending upon where they are exposed.

(*e*) *Bed III*

Reck described Bed III as 'a sub-aerial land deposit—the contents of which were brought down mainly by water action from the volcanic highlands to the east'. It is, indeed, a deposit which is in striking contrast to the beds which are above and below it and from both of which it is separated by marked discordances.

Bed III is dominantly red in colour in the eastern and central areas but to the west it ceases to be red and is sometimes difficult to identify. It is partly of terrestrial origin and partly fluviatile. In some localities it contains torrential river gravels and boulder beds. It seems to be indicative of a marked change of climate from that which preceded it in upper Bed II times. It is also very distinct from the mainly fluviatile and partly lacustrine deposits of the lower part of Bed IV.

To the west of site PLK, in the main gorge, Bed III ceases to be red in colour but can usually be traced because it is different in texture and content from the beds above and below it.

The degree of discordance between Bed IV and Bed III is much greater than Reck believed it to be. Although the contact is almost horizontal in many places, there are also major unconformities.

In the main, Bed III is not fossiliferous, nor does it contain many living sites of early man. At one or two places a few fossils and stone tools have been found in fluviatile parts of the deposit.

(*f*) *Bed IV*

Bed IV is also a much more complex deposit than was appreciated by Reck. The lower part is mainly of fluviatile origin but there are some deposits formed in shallow water. At the top there are aeolian sands that are almost desertic in character. A detailed study of Bed IV has not yet been made.

In contrast to Bed III the lower part of Bed IV has abundant remains of *Unio* as well as of fish and crocodile. The lower and middle parts of Bed IV have numerous ancient land surfaces upon which remains of large mammals occur, usually in association with cultural material. This fauna has some elements in common with that of the upper part of Bed II and others which are new. In contrast, the aeolian deposits at the top are practically devoid of fauna or culture. This dry period was marked by the disappearance and extinction of a very large number of mammalian genera and species which had been common in Bed IV.

(*g*) *Bed Va*

After Bed IV had been deposited there was a period of intense tectonic activity and faulting (see below), as a result of which the drainage of the whole area was reversed. A wide mature valley was cut from the west to the east, draining into the Balbal depression. The various terrace and river gravels and the fluviatile beds of this period of valley cutting are known as 'Va'. The gravels of this time invariably contain many pebbles derived from the basement complex area to the west and they are always very intensely rolled. These gravels are to be seen not only in the western and central regions but also right down to the edge of the Balbal in the first fault scarp, demonstrating the strength of the river that cut this wide mature valley.

(*h*) *Bed V*

Bed V is a fawn-coloured, windblown deposit, mainly consisting of volcanic ash. Locally, in a few places, it contains mammalian fossils, while it also contains rare molluscs of a species of *Limnocolaria*, which is today found living in very arid regions of Kenya. Bed V filled in the mature valley which followed the faulting and often lies unconformably upon Beds I to IV and even in places on the lava.

(*i*) *The steppe limestone*

Although there are some deposits of steppe limestone (or caliche) in other horizons at Olduvai, the main deposit was formed after Bed V.

(*j*) *Bed VI*

In some places, mainly on the east side of side gullies, there is a black volcanic ash, rich in biotite, over Bed V, as well as over the steppe limestone. This is referred to as Bed VI.

THE FAULTING

We have already noted that after the deposition of Beds I to IV (including the closing arid deposits at the top of Bed IV), a severe tectonic movement occurred that altered the whole drainage pattern and initiated the cutting of a wide mature valley from west to east, as well as causing some peneplanation and sheet erosion. This faulting was described by Professor Reck, who estimated that the total throw, caused by the five faults which he recognised, was of the order of 440 ft. There are actually more faults than were originally recognised, and the total throw is much greater.

The first fault forms the western scarp of the Balbal depression. It is not possible to estimate the depth of the throw of this fault since a large accumulation of silt has formed on the floor of the Balbal (especially in the area round the mouth of the Olduvai Gorge), which masks the whole surface. The block that lies between this first fault and the second fault appears to be tilted to the west, before being bent upwards near the second fault. The second fault does not seem to have a throw of much more than 40 ft. It can be seen to affect the underlying lavas as well as the main sedimentary series. It cuts the gorge near where the side gully, known as CK, meets the main gorge. The deposits are bent towards the fracture at the fault line, but to the east they dip very markedly so that Bed III and all the deposits beneath it disappear out of sight below the present stream bed. They reappear near the Balbal as a result of the westward tilt of the fault block in that area.

The third fault crosses the gorge in the region of Elephant Karongo and what is known as 'Reck's Man site'. The fault through the lava is particularly clearly seen as a perpendicular face as one descends into the valley by the footpath on the north side. The throw of this fault is of the order of 100 ft. The displacement was remarkably horizontal. Immediately to the west of the fault line Bed III is visible at the top of the cliff, covered by a thin layer of Bed V and then by steppe lime. The whole of Bed IV has been removed by sheet erosion and peneplanation. To the east of the fault line a large part of Bed IV is preserved.

The fourth fault consists of a double fault with a horst in the middle having a throw of about 40 ft. both to the east and to the west.

What Professor Reck termed the fifth fault lies some eight miles to the west of the fourth fault. This has a throw of over 100 ft. and forms a long bank which looks as though it might be somewhat younger than the other faults described.

In addition to these five main faults there are several others which have not yet been fully studied. One of these affects the FLK area, with a throw to the west. This, or a small parallel fault, can also be seen at site MNK in the side gorge. Other small faults can be seen both in the main and side gorges.

A careful level section was made by Richard Leakey from the floor of the Balbal, near the point where the gorge cuts down through the cliffs of the first fault, to a point above the fifth fault, which gave a displacement of 638 ± ft. Since the floor of the Balbal, at the point measured from, is covered by a thick mantle of recent deposits the total displacement must be somewhat greater than this figure.

CHAPTER II

REVIEW OF EARLIER REPORTS ON THE FAUNA

In the chapter which Dr A. T. Hopwood wrote for my Olduvai book, published in 1951, he dealt with certain aspects of the fauna of Olduvai. He gave his interpretation of the fossil evidence then available in respect of the age of Beds I to IV. He concluded that these four beds formed 'a single unit' and that, in his own words, 'there is no faunistic evidence to suggest that the lower part of the Olduvai series is of Lower Pleistocene age'.

In 1935 (Leakey, 1935) I had published the view that Bed I of Olduvai was of Lower Pleistocene age; but because of Dr Hopwood's categorical statement, based upon his study of the fauna, I later retracted this opinion, and in the 1951 book I accepted his suggestion and treated the whole of the main Olduvai sequence as belonging to the Middle Pleistocene. In spite of this, however, Dr D. G. MacInnes and I felt constrained to point out that we considered the break between Beds I and II and Bed IV was more significant than Dr Hopwood had suggested. However, we accepted the fossils of Beds I and II as representing a single faunal unit, for we did not, in those days, see any reason to challenge the identifications which were given in Dr Hopwood's list.

As a result of studying the Bed I fauna obtained during 1959–60 and of comparing it with the large series of fossil mammals from Bed II, found at sites BK II and SHK II, we began to doubt whether all the identifications made by Dr Hopwood in respect of the fauna of Bed I were really justified. Consequently, I re-examined (in the British Museum of Natural History) the collections upon which Dr Hopwood's conclusions were based.

BED I FAUNA

Dr Hopwood listed twenty-three genera as being represented in the Olduvai fauna of Bed I. Of these, one genus, *Strepsiceros*, was supposedly represented by two species, the others by one each. His suggested list for Bed I thus contained twenty-four species. Re-examination of the material shows that a great many of these identifications were founded upon inadequate data. In some cases they seem to have been based upon material which is quite incapable of sound specific identification and only doubtfully of even generic classification. In other cases, the identifications seem to have been based upon specimens which had been found resting upon the surface of Bed I. Some of these may never have belonged to a Bed I context but may have been washed down the scree slope from higher levels, to come to rest upon it. It seems that generic and specific identification of well-preserved material from Bed II sometimes led Dr Hopwood to include the same genera and species in Bed I, upon the basis of much more fragmentary material obtained from that deposit.

Of the twenty-four species which Dr Hopwood listed from Bed I, he considered that no less than ten—the porcupine *Hystrix galatea*, the spotted hyaena *Crocuta crocuta*, the lion *Felis leo*, the leopard *Panthera pardus*, the giraffe *Giraffa* cf. *capensis*, the eland *Taurotragus oryx*, the greater kudu *Strepsiceros strepsiceros*, the lesser kudu *S. imberbis*, Hunter's antelope *Beatragus hunteri*, the wildebeest *Gorgon taurinus*—were the same as the species which are still living in Africa today. If this conclusion was correct, then my suggestion of a Lower Pleistocene age for Bed I, put forward in 1935, was most improbable.

The re-examination which I have made shows that the following animals are genuinely represented in Bed I: *Deinotherium bozasi*, *Stylohipparion albertense*, *Equus oldowayensis*, *Metaschizatherium* cf. *hennigi*, *Libytherium oldowayensis*,[1] and *Parmularius altidens*. Thus, only six out

[1] Hopwood referred to this as *Sivatherium oldowayensis*.

of the twenty-four species which were listed by Dr Hopwood can be identified with any degree of certainty, and none of them is a living species. Only one, the horse, is a living genus. As we shall see later, the new excavations have added numerous new genera and species to the Bed I fauna and so has the re-examination of the earlier material. Before discussing this material in the next two chapters, we must look briefly at that part of Dr Hopwood's list which is rejected, in order to make it clear why it is not considered to be admissible.

1. *Simopithecus leakeyi.* The identification of this genus and species in Bed I was based upon a single specimen. It carries a field mark 'Bed I', but it was not found *in situ.* The colour and condition of preservation make it almost certain, in the light of our much greater knowledge today, that it was derived from the lower part of Bed II. Extensive digging in Bed I has so far not yielded any examples which can with certainty be assigned to *Simopithecus leakeyi*, but this primate occurs commonly in Bed II. The single surface-find does not, at present, justify inclusion of the species, *S. leakeyi*, in the Bed I faunal list.[1] There are, however, bones and teeth of general *Simopithecus* affinities at this level which have not yet been identified as to species.

2. *Hystrix galatea.* The material which led Dr Hopwood to include this living species of porcupine in Bed I apparently consisted of three specimens in the British Museum of Natural History. Two of these are from the site MK I and belong unquestionably to Bed I. At this site, as well as at site FLK N I, further examples of porcupine have since been found. The new finds compare well with the two original specimens. They do *not*, however, represent the living porcupine, but a quite different animal of much larger size. It seems to be comparable to the giant porcupine of the Australopithecine deposits in the Transvaal, which has been described by Mrs Marjorie Greenwood (1955). The third specimen was found lying upon Bed I at site HWK. It resembles the living species fairly closely in size, and it was probably correctly identified as *Hystrix galatea.* But since it is a surface specimen the species cannot be included in the Bed I faunal list. The condition of this last specimen, moreover, suggests that it was probably derived from Bed V.

3, 4, 5 and 6. *Carnivora.* Dr Hopwood listed four carnivores in Bed I. These are, *Crocuta crocuta*, the spotted hyaena, *Panthera pardus*, the leopard, *Felis leo*, the lion, and *Acinonyx*, the cheetah. Re-examination of the carnivore material by Dr Ewer does not confirm any of these identifications. There is clear evidence of the presence, in Bed I, of some species of hyaena, several large felids and some other carnivores, but not of any living species, except perhaps a jackal.

7. *Ceratotherium simum.* Dr Hopwood included the white rhinoceros, *Ceratotherium simum*, in the Bed I fauna. While there is some evidence to suggest that the white rhinoceros occurs in the higher levels at Olduvai, the Bed I rhinoceros appears to be different.

8. *Potamochoerus* (*Koiropotamus*) *majus.* The material which was originally cited by Dr Hopwood as the paratype of the species *Potamochoerus majus* consisted of two upper canines from Bed I. Although we now know a great deal more about *P. majus*, we have still not found any *upper* canines of this pig in direct association with molars which can be proved to belong to the species. Indeed, the two tusks which formed the paratype closely resemble those of *Notochoerus* and *Potamochoerus intermedius*. The listing of *P. majus* in Bed I, therefore, is not justifiable in the present state of our knowledge. A pig ancestral to *P. majus* was, however, present at that level (see chapter III). In my subsequent monograph on the fossil Suidae of East Africa I also identified certain molar teeth from Bed I as belonging to *P. majus*. It is now apparent (see chapter III) that these molar teeth were wrongly identified and that the pigs in Bed I, which have molars like those of *P. majus*, belong to a quite different genus, *Ectopotamochoerus*.

9. *Mesochoerus olduvaiensis.* This is another giant pig. The holotype was found in Bed II, while the paratype was listed by me as coming from Bed I. This would seem to justify including this species

[1] Drs Leakey and Whitworth (1958) have suggested that *Simopithecus leakeyi* is really only a subspecies of *S. oswaldi* (see chapter III).

of *Mesochoerus* in the Bed I fauna. New evidence has since come to light which shows that without the canine teeth it is impossible to identify certain groups of fossil pigs, since the molar teeth of several different genera are often very similar (see chapter III). There is no evidence of *Mesochoerus olduvaiensis* in Bed I, but only of *M. heseloni* and of a new genus *Promesochoerus*.

10. *Hippopotamus gorgops*. While whole skulls of this species are known from Beds II and IV, only fragments of hippopotamus have so far been studied from Bed I. It is possible that they may represent the species *gorgops* but this is not yet established. No skull with the basic characters of this species has yet been found in Bed I. The presence of *Hippopotamus gorgops* in the faunal list for Bed I is therefore still doubtful.[1]

11. *Giraffa* cf. *capensis*. Specimens representing a giraffe are certainly present in the deposits of Bed I. There is, however, no specimen sufficiently diagnostic to justify specific identification. Since neither of the giraffes from Bed II represent living species, it is most unlikely that the earlier giraffe was of this type.

12. *Taurotragus oryx*. The Bed I specimens identified as the living species are very fragmentary and do not justify specific identification, while even the genus is not completely certain. Members of the genus are certainly present in the higher levels of Olduvai, but even in Bed IV, from which there is a nearly complete skull, the species is quite distinct from either of those now living in Africa.

13 and 14. *Strepsiceros strepsiceros* and *S. imberbis*. Dr Hopwood's inclusion of the two living species of kudu in the fauna of Bed I seems to have been based upon fragments, some of which resemble the horn cores of these species. We now possess two skulls, with the horn cores intact, which have been found *in situ* in Bed I. They clearly represent a wholly different species with quite different skull structure, although the horn cores resemble, in some respects, those of the living greater kudu. There is nothing in the fossil material from Olduvai in the British Museum of Natural History that would justify the inclusion of *Strepsiceros imberbis* in the lists for Bed I. Both the identifications of living kudu, as part of the Bed I fauna, must therefore be abandoned.

15. *Damaliscus angusticornis*. This alcelaphine, which Dr Schwarz described from material found in the upper beds of Olduvai, was listed by Dr Hopwood as also represented in Bed I. Re-examination of his material shows that this identification is doubtful. The specimens from Bed I were too fragmentary to be assigned to the species *angusticornis* with any degree of certainty. The *Damaliscus* specimens subsequently found in Bed I are of primitive type and belong to a distinct species, *antiquus*.

16. *Beatragus hunteri*. An antelope which is similar in some ways to Hunter's antelope occurs in Bed I, but it is a distinct species. It was, therefore, incorrect to list the living species as occurring in Bed I.

17. *Gorgon taurinus*. The wildebeest of Bed I, Olduvai, is not the living species. There is nothing identifiable as *Gorgon taurinus* in the Bed I collections in the British Museum of Natural History. Even the specimens from Beds II and IV are distinct and belong to the species *olduvaiensis*.

18. *Elephas* (*antiquus*) *recki*. Relatively few elephant remains have so far been found *in situ* in Bed I and these cannot be regarded as belonging to the species *recki*. They represent a more primitive elephant with low-crowned molars, which is provisionally identified as *Elephas* cf. *africanavus*. (This is a species which has been reported from North Africa in Villafranchian deposits.)

From the foregoing pages it is clear that the suggestion made by Dr Hopwood that Bed I Olduvai contains many living species is wholly unfounded. The various conclusions which have been based upon this assessment must now be disregarded. Similarly, a number of the earlier identifications of the fauna of Beds II, III and IV, prove to be untenable in the light of new discoveries.

BED II FAUNA

Dr Hopwood's list showed thirty genera for Bed II. One of these, *Hippotragus*, was reported to be represented by two species, both the same as those

[1] The only specimen from Bed I so far examined seems to differ in many important respects from *Hippopotamus gorgops*.

living today, namely, the sable and roan antelopes. Re-examination of the material that was available to Dr Hopwood in the British Museum of Natural History confirms the presence of the following twenty-six genera: *Simopithecus*, *Canis*, *Aonyx*, *Elephas*, *Deinotherium*, *Stylohipparion*, *Hippotigris*, *Equus*, *Ceratotherium*, *Diceros*, *Mesochoerus*, *Phacochoerus*, *Hippopotamus*, *Giraffa*, *Libytherium*, *Taurotragus*, *Strepsiceros*, *Bularchus*, *Hippotragus*, *Damaliscus*, *Alcelaphus*, *Beatragus*, *Gorgon*, *Gazella*, *Phenacotragus* and *Pultiphagonides*. The following four do not seem to be present: *Felis* (*Leo*), *Anancus*, *Notochoerus*, and *Tragelaphus*. Of the twenty-six valid genera the specific identification in fifteen cases seems well founded, while in ten cases it cannot at present be sustained. The *Simopithecus* is *oswaldi* not *leakeyi*, the *Ceratotherium* is almost certainly not *simum*, the *Phacochoerus* is not *africanus*, the *Giraffa* is not *capensis*, the *Taurotragus* is not *oryx*, the *Strepsiceros* is not *S. strepsiceros* but *grandis*, the two *Hippotragines* are similar to *equinus* and *niger* but not necessarily identical, the *Beatragus* is not *hunteri*, while the *Gazella* is not *G. gazella*.

Dr Hopwood stated that in his opinion there were a number of living species represented in Bed II. He listed these as the lion, the zebra, the white rhino, the black rhino, the wart-hog, the eland, the greater kudu, the roan, the Hunter's antelope and the gazelle. Had this been true, it would be difficult to place Bed II in the lower half of the Middle Pleistocene. It is possible that the zebra and the black rhinoceros are the same as those living today. None of the others seem to represent the living species.

As will be seen in the next three chapters, a number of additional genera and species have been added to the faunal list for Bed II since Dr Hopwood's report, and a review of the present state of our knowledge of the Bed II fauna will be given in chapter VI.

BED III FAUNA

Dr Hopwood listed eleven genera as being represented among the fossils from Bed III. There is some doubt whether any of them were represented by *in situ* material.

BED IV FAUNA

When we come to Bed IV, Dr Hopwood's list contained twenty-nine genera, two of which are each represented by two species. Re-examination of the material confirms the identification of twenty-seven genera. The only doubt is in respect of *Felis* (*Leo*) and *Tragelaphus*. Of the listed species, however, twelve are doubtful. For example, the *Giraffa* is *jumae* not *capensis*, the *Gorgon* is *olduvaiensis* not *taurinus*, the *Phacochoerus* is *altidens* not *africanus*. The *Taurotragus* is *arkelli* not *oryx*, the *Hippotragus* is probably not *equinus*, the *Beatragus* is not *hunteri* and the *Gazella* is not *G. gazella*.

A few of the species which we can accept as certain in Dr Hopwood's list for Bed IV represent living forms. These are the black and white rhinoceros, the zebra and possibly the *Nesotragus* identified by Schwarz. Relatively little has been added to our knowledge of the Bed IV fauna since 1931.

The fauna of Bed IV seems to fit in very closely with that from the deposits of Olorgesailie and Kanjera.

DISCUSSION OF THE OMO FAUNA

Reports about the fossil fauna of different sites all too often cause confusion among non-specialists. Even my friend Professor Arambourg's excellent study of the fauna of Omo (1947) can be misleading to those who do not examine the published data very critically. A certain number of genera and species are represented at Omo by such well-preserved specimens that they are beyond dispute. Others are represented by much less complete material. Professor Arambourg himself made this very clear but he has often been misquoted. A brief review of the Omo fauna is, therefore, given here.

Professor Arambourg has stated that, in his opinion, the fauna of Omo should be regarded as belonging to the Upper Villafranchian or the closing stages of the Lower Pleistocene. Recently, some scientists, who do not seem to have studied Professor Arambourg's reports carefully, have treated Omo as the equivalent of the Kanam East and Kanam West deposits. Others, including myself, have equated it with the Kaiso sequence.

Considering the Proboscidea first: *Deinotherium bozasi* and *Elephas recki* are certainly present at

Omo, while another elephant, which was included as *E.* aff. *meridionalis*, seems to be the same as Dr Dietrich's *E. exoptatus*.

Turning next to the perissodactyls: a *Metaschizatherium* (or *chalichothere*) is certainly present, but its specific identification is not possible. So far as the rhinoceros group is concerned, there is a large rhinoceros which has clear affinities to the white rhinoceros rather than to the black. But on the basis of the material available it would be unwise to give generic classification, let alone specific. The specimens could represent either *Atelodus* (*Ceratotherium*), or *Serengeticeros*. As Professor Arambourg himself explained, the Equidae are noticeably scarce at Omo. Although he lists as being present *Stylohipparion albertense* and a large *Equus*, he himself states, categorically, that the specific identifications are both purely provisional.

As far as the carnivores are concerned, a sabre-toothed felid is represented by a single fragmentary specimen. Although it was given generic and specific rank by Professor Arambourg, the identification as to genus and species is most doubtful.

Professor Arambourg also listed a primate, which he referred to the South African genus *Dinopithecus*. This is also represented by very fragmentary material, and the most that can be said is that a large baboon allied to *Simopithecus* is present at Omo.

With regard to the Artiodactyls, the evidence is a little more complete, but only in respect of a few genera. A *Libytherium* is certainly present and it is probably the same as that from Olduvai. Arambourg's *Giraffa gracilis*, a large but very slenderly built giraffe, is common, while there is some evidence of a more massive giraffe which cannot be identified specifically. The genus *Hippopotamus* is very well represented at Omo. It is not, however, of the species *gorgops*, which is found at Olduvai from Bed II upwards, nor is it *amphibius*, nor *imaguncula*. Professor Arambourg has placed it in a new species *protamphibius*, on account of a number of special characters, which he has discussed in detail.

As far as the Suidae are concerned, true pigs are represented by *Mesochoerus* (*Omochoerus*) *heseloni*, *Pronotochoerus jacksoni* and by *Notochoerus euilus*. The *Phacochoerus* specimen, included in his list, came from undated beds near Todenyang and not from Omo.

Finally, we come to the Bovidae. This family has some remarkable representatives in the Omo deposits. It is notable for the relative abundance of a few limited forms and the scarcity of all others. The three commonest Bovidae are *Aepyceros melampus*, an impala; *Tragelaphus nakuae*, an extinct antelope allied to the bongo and of about the same size; and *Menelikia lyocera*, an antelope which shows affinities with many different existing subfamilies, but which stands alone and is unrecorded elsewhere. A water buck, and a reed buck are also clearly represented. Other Bovidae reported from Omo are *Oryx* sp., *Gazella praethomsoni*, *Synceros* sp., *Strepsiceros* sp. and *Antidorcas*. The existence of these is problematical since the identifications were based upon very fragmentary specimens which are not really diagnostic. There are also indications of the presence of some form of alcelaphine.

If, then, we list the fauna of Omo, we find that the identifiable elements consist of:

Elephas aff. *meridionalis* (*exoptatus*)	A large extinct baboon
Elephas recki	*Libytherium*
Deinotherium bozasi	*Mesochoerus heseloni*
Hippopotamus protamphibius	*Pronotochoerus jacksoni*
Metaschizatherium sp.	*Notochoerus euilus*
Stylohipparion sp.	*Tragelaphus nakuae*
Equus sp.	*Menelikia lyocera*
A large rhinoceros	*Aepyceros melampus*
Giraffa gracilis	*Kobus* sp.
A large giraffe	*Redunca* sp.
	A sabre-toothed feline

The presence of the remains of water buck, reed buck, impala and an antelope allied to the bongo, together with many examples of hippopotamus, all seem to indicate somewhat swampy wet conditions. The remains of *Deinotherium* and *Notochoerus euilus* tend to confirm this. The scarcity of alcelaphines, gazelles and other plains fauna, as well as of the Equidae, tends to confirm the view that the ecological setting may not have been suited to the type of fauna one associates today with savannah and open plains. However, as we shall see in chapter VI, the use of large mammals for interpreting ecological setting is not to be recommended.

CHAPTER III

MAMMALIAN FAUNA: OTHER THAN BOVIDAE

The fauna of Olduvai is now represented by very extensive collections found *in situ* in Beds I and II. These are supplemented by material collected on the surface or excavated from exposures when the fossils had become visible as a result of erosion. These collections will require many years of work before the final reports can be published. This is partly because many specimens have to be sent to specialists all over the world and partly because the quantity of material in certain groups is very great indeed. In the present volume it is possible to give only a summary, but this will present a clearer picture of the fauna of Olduvai than has been possible in the past.

In this and the next two chapters, only the families, subfamilies and genera are given in some cases, since the detailed work has not yet been carried out. In other groups, such as the Bovidae and the Suidae, a more complete picture is given, with diagnoses and descriptions of new genera and species, where possible. Material which had previously been inadequately described is dealt with a little more fully and is illustrated. In a few cases the quantity of material is so great, as for example in the genera *Pelorovis* and *Bularchus*, that no attempt has been made to discuss it in detail for the time being.

In this chapter, various scientists have collaborated to present a preliminary statement, and I am most grateful to them.

The following is a list of the mammalian families, subfamilies, genera and species, other than the Bovidae, dealt with.

Order: **INSECTIVORA**

Family: ERINACEIDAE

1. *Erinaceus* cf. *major*

Family: SORICIDAE

2. *Myosorex* cf. *robinsoni*
3. *Crocidura* cf. *hindei*
4. *Suncus* sp.
5. *Suncus* cf. *lixus*
6. *Suncus* cf. *orangiae*

Family: MACROSCELIDIDAE

7. *Nasilio* sp.indet.
8. ? *Elephantulus* sp.indet.

Order: **CHIROPTERA**

Family: MEGADERMIDAE

9. Gen. et sp.indet.

Family: VESPERTILIONIDAE

10. Cf. *Nycticeius* (*Scoteinus*) *schlieffeni*
11. Cf. *Pipistrellus* (*Scotozous*) *rueppelei*

Order: **PRIMATES**

Family: GALAGIDAE

12. *Galago senegalensis*

Family: CERCOPITHECIDAE

13. *Simopithecus oswaldi*
14. *S. jonathani*
15. *Papio* sp.indet.

Order: **RODENTIA**

Family: MURIDAE

16. *Otomys kempi*
17. *Tatera* sp.indet.
18. *Gerbillus* sp.indet.
19. *Saccostomus* sp.indet.
20. Gen. et sp.indet.
21. Gen. et sp.indet.
22. Gen. et sp.indet.
23. Gen. et sp.indet.
24. Gen. et sp.indet.
25. Gen. et sp.indet.
26. Cf. *Grammomys*
27. Gen. et sp.indet.
28. Gen. et sp.indet.
29. Cf. *Steatomys*
30. *Dendromys* sp.
31. *Dendromys* sp.

Family: SCIURIDAE

32. *Xerus* sp.

Family: PEDETIDAE

33. *Pedetes* indet.

MAMMALIAN FAUNA: OTHER THAN BOVIDAE

Family: HYSTRICIDAE

34. *Hystrix* indet.

Family: BATHYERGIDAE

35. *Heterocephalus* sp.

Order: **LAGOMORPHA**

Family: LEPORIDAE

36. *Lepus* sp.
37. *Serengetilagus* sp.

Order: **CARNIVORA**

Subfamily: *MACHAIRODONTINAE*

38. Gen. et sp.indet.
39. Gen. et sp.indet.

Family: FELIDAE

40. *Panthera* cf. *tigris*
41. *Panthera* cf. *crassidens*

Family: HYAENIDAE

42. *Crocuta* aff. *ultra*

Family: CANIDAE

43. *Canis mesomelas*
44. *Otocyon recki*
45. *Canis africanus*

Family: MUSTELIDAE

46. *Lutra* indet.
47. *Aonyx* indet.

Family: VIVERRIDAE

48. Herpestinae gen. et sp.indet.
49. *Herpestes* sp.
50. Cf. *Herpestes*
51. Cf. *Herpestes*
52. *Genetta* sp.
53. Herpestinae gen. et sp.indet.
54. *Genetta* sp.
55. *Ichneumia* sp.
56. *Genetta* sp.
57. Gen. et sp.indet.
58. Gen. et sp.indet.
59. Gen. et sp.indet.
60. Gen. et sp.indet.
61. *Crossarchus* sp.

Order: **PROBOSCIDEA**

Family: ELEPHANTIDAE

62. *Elephas* cf. *africanavus*
63. *E. recki* (a primitive stage)
64. *E. recki*

Family: DEINOTHERIDAE

65. *Deinotherium bozasi*

Order: **PERISSODACTYLA**

Family: RHINOCEROTIDAE

66. *Ceratotherium efficax*
67. *C. simum*
68. *Diceros bicornis*

Family: CHALICOTHERIIDAE

69. *Ancylotherium* cf. *hennigi*

Family: EQUIDAE

70. *Equus oldowayensis*
71. *Equus* sp.
72. *Equus* sp.
73. *Stylohipparion albertense*
74. *Stylohipparion* sp.
75. Gen. et sp.indet.

Order: **ARTIODACTYLA**

Family: SUIDAE

76. *Mesochoerus* cf. *heseloni*
77. *Mesochoerus olduvaiensis*
78. *Potamochoerus majus*
79. *P. intermedius*
80. *Promesochoerus mukiri*
81. *Ectopotamochoerus dubius*
82. *Pronotochoerus* cf. *jacksoni*
83. *Notochoerus euilus*
84. *N. compactus*
85. *N. hopwoodi*
86. *Tapinochoerus meadowsi*
87. *T. minutus*
88. *Tapinochoerus* sp.
89. *Phacochoerus altidens robustus*
90. *P. altidens altidens*
91. *Afrochoerus nicoli*
92. *Orthostonyx brachyops*

Family: HIPPOPOTAMIDAE

93. *Hippopotamus gorgops*
94. *Hippopotamus* sp.

Family: GIRAFFIDAE

95. *Giraffa gracilis*
96. *G. jumae*
97. *Okapia stillei*
98. *Libytherium olduvaiensis*

Order: **INSECTIVORA**

The Insectivora of Olduvai are in process of being studied in detail by Professor P. M. Butler and Mrs M. Greenwood. The detailed report will be published in due course in the series entitled 'Fossil Mammals of Africa'. The Insectivora recorded include shrews, elephant shrews and hedgehogs. The following preliminary note has been submitted:

Study of the numerous insectivore remains from Olduvai is still at an early stage. It is possible at present to give only tentative and approximate identifications, based mainly upon the lower jaws which, together with limb bones (as yet unstudied), make up the bulk of the material.

Insectivores were obtained from the following levels in Bed I:

FLK N I	Layer 1	FLK I	*Zinjanthropus* level
	Layer 2		
	Layer 3	FLK NN I	Layer 1
	Layer 4		Layer 2
	Layer 5		Layer 3
	Layer 6		

Family: ERINACEIDAE

Erinaceus cf. *major* Broom (1948 *a*). A hedgehog larger than *E. frontalis* Smith and apparently similar to *E.* (*Atelerix*) *major*, a species described from Bolt's Farm. It first appears in FLK N I, layer 5, and is common in layers 4 and 3. It is similar in size to *E. algirus* and may link this North African species with the South African *E. frontalis.*

Family: SORICIDAE

Myosorex cf. *robinsoni* Meester. This species can be recognised by the presence of an additional lower pre-molar anterior to P_4. It differs only in minor aspects from *M. robinsoni*, which occurs at Swartkrans, Makapansgat, Sterkfontein and Bolt's Farm. It was found in FLK NN I, layers 2 and 3, FLK I, and in all the layers of FLK N I. In FLK N I, layer 5, it is the commonest shrew (fourteen specimens out of twenty-two).

Crocidura cf. *hindei* Thomas. This is a rare species, represented by only ten specimens, obtained from FLK NN I, layers 3 and 1, and from FLK N I, layers 5, 4, and 3. It does not agree with any of the species from the South African Australopithecine sites distinguished by Meester (1955).

Suncus sp. 1. This small species, of the size of *S. gracilis* Blainville, first appears in FLK N I, layer 5, and persists till layer 1. It is fairly common. It shows distinctive features which seem to exclude it from all the living species, but it resembles *S. gracilis* more closely than species of *Crocidura* of the same size. It is smaller than the *Suncus* sp. of Meester (1955).

Suncus cf. *lixus* Thomas. This shrew is abundant in layers 1, 2 and 3 of FLK NN I, from which 242 lower jaw fragments have been obtained. It does not occur at all in the higher layers of Bed I and has not been recognised from the South African Australopithecine sites. The lower jaws and teeth differ in only minor respects from living specimens of this species.

Suncus cf. *orangiae* Roberts. Most specimens from FLK N I appear to belong to a species of *Suncus*, as there are two palates, definitely referable to this genus, with teeth that fit those in the mandibles. There is only one specimen from layers below FLK N I, layer 6; this is from FLK I, the *Zinjanthropus* level. The unnamed species of *Suncus* from Swartkrans, Sterkfontein, Makapansgat and Kromdraai agrees in size with the Olduvai form.

No members of the Chrysochloridae have been found at Olduvai, though two species are so far known from the South African Australopithecine sites.

Family: MACROSCELIDIDAE

Nasilio sp. Specimens of this genus, recognised by the presence of a small third molar in the lower jaw, are abundant in FLK N I from layer 5 upwards. It is particularly abundant in layers 3 and 2. Maxillary specimens agree in pre-molar structure with *Elephantulus langi* Broom (1937), which may belong to *Nasilio.*

? *Elephantulus* sp. A small number of specimens from FLK N I, occurring in all the layers from layer 5 upwards, do not belong to *Nasilio* because of the absence of M_3. It is possible that the lower jaws lacking M_3, which Broom (1937) referred to *Elephantulus langi*, correspond to this Olduvai species. Two juvenile specimens of an unidentified macroscelid occur in deeper layers of Bed I, one from FLK NN I, layer 2 or 3, and the other from FLK I.

CONCLUSIONS

Two insectivore faunas can be distinguished. The earlier fauna, from FLK NN I, layers 3 to 1, contains the following species:

Myosorex cf. *robinsoni*
Crocidura cf. *hindei*
Suncus cf. *lixus* (abundant)
Macroscelididae indet.

In FLK N I this is replaced by the later fauna:

Erinaceus cf. *major* (common)
Myosorex cf. *robinsoni*
Crocidura cf. *hindei*
Suncus sp. 1 (common)
Suncus cf. *orangiae* (abundant)
Nasilio sp. (abundant)
? *Elephantulus* sp.

The levels FLK N I, layer 6, and FLK I, contain *Suncus* cf. *orangiae*, but not *Nasilio* or *Suncus* sp. 1. Unfortunately, the insectivores of these layers are very poorly known.

It may be significant that the later fauna shows some resemblance to the insectivore fauna of the Australopithecine sites of South Africa. In a number of cases the species are probably not the same, but this might be attributed to the geographical distance between the sites.

Order: CHIROPTERA

Remains of bats are very rare in the fossil beds of Olduvai but a few occur. The following preliminary note has been submitted by Professor Butler and Mrs Greenwood, who are studying the material:

Family: MEGADERMIDAE

Two specimens from FLK N I—a mandible from layer 3 and a maxilla from layer 2—appear to represent an extinct species, perhaps an extinct genus, of megadermid.

Family: VESPERTILIONIDAE

There are two mandibles from FLK N I, one from layer 1, and the others from layers 1–3, which resemble *Nycticeius* (*Scoteinus*) *schlieffeni* Peters. A mandible from FLK NN I, layer 3, resembles *Pipistrellus* (*Scotozous*) *rueppelei* Fischer, but is somewhat larger, about the size of *P.* (*Scotozous*) *dormeri* from India.

Order: PRIMATES

The following preliminary note has been received from Professor George Gaylord Simpson.

Family: GALAGIDAE

Genus: *Galago*

Species: *senegalensis* E. Geoffroy, 1796

Specimens of a fossil Galago from Bed I, Olduvai, Tanganyika, are indistinguishable from living *Galago senegalensis* and are of interest in indicating the presence, at a remote Pleistocene date, of a lesser Galago, or bush-baby, nearly if not quite identical with those still living in the same general area.

Primates Linnaeus, 1758
Lorisiformes Gregory, 1915
Lorisidae Gregory, 1951
Galaginae Mivart, 1864
Galago E. Geoffroy, 1796
Galago senegalensis E. Geoffroy, 1796

HORIZON AND LOCALITY

All specimens in hand are from FLK I (the *Zinjanthropus* level) in Bed I at Olduvai.

MATERIALS

In addition to a number of skeletal fragments, which do not add significantly to the taxonomic information, the following more precisely identifiable jaw fragments are in hand:

No. FLK I, E–G 391, lower jaw with parts of both rami and crowns of left P_2–M_2 and right P_3–M_1.

No. FLK I, E–G 392, partial left ramus with crowns of P_2–M_2.

No. FLK I, E–G 393, partial left ramus with crowns of P_3–M_2.

No. FLK I, E–G 394, partial left ramus with crowns of P_4–M_1.

DISCUSSION

The fossils show expected minor variation in size and structure, but they clearly represent a single species. In all respects they seem to be within the observed or probable range of the living species *Galago senegalensis*, a steppe and savannah species that is the most widely distributed of all African prosimians and exhibits marked variation both regional and local. Of the subspecies recognised by Hill,[1] *G. s. braccatus* occurs nearest to Olduvai, but of specimens available to me for first-hand comparison the next subspecies to the south, *G. s. moholi*, is geographically closest to Olduvai. The dimensions of P_4–M_2 in the fossils average slightly larger, about 7 per cent, than in our specimens of *moholi*, but the ranges overlap. I do not have available dental measurements of *braccatus*, but the skull length is said (Hill after Schwarz) to average about 8 per cent larger in recent *braccatus*, which is thus probably of almost exactly the same size as the Olduvai fossils. I do not mean to suggest that the fossils belong to the living subspecies of the same region, or even that such subspecies are definable and valid as now recognised. In fact they are plainly highly arbitrary and their supposed distributions are evidently more conventional than realistic. For example, Hill's map (*op. cit.* fig. 75) does not show *Galago senegalensis braccatus* as occurring nearer than about 150 miles to its type locality!

The most nearly distinctive character of the fossils is that

Table 1. *Measurements in millimetres of teeth of fossil and recent* Galago senegalensis

(L = length, and W = width, of crowns as defined in the text)

	P_3		P_4		M_1		M_2	
	L	W	L	W	L	W	L	W
Olduvai, Bed I								
No. FLK I E–G 391	2·6*	1·4*	2·2*	1·6*	2·4*	2·0*	2·2	2·1
No. FLK I E–G 392	2·3	1·5	2·2	1·6	2·5	2·1	—	—
No. FLK I E–G 393	2·4	1·4	2·2	1·5	2·3	2·1	2·3	2·1
No. FLK I E–G 394	—	—	2·3	1·7	2·4	2·1	—	—
Recent *G. s. moholi*								
Male	2·2	1·4	2·1	1·5	2·2	1·8	2·1	1·8
Female	1·9	1·7	2·0	1·6	2·2	2·0	2·2	2·0

* Mean of right and left sides.

[1] W. C. Hill, *Primates. Comparative Anatomy and Taxonomy.* Vol. I. *Stepsirhini* (Interscience, New York, and Edinburgh University Press, 1953).

P_2 and P_3 average distinctly greater in longitudinal dimensions, both absolutely and in proportion to P_4–M_2, than in compared specimens of recent *moholi*, but even in this respect the ranges probably overlapped for that subspecies and surely for the species as a whole. It should also be noted that the peculiar shape and implantation of P_{2-3} make it difficult to obtain reliably comparable measurements and that slight distortion of fossil specimens can change conventional measurements of these two teeth considerably. The fact that the fossils cannot be definitely distinguished from recent *Galago senegalensis* does not amount to absolute determination that they are of that species. Larger series, especially with upper dentitions and skulls, might well demonstrate a distinction, but the parts now known in the Olduvai Bed I form do suffice to indicate that it is in any event very closely related to the living lesser Galago of the same general region.

Measurement of the teeth of Olduvai specimens and of two more or less characteristic recent specimens are given in Table 1. These measurements were taken optically with uniform projection on to an occlusal plane. Length is the longitudinal or mesiodistal distance between parallel transverse tangents to the crown as thus projected, and width is the correspondingly defined transverse or bucco-lingual distance between longitudinal tangents.

Family: CERCOPITHECIDAE
Subfamily: *CYNOCEPHALINAE*
Genus: ***Simopithecus*** Andrews
Species: (*a*) ***oswaldi*** Andrews
Subspecies: ***leakeyi*** Hopwood
Species: (*b*) ***jonathani*** Leakey and Whitworth

Andrews based the genus and species upon material collected near Homa Mountain by Dr Felix Oswald in 1913. This site, which is now known as Kanjera, lies on the southern shore of the Kavirondo Gulf of Lake Victoria in Kenya.

Simopithecus is an extinct genus of baboon, some species of which attained gigantic size. The genus is generally regarded as in some respects related to the gelada baboons of southern Ethiopia and Somaliland.

Dr Hopwood considered that the *Simopithecus* which occurs at Olduvai in Beds II and IV was distinct from the species *oswaldi* on account of its size, but Dr Leakey and Dr Whitworth (1958) in their revision of the genus reduced Dr Hopwood's 'leakeyi' to subspecific rank. They also gave revised generic and specific diagnoses as follows.

DIAGNOSIS OF GENUS

'Extinct, broad-faced, short-muzzled baboons, exhibiting pronounced sexual dimorphism in respect of size, as well as in the morphology of canines and lower anterior premolars. Maxillary fossae absent or very poorly developed (unlike *Gorgopithecus*); distinct, projecting mastoid process; marked post-orbital constriction. Well-defined temporal and nuchal crests, and, in adult males, powerful sagittal crest, all associated with generally massive bone structure of adult skull. Long, robust post-glenoid process. *Eminentia articularis* restricted in antero-posterior direction, and in part slopes backward towards well-defined glenoid fossa. Basi-occipital region of calvarium placed high above level of alveolar margin, so that backward prolongation of line of alveolar margin would fall well below occipital condyles.

'Mandible possesses very robust corpus. Ascending rami tall relative to length of mandible and almost vertical. Mandibular fossae absent or poorly developed; but channel behind M_3 for passage of buccinator muscle from inner side of ascending ramus to outer surface of mandibular corpus, is enormously developed (as in *Gorilla*, but unlike *Papio*). Symphysial region possesses strong backward projection from genial fossa to form marked simian shelf. Region of symphysis lying between pre-molars flat or only very shallowly excavated (unlike *Papio* and *Theropithecus*, where top of symphysis in this region deeply excavated). Articular surfaces of condyles antero-posteriorly narrow, and face wholly upward. Condyles tend to be defined behind by well-marked lip, below which posteriorly is a depression of inverted triangular shape to accommodate large post-glenoid process, when mandible is articulated.

'Molar and pre-molar teeth characterized by tall, rather angular cusps. Dental arcades of both jaws decrease in breadth anteriorly, appreciably in females, less markedly in males, a feature related to small size of incisor teeth. In unworn specimens, the two lingual cusps of upper molars and the two buccal cusps of lower molars are linked by a high longitudinal crest. Posterior part of M^3 subequal to anterior part, and M_3 possesses strong talonid. Incisors very small and rather low crowned. Usually no diastema between upper lateral incisor and canine (unlike *Gorgopithecus*), but, if present, exceedingly small.'

REVISED DIAGNOSIS OF SPECIES 'OSWALDI'

'A *Simopithecus* with skull approximately as large as, or slightly larger than, that of modern *Papio*. Muzzle short, length in females (glabella to alveolar point) about 75 per cent length of calvarium (glabella to inion); in males about 90 per cent. Strong post-orbital constriction of cranium which is reduced there to a breadth about 40 per cent of bi-zygomatic breadth. Temporal crests (actually situated on frontal bones) unite posteriorly at bregma. Well-defined supra-orbital notch in both sexes.

'Ascending ramus of mandible high and approaching vertical, height being about 60 per cent of mandibular length. Only small part of articular condyle extends laterally outside plane of sigmoid notch.

'Upper canines sexually dimorphic; in males tusk-like, but stouter and lower crowned than in *Papio*. M_3 possesses well-developed and large talonid.'

DIAGNOSIS OF SPECIES 'JONATHANI'

'A *Simopithecus* of gigantic size, the female mandible being as large as, and slightly more massive than, the mandible of an average adult male gorilla. Ascending ramus more nearly vertical than in *S. oswaldi* and appreciably higher relative to length of mandible which resembles that of female *Gorilla* in general shape. Mandibular condyles larger transversely than in *S. oswaldi*, owing to lateral extension external to the plane of the sigmoid notch. Cheek teeth smaller in relation to mandible than in *S. oswaldi* and talonid of M_3 reduced. Exceptionally deep channel behind M_3 for buccinator muscle, very similar to that seen in male *Paranthropus crassidens*.'

Whereas there is some material which seems to represent a *Simopithecus* of about the same size as *oswaldi* in Bed I, none of the specimens so far found *in situ* are sufficiently diagnostic to make it certain whether they represent the same species as that which is so common in Beds II and IV. For the moment, the Bed I material can only be regarded as *Simopithecus* sp. without being more definite.

Genus: ***Papio***

Species: indet.

A baboon of the genus *Papio* was described from Olduvai by Dr Remane (1925) on the basis of Professor Reck's 1913 collection. It is the skull of a juvenile but it does seem to be that of a true *Papio*. It is also clearly not the same as the juvenile *Simopithecus* material. There is no clear indication as to which bed this specimen came from. Its preservation suggests that it may have been found in Bed II. During 1962, a very crushed primate skull was found in Bed I, at FLK NN I. It probably represents a *Papio*.

Order: **RODENTIA** and **LAGOMORPHA**

The following preliminary note has been supplied by l'Abbé R. Lavocat:

The rodents which have been studied come from the following sites and levels:

FLK NN I	Level 1—very few.
FLK NN I	Level 2—few.
FLK NN I	Level 3—only a modest number, but of importance.
FLK I	The main site at the *Zinjanthropus* level —few.
FLK N I	Levels 1, 2, 3 and 4—large or very large numbers.
FLK N I	Level 5—few.
FLK N I	Level 6—very few.

The material consists of jaws, mostly with teeth in position, isolated teeth—both molars and incisors—and limb bones. Only jaws with teeth in them and the isolated molars have been used in the present preliminary study.

In view of the urgent need to prepare a preliminary note, only a portion of each of these elements has been sampled in those cases where there was a large quantity of material. In such cases the sample was, however, made of a sufficiently large part of the whole—about a quarter or a fifth in each case—and was never less than several hundred specimens. This was done in order to have a clear picture of the assemblage and the result probably only omits a few varieties which are rare and which may have been overlooked by such sampling.

The essential results of the preliminary examination are summarised in Table 2, and my report consists of a brief commentary.

Otomys kempi (abundant). The *Otomys* is attributed to the species *kempi* on the basis of the number of plates of the highly characteristic molars as well as on the presence of a single groove on the incisors. In respect of this species there are, in the collection, some cranial fragments which it has not yet been possible to examine. It is, therefore, impossible to exclude *a priori* the possibility that these fragments of skull may exhibit characters which will distinguish the material as a new species. In respect, however, of the dentition, the material is very constant from all the sites and all the levels. The plate formula of the teeth appears to be different from the available material from the Australopithecine caves of South Africa. Taken in conjunction with the *Tatera* and the Muridae nos. 2–3 in the list, *Otomys* constitutes one of the main elements of the fauna, being abundant at all sites and in all levels.

Tatera sp. (fairly common). It is well known that it is difficult, not to say impossible, to recognise specific characters in the teeth of *Tatera*. One might be tempted to attribute special importance to the 'fossette', which is present in the anterior lobe of the lower first molar. In fact, the specimens show that, although in most cases the 'fossette' represents the remains of a median constriction open towards the back of the lobe, it may sometimes be open towards the front. Occasionally, however, it represents the last trace of a longitudinal valley, which is sometimes to be seen in unworn teeth, separating two entirely distinct lobes each of which is surrounded by a complete border of enamel. I have not succeeded, so far, in finding any constant and clear differences between the teeth of *Tatera* from Olduvai and those from the Australopithecine caves in South Africa.

Gerbillus sp. This form, which is quite abundant in FLK N I, seems, so far as my present data go, to be totally absent in FLK I and FLK NN I, both of which sites are lower down in the geological sequence than FLK N I and are at approximately the same level. Unfortunately, the total number of specimens of other rodents from the last two sites is not sufficient to enable one to say that the absence of *Gerbillus* is really significant. On the other hand, it may be noted that *Gerbillus* is present in levels 5 and 6 FLK N I, even though in those two levels the total number of rodent specimens is also small. In any case, it must be assumed that, even if the species did exist at those levels where it has not so far been found, it was relatively rare. The species is clearly different from that represented in the material from the Australopithecine caves which I have available for study.

Saccostomus. This genus is relatively well represented by a form which is not easy to distinguish from the living species. The lack of available material and the limited time at my disposal has made it impossible for me to be quite sure that Muridae 1 represents the lower dentition of *Saccostomus*, which, however, I believe is likely.

Muridae 2 and 3: these represent, in all probability, the upper and lower dentition of one and the same species. It represents a creature of notable size and with a dental formula similar to *Rhabdomys*, but the exact genus is not yet

Table 2. *The rodents*

	FLK NN I			FLK I	FLK N I					
	1	2	3		1	2	3	4	5	6
Otomys kempi	×	×	×	×	×	×	×	×	×	×
Tatera	×	×	×	×	×	×	×	×	×	×
Gerbillus	.	.	.	.	×	×	×	×	×	×
Muridae 1 (inf.)	.	.	×	×	×	×	×	×	×	×
Saccostomus (sup.)	.	.	.	×	×	×	×	×	×	×
Muridae 2 (inf.)	.	×	×	×	×	×	×	×	×	×
Muridae 3 (sup.)	.	×	×	×	×	×	×	×	×	×
Muridae 4 (inf.)	.	×	×	.	×	×	×	×	×	×
Muridae 5	.	×	×	×	×	×	.	×	×	×
Muridae 6 (sup.)	.	.	×	×	×	×	.	×	×	.
Muridae cf. 3 (sup.)	×	.	.	.	.	.	×	×	.	.
Muridae 7	.	.	.	.	.	×	×	.	.	.
Muridae 8	.	.	.	.	.	.	×	.	.	.
Muridae 9	.	×	.	.	.	.	.	.	.	.
Muridae 10 (sup.)	.	.	×	×	.	.	.	.	.	.
Muridae X (inf.)	.	×	×	×	.	.	.	.	.	.
Muridae 11 (*Grammomys*)	.	.	×	.	.	.	.	.	.	.
Muridae 12	.	.	×	×	.	.	.	.	.	.
Muridae 13	.	.	.	×	.	.	.	.	.	.
Steatomys	×	×	×	×	×	×	×	×	×	×
Dendromys 1	.	.	×	.	×	×	×	×	×	×
Dendromys 2	.	.	.	.	.	×	.	×	.	.
Sciuridae	.	.	.	.	×	×	×	×	×	×
Pedetes	×	.	.	.	.	.	.	.	.	.
Hystrix	×	.	.	×	×	.	.	.	.	×
Lagomorpha	.	.	.	×	.	×	×	×	.	.
Heterocephalus	.	.	.	.	.	.	×	×	×	×

clear. This is a very common form and some indications of its presence are found in practically all levels.

Muridae 4: a form represented by lower dentitions of an animal smaller than the preceding one is also present in some abundance at all sites.

Muridae 5: a form which is present, more or less, in all places, but represented by only a few specimens (possibly because of its small size). It is an animal about the size of *Mus musculus* or *Leggada.* The rare examples from FLK I and FLK NN I are on the whole of rather smaller size. Only one upper dentition of this type has so far been seen from level 2 of FLK NN I. The characters of the upper first molar are much closer to *Mus musculus* than to *Leggada.*

Muridae 6: this rare form (only a few upper molars are present) approaches the large *Rattus* group. In this group the teeth of the living genera are not of much value as diagnostic characters. Except for the Muridae nos. 10, 11 of my list the other Muridae are based upon a few isolated specimens and are not yet clearly identifiable. There is an upper molar which is probably of *Grammomys,* which is not easy to separate from *Grammomys surdaster* of the present day. It comes from FLK NN I, layer 3. Unless proved otherwise, I consider Muridae 10 and X of my list as representing a single large-sized species (upper molars 1–3 approximately 7 mm.), which I cannot at present identify. It seems to be related to *Rhabdomys* but is, nevertheless, quite distinct and different from the Muridae represented in the list as Muridae nos. 2–3. The species Muridae 10 and X are relatively common in FLK NN I, layer 3.

Steatomys. A very constant type but never abundant.

Dendromys sp. Less common than *Steatomys* and probably represented by two species.

Xerus sp. It is probably to the genus *Xerus* that one must look for the relationships of the only Sciuridae in the collection. It is not common and is only from site FLK N I.

Pedetes sp. This genus is certainly present in layer 1 of FLK N I, although only represented by a few specimens. This species is interesting because of its being typical of open country.

Hystrix. This genus is present only in certain levels but at each of the three sites. As it is always a rare animal, its absence in certain levels may not be significant.

Heterocephalus sp. It is noted with interest that this genus is present in the collection. It is rare in layer 3 of FLK N I which is, nevertheless, rich in other material. It is fairly abundant on the other hand in levels 4 and 5 of FLK N I. It is too early to interpret these facts, although its total absence from FLK NN I and FLK I, as well as levels 1 and 2 of FLK N I, is to be noted.

To summarise, the rodent fauna of the different Olduvai Bed I levels seems to be very homogeneous and made up of species seemingly identical, or very near, to those of the present day. The relative population differences as between one level and another are often marked by examples which are in any case rare. The absence of these rare species in the material from other levels is not necessarily important. On the other hand, the comparison between FLK N I and the two sites FLK I and FLK NN I is made difficult by the rarity of many of the forms available. It seems possible to see certain differences between these two groups of deposits, but the certainty of this difference cannot be established without much more detailed work, and not at the present time. A certain number of forms in this fauna from FLK N I—and they are the most numerous, moreover—seem to indicate a steppe, or sub-desertic environment.

Lagomorpha: these are present but not common.

(A very large lagomorph, apparently belonging to the genus *Lepus*, as well as a lagomorph closely resembling *Serengetilagus*, occurred at site FLK N I (pl. 13). L.S.B.L.)

Order: CARNIVORA

Dr R. F. Ewer has supplied the following preliminary note on the large carnivora of Beds I and II;

Earlier accounts of the Olduvai carnivore fauna (see, for example, Hopwood, 1951) tend to be somewhat misleading. A number of extant species are listed as occurring in the deposits and the impression conveyed is that the fauna did not differ very greatly from that of present-day Africa. Some of the identifications are incorrect, while others are based on fragments too small to be of much value. The large carnivore remains from Beds I and II include a number of limb bones and various skull and jaw fragments, but unfortunately very few of the specimens are sufficient for an unequivocal specific identification to be possible. An examination of all the carnivore material at present available shows that in fact only one extant species, the black-backed jackal, *Canis mesomelas* Schreber, is certainly present in the Olduvai fauna and even this is probably subspecifically distinct from the present-day form.

A description of the specimens on which this statement is based is given below. Although only a few certain identifications can be made, the remains are sufficient to give a picture of the general character of the large carnivore fauna. The assemblage is of the same type as occurs in Africa today, comprising large predators, scavenging hyaenas and jackals, a wolf-like animal and a relative of the insectivorous bat-eared fox. Most of the species are, however, not identical with their present-day ecological counterparts: some, such as the *Crocuta* and the bat-eared fox, are more primitive than their extant relatives, the 'lion' resembles a tiger more closely than it does the extant *Panthera leo* (Linn.) and at least one, the sabre-tooth, belongs to an extinct genus. It thus seems that, although most of the carnivores of these extinct assemblages must have played much the same roles as their modern counterparts do today, the conclusion that extant species were relatively common is not justified.

LARGE CARNIVORA OF BED I

Subfamily: *MACHAIRODONTINAE*

A single specimen from site FLK NN I consists of the anterior part of one ramus of a lower jaw. The presence of a distinct mental process, the raised level of the incisor row, small canine, large diastema and sharp angle between the anterior and lateral faces of the mandible all show that the specimen belongs to an advanced sabre-toothed feline, showing a high degree of specialisation. The species was relatively small, the width of the mandible across the symphysis being about 35 mm. Without further material not even a generic identification can be made with certainty.

Family: HYAENIDAE

Genus: *Crocuta*

Species: aff. *ultra*

Specimen no. 263, 1959, a mandible with cheek teeth well preserved. The dimensions of the teeth are as follows:

		mm.
P_2	length	13·2
	breadth	9·0
P_3	length	18·6
	breadth	12·3
P_4	length	20·0
	breadth	11·4
M_1	length	26·7
	breadth	11·5

M_1 length: P_4 length 133%

P_3 width: P_4 width 108%

The mandible is relatively light, and not deepened below M_1. The carnassial possesses a small metaconid, but the talonid is completely unworn, showing that the upper molar must have been too reduced to make a functional occlusion with the lower carnassial. The carnassial is considerably longer, relative to P_4, than is the case even in the most advanced members of the genus *Hyaena*, while P_3, although not as broad relative to P_4 as in the extant *Crocuta*, is broader than in any *Hyaena* (see Ewer, 1954). Clearly the specimen belongs to the genus *Crocuta*, but to a species more primitive than the extant one, since P_3 is relatively narrower and a small metaconid is present on M_1 (Ewer, 1954). Although it cannot be certainly identified with either, it closely resembles *Crocuta ultra* Ewer, and *Crocuta venustula* Ewer, from the Transvaal Cave deposits of Swartkrans and Kromdraai respectively. Possibly the latter two and the Olduvai form should not be regarded as different species, but only as subspecifically distinct.

A number of other specimens from Bed I, which belong to a large hyaenid, probably belong to the same species as the mandible. These are as follows:

No. M. 20232, a toothless mandible.

No. M. 14675, a basi-cranium.

No. FLK I 506, 1960, an associated humerus, radius and ulna.

Nos. FLK N I 6140 and 6307, 1960, the distal end of a humerus and proximal end of an ulna, much crushed.

Specimen no. 1510, 1957, is a hyaenid mandibular fragment with P_3 and P_4. The teeth are a trifle larger than those of specimen 263 from Bed I, but their relative lengths are almost identical and it therefore seems reasonable to refer the specimen to the same species, *Crocuta* aff. *ultra*. Some limb bones belonging to a juvenile individual are also present and these too may belong to the same species.

Family: **CANIDAE**

Genus: ***Canis***

Species: ***mesomelas latirostris*** (Pohle)

Jackal remains from Bed I comprise no. 358, 1960, a damaged skull with associated mandible; no. FLK N I 6252, 1960, a much-crushed skull; no. K 14, 1960, a maxillary fragment and no. FLK N I 6237, 1960, a pair of mandibular rami, broken off behind the carnassials.

The dimensions of the teeth of these specimens in millimetres are as follows:

		No. 358			
		Right	Left	No. 6252	No. K 14
P^1	length	6·8	—	—	—
	breadth	3·2	—	—	—
P^2	length	9·3	—	—	—
	breadth	3·3	—	—	—
P^3	length	10·6	10·1	—	—
	breadth	3·8	—	—	—
P^4	length	—	17·8	18·0	18·5
	breadth	7·4	7·3	7·6	8·1
M^1	length	11·2	11·1	11·3	13·2
	breadth	13·2	13·1	13·0	14·5
M^2	length	6·4	6·6	*c.* 5·7	6·7
	breadth	*c.* 9·1	*c.* 8·0	9·6	9·0
Carnassial: molar ratio (= length of P^4: length of M^1)		—	1·60	1·60	1·42
Breadth of palate at back of P^4		57·5	—	—	—

		No. 358		No. 6237	
		Right	Left	Right	Left
P_1	length	4·3	4·1	4·6	—
	breadth	3·2	3·1	3·1	—
P_2	length	9·0	8·8	8·2	8·5
	breadth	3·8	4·0	3·4	3·5
P_3	length	9·5	9·7	9·1	9·2
	breadth	4·1	4·1	3·5	3·8
P_4	length	11·2	11·4	10·7	10·7
	breadth	5·2	5·1	4·6	4·7
M_1	length	19·0	18·3	19·5	18·5
	breadth	7·3	7·3	7·1	7·3
M_2	length	9·0	8·9	—	*c.* 8·0 (much worn)
	breadth	6·2	6·1	—	5·8
M_3	length	4·3	4·2	—	—
	breadth	3·6	3·5	—	—
Carnassial: molar ratio (= M_1 length: M_2 length)		2·11	2·05	—	*c.* 2·30

The carnassial: molar ratios (see Ewer, 1956) show that these specimens closely resemble *Canis mesomelas* Schreber, and differ from *C. adustus* Sundevall. They also agree with the measurements given by Pohle (1928) for the specimen from Olduvai which he made the type of a new subspecies, *C. mesomelas latirostris*, said to differ from the extant form mainly in its broader palate. The width of the palate of Pohle's specimen at the back of P^4 is 58 mm. This is very close to the value of 57·5 mm. found for specimen no. 358, 1960, the only one of the present series in which this measurement can be made. It therefore seems reasonable to refer these specimens to Pohle's subspecies, although clearly a larger series of specimens would be desirable to make certain that the subspecific distinction from the extant form is in fact warranted.

Genus: ***Otocyon***

Species: ***recki*** (Pohle)

In his monograph (1928) Pohle described a skull from Olduvai closely resembling the extant *Otocyon megalotis* (Desmarest) except for the fact that it possessed only two upper molars. This he referred to a new genus, *Prototocyon*, which he regarded as more primitive than, and probably ancestral to, the extant form. Further remains of three specimens of an *Otocyon* have now been recovered from Bed I, but they differ from Pohle's specimen in having the three upper molars characteristic of the living species. In two of the specimens the tooth itself has been lost, but the alveoli are present, while in the last both left and right third molars are preserved. As will be seen from the measurements given below, the tooth is smaller in comparison with the first molar than in any of the nineteen specimens of the extant species which were measured for comparison. From Pohle's photograph of his specimen it is impossible to be certain that the posterior end of the maxilla is complete and undamaged. A small portion, bearing a rudimentary M^3, might possibly have been lost from the specimen. It thus seemed essential to re-examine Pohle's specimen. Through the kindness of Dr W. O. Dietrich I have been able to do this, and there is no doubt that, although the right side of the maxilla is broken off behind P^4, on the left side it is complete and undamaged, and Pohle was perfectly correct in saying that only two upper molars were present.

Since the more recently discovered specimens come from Bed I, they cannot be significantly younger than Pohle's specimen and it seems very unlikely that the material represents two different species. It seems more probable that all belong to a single species in which the evolution of the extra upper molars had reached a stage where a small third upper molar was generally present, but individuals lacking it were not uncommon in the population. In an examination of twenty-seven skulls of the extant species in the collections of the British Museum of Natural History and the Coryndon Memorial Museum I have found no specimen with only two upper molars, but one in which a rudimentary fourth molar was present. The number of molars is thus not fully constant in the extant species, and there is no reason to believe that it must have been so in the past. In fact, if the number of molars has increased from the normal canid two, to the three characteristic of the extant species, it is virtually certain that a stage must have existed in which the presence of the extra tooth was a variable character. Populations of the immediate precursor of the extant species would be expected to include individuals in which the 'extra' tooth was absent, and it therefore does not seem desirable to distinguish this evolutionary stage as a separate genus. I have therefore referred all the fossil specimens to a single species, to which Pohle's species name *recki* must be applied, but have not accorded it separate generic status. *Otocyon recki* is regarded as a species differing from the extant *O. megalotis* in that the third upper molar is at an earlier stage of development; it may be absent, or if present is relatively small, its length not being more than three-quarters of that of the first molar.

		No. 308		No. 6275	
		Right mm.	Left mm.	Right mm.	Left mm.
P^4	length	6·4	6·3	*c.* 6·4	6·3
	breadth	4·5	4·3	4·8	5·0
M^1	length	5·7	6·0	6·4	6·4
	breadth	7·5	7·3	7·8	8·0
M^2	length	5·5	5·4	5·7	5·7
	breadth	6·7	6·9	7·5	7·3
M^3	length	—	—	3·3	3·4
	breadth	—	—	5·3	5·3
M^3	length as % M^1 length	—	—	57·9%	59·6%

In a sample of nineteen specimens of the extant *O. megalotis* measured in the British Museum of Natural History collection the length of M^3 ranged from 69·6 per cent to 89·8 per cent of that of M^1. The mean was 80·17 per cent, with a standard deviation of 5·83. For this percentage the fossil thus differs from the extant sample by more than three standard deviations.

Family: **MUSTELIDAE**

Specimen no. 24, 1959, is the first upper molar of *Lutra* sp.[1]

LARGE CARNIVORA OF BED II

Family: **FELIDAE**

Genus: ***Panthera***

Species: sp. indet.

Specimen no. 1273, 1957, is an almost complete mandible of a large felid. The dimensions of the teeth do not show any significant differences from those of the extant lion, but the structure of the posterior part of the jaw is utterly different and closely resembles that of the extant tiger (pl. 14). In the extant lion the lower border is convex and from the point directly below P_4 its margin rises slightly to form the angular process, whereas in the tiger and in the fossil specimen the lower border is concave and from the point below P_4 the margin descends to form the angular process. Specimens no. 301, 1955, an upper canine, no. 1033, 1957, a damaged P^4, and no. 302, 1955, a damaged canine, probably belong to the same species. Although the mandible cannot be distinguished from that of the extant tiger it would be premature to conclude that the latter species was present at Olduvai. In its behaviour and mode of life the extant lion is the most atypical of the large cats: it may well be that the structure of its jaw, which must reflect the details of the arrangement and relative sizes of jaw muscles, is also atypical and has been derived from a condition more like that of the extant tiger. Thus, while the fossil mandible may belong to the extant tiger, it might equally well belong to an ancestor of the extant lion.

		No. 1273	
		Right mm.	Left mm.
C_1	length (at alveolar margin)	—	24·5
	breadth	—	18·5
P_3	length	18·9	18·7
	breadth	10·2	10·5
P_4	length	27·3	27·4
	breadth	13·7	13·7
M_1	length	27·7	27·9
	breadth	14·7	14·7

Subfamily: ? ***MACHAIRODONTINAE***

Specimen no. 791, 1957, is a damaged root of a very large laterally compressed tooth, which may possibly be the upper canine of a large species of sabre-toothed feline. It is far too large to belong to the small species which occurs in Bed I.

Family: **CANIDAE**

Genus: ***Canis***

Species: ***africanus*** Pohle

Pohle (1928) described from Olduvai a large wolf-like animal, which he named *Canis africanus*. Subsequently Dr Hopwood (1951) tentatively suggested that this animal was really the extant *Lycaon pictus* (Temminck). This is certainly incorrect. Both the upper molars and the lower carnassials of *Lycaon* are easily distinguished from those of *Canis*, and in Pohle's specimen the upper molars clearly belong to the latter genus. The more recently discovered specimens from Bed II consist of three mandibular fragments, nos. M. 15017, M. 15018 and M. 15019. In the last two specimens the teeth are too heavily worn to show any distinctive characters, but in M. 15017 the carnassial

[1] A skull and mandible are now known but have not yet been studied.

certainly belongs to *Canis*, not *Lycaon*. It is thus clear that there is no justification for identifying the specimens as *Lycaon* and they are therefore referred to Pohle's *Canis africanus*.

Incertae sedis

In addition to these specimens found *in situ* in Bed II a mandible, no. M. 20233, was found on the surface. Its horizon is therefore uncertain, it may belong to Bed II or to some later period. The teeth have been lost, but the alveoli show that only two pre-molars were present and that the diastema was not very large. Its general structure suggests that it is felid and its size is that of a large leopard. The carnassial alveolus, 25·4 mm. long, shows that this tooth was much larger than that of the extant leopard. The specimen does not belong to, nor even very closely resembles, any extant species, but it might possibly belong to *Felis crassidens* Broom, which occurs at Kromdraai (Broom, 1948*a*). A damaged distal end of a humerus and a nearly complete ulna which belong to a felid also come from Bed II; they closely resemble those of a leopard but are considerably more massive (pl. 15). It seems likely that they belong to the same species as the jaw fragment. Without better material the suggested attribution of these remains to *Felis crassidens* cannot be more than tentative, but it is certain that they do not belong to any extant species of felid.

The table below summarises the provisional identifications which have been made for material from Beds I and II. Those which are certainly extinct are marked *. It will be seen that out of ten carnivores recorded only one, *Canis mesomelas*, is certainly extant. This gives a maximum estimate of nine out of ten extinct carnivores. If we allow that the otter and the 'tiger' could possibly belong to extant species then we arrive at a minimum estimate of seven out of ten carnivora which are extinct. The number of large carnivores identified is, of course, so small that little value can be attached to these figures in isolation and they must be combined with those derived from the study of other orders before a reliable estimate of the proportion of extant species present in the deposits can be made.

Bed I	Bed II
*Small sabre-toothed feline	*Panthera* aff. *tigris*
**Crocuta* aff. *ultra*	**Crocuta* aff. *ultra*
**Otocyon recki*	**Canis africanus*
Canis mesomelas latirostris	**Felis* aff. *crassidens*
Lutra sp.	*? Large sabre-toothed feline

I should like to express my thanks to the authorities of the British Museum of Natural History for allowing me to study material in their collections, and to Dr W. O. Dietrich for allowing me to examine Pohle's type specimens in the Palaeontological Institute of the University in Berlin.

In addition to the larger carnivores listed above by Dr Ewer from Olduvai, reference must be made to a few other specimens of considerable importance but as yet unidentified.

Subfamily: *LUTRINAE*

Incertae sedis

In the British Museum of Natural History collections there is an incomplete cranium (M. 14690) which was referred to *Aonyx* by Dr Hopwood. While it is not possible to give any detailed description of this somewhat weathered specimen it is more massive than the present-day African *Aonyx*. It also differs morphologically from the modern *Aonyx* in the basi-occipital region and the structure of the occiput. It is also clearly different from the African *Mellivora*.

Subfamily: *FELINAE*

Incertae sedis

In the British Museum of Natural History collections from Olduvai there are two distal ends of tibiae of large size, nos. M. 20231 and M. 20230, the latter being slightly damaged in the articular region. These tibiae are unlike those of any of the larger living felids of Africa. They are, moreover, too small to be associated with the felid described by Dr Ewer, which has affinities with the tiger. It is possible that they represent a member of the subfamily Machairodontinae, since it occurs in both Beds I and II at Olduvai. There are, however, certain resemblances to *Panthera pardus*, although the specimens are more massive. In view of the fact that Dr Ewer has recorded the presence of *Panthera* cf. *crassidens*, they could, perhaps, belong to this species.

The distal end of a humerus, no. M. 14676, in the collections from Olduvai at the British Museum of Natural History, is clearly of a large felid allied to the lion and tiger rather than to the leopard. But it is noticeably smaller than in the East African lion and also somewhat different. Morphologically it is unlike and also larger than in the leopard.

Family: VIVERRIDAE

Mme Germaine Petter has supplied the following preliminary note on the Viverridae.

In the course of the excavations at Olduvai Gorge at the FLK group of sites, Dr and Mrs Leakey brought to light a viverrid fauna which they have been kind enough to entrust to me to study.

This material deserves a very extensive examination and the brief summary which is given here is only intended to give an indication of the importance of the material. It is represented by isolated teeth, of which a number are milk dentitions, by various mandibular fragments, which in many cases carry dentition in an excellent state of preservation, and by a reasonable number of bones of the skeleton.

The wide range represented by the specimens makes their study a delicate operation, and a very careful comparison with living species will clearly be necessary for a proper understanding. Consequently, the preliminary study which has so far been carried out allows only a very general indication of the nature of the material to be given.

Since the interpretation of the post-cranial skeletal material will be particularly difficult, the indications which it will be possible to draw from this material cannot be given until a later date. On the other hand, the examination of the teeth makes it possible, even now, to indicate the existence of a number of genera of Viverridae in the fauna of Olduvai.

SITE FLK N I

Layer 1

The few isolated teeth found in this level do not allow, for the moment, an exact generic determination. One can merely mention the existence of a 4th upper pre-molar from the left side of a small species of the subfamily Herpestinae.

Layer 3

Among the six teeth which were recovered from this level, over and above skeletal fragments, it is possible to recognise the following: an upper first molar belonging to the genus *Herpestes*, and two well-preserved upper molars, which will certainly be identifiable but which, for the moment, I cannot place, exactly, in any genus.

Layer 4

This is the level which appears to be the richest in Viverridae remains at the FLK N I site. From this level one can recognise the following:

(1) A right mandible with the 1st, 2nd and 4th pre-molars and the first molar, which belongs to something allied to *Herpestes* or *Bdeogale*.

(2) Two 4th upper pre-molars of a milk dentition of a very large species: Incertae sedis.

(3) An upper 1st molar and 2nd molar belonging to a new species of *Mungos*.

(4) A 1st upper molar which can perhaps be regarded as belonging to one species of *Herpestes*.

(5) A 1st upper molar of a genus which I cannot, as yet, identify, and which differs in its structure from corresponding teeth of all living African genera.

(6) A right mandible with lower 4th pre-molar and a 4th lower pre-molar of the milk dentition of a small species of the subfamily Herpestinae.

(7) A right mandible with lower pre-molars 2, 3, 4 and molar 1 and the sockets of molar 2, very well preserved and representing a species of the genus *Genetta*, which is, however, of a smaller build than the one mentioned above as referring to *Genetta*.

(8) A right mandible with a complete dental series very well preserved and found in association with some upper teeth. It is referable to a new species of *Mungos*.

SITE FLK NN I

This site has yielded a number of specimens of very special interest.

Layer 1

A right mandible with 4th milk pre-molar and with a canine of a small species of the subfamily Herpestinae. It is a much smaller creature than that represented in layer 4 of site FLK N I above.

Layers 2 *and* 3

(1) An upper molar of a species of *Genetta* smaller than those found in layer 4 of site FLK N I above.

(2) An upper molar of similar structure to that found in layer 4 of site FLK N I above, and which, as I stated above, cannot yet be fitted into any known genus.

(3) A well-preserved upper 1st milk molar of great size which probably belongs to the same species of *Civettictis* as the teeth mentioned further on.

(4) A lower 1st molar of the same species.

(5) A left mandible with milk pre-molars and the 4th definitive pre-molar, belonging to a new species of *Civettictis*, larger than the living one.

Layer 3

(1) A left and a right mandible with CP, P_2, P_3, P_4, M_1, M_2 and socket of M_3 belonging to *Prototocyon*, which is a canidae, not a viverridae.

(2) A right mandible with pre-molars 2, 3 and 4 and molars 1 and 2 in excellent preservation and belonging to a small species with resemblances to the genus *Crossarchus*.

(3) A left mandible with the lower 3rd pre-molar and the 4th milk pre-molar of a species which probably belongs to the genus *Genetta*.

CONCLUSION

This preliminary study makes it possible to indicate the very great and certain interest of the Viverridae fauna from Olduvai. It makes it possible to show the co-existence, side by side, of species belonging to living genera and of forms which are today extinct throughout the continent and probably not known to science. It makes it possible to envisage that the detailed study of the material will bring very interesting information to light concerning stratigraphy and biogeography.

Order: **PROBOSCIDEA**

Family: **ELEPHANTIDAE**

Genus: ***Elephas***

Species: cf. ***africanavus*** Arambourg

The fossil elephant material from Olduvai has generally been regarded as belonging to the species

Elephas recki (formerly known as *E. antiquus recki*, and sometimes as *Palaeoloxodon recki*). Dr Leakey and Dr MacInnes, however, in 1951 recorded their view that there was another more primitive elephant in Bed I. They suggested that it was *Elephas exoptatus*, described by Dr Dietrich (1941) from Vogelflüss (Laetolil). This appears to be the same species as that which Professor Arambourg termed *E.* aff. *meridionalis*.

Professor Arambourg (1949) has re-examined a quantity of fossil elephant material from North Africa. He has indicated that, in his opinion, there are two primitive elephants in the older part of the Pleistocene of Africa. The first and more primitive of these he has named *E. africanavus* and he regards the second as closely related to *E. meridionalis*.

During the recent re-examination of the fossil material from Olduvai, now housed in the British Museum of Natural History, it became clear that there are three distinct forms of elephant present in the Olduvai deposits. The majority of the specimens represent *E. recki* and come from Beds II and IV. A few of the specimens which came from lower Bed II seem to correspond to Dr Dietrich's *E. exoptatus* (Professor Arambourg's *E.* aff. *meridionalis*), while the elephant from Bed I seems to be in an evolutionary stage similar to *E. africanavus* (pl. 16).

I could find only one specimen of an elephant from Bed I in the collections at the British Museum of Natural History. This consists of a large part of a skeleton, together with a lower jaw and some of the upper dentition, excavated by Dr and Mrs Leakey from site MK I in 1935. It is embedded in matrix, but a preliminary examination suggests that it is *E.* cf. *africanavus* (pl. 17). The molars are very low-crowned and the individual unworn plates have the characteristic shape and proportions of *E. africanavus*. There are, in addition to this specimen, several others which were found on the surface of Bed I but which probably come from Bed II. There are also some examples marked as having been excavated from Bed I, which belong to the base of Bed II. These were excavated in 1931–2 when the 'marker bed', now used to separate Bed I from Bed II, had not been sufficiently defined. In the Coryndon Museum Centre for Prehistory and Palaeontology there is part of a molar, collected from Bed I, which resembles *E.* cf. *africanavus* (pl. 16).

Species: ***recki***, early form
= *Elephas exoptatus* Dietrich
= (*Archidiskodon exoptatus* Dietrich and *Elephas* aff. *meridionalis* Arambourg)

Among the fossil remains excavated from lower Bed II at Olduvai there are elephant jaws and teeth which are lower-crowned and more primitive than those of typical *Elephas recki*, but also more evolved than *E.* cf. *africanavus* (pl. 18).

The teeth correspond closely in size and morphology to those found by Professor Arambourg at Ain Hanech, which he described as *E.* aff. *meridionalis* (Arambourg, 1949). He now regards this elephant as characteristic of the Upper Villafranchian in North Africa.

Species: ***recki*** (Dietrich)
= *Elephas antiquus recki*, *Palaeoloxodon recki* and *Archidiskodon recki*

This species is now regarded by most authors as distinct from *E. antiquus* and is given full specific rank. It appears in its typical form for the first time at Olduvai in Bed II and continues to the end of Bed IV. The teeth of this species are higher-crowned and have plates of different shape from those of either *africanavus* or *exoptatus*. *Elephas recki* is the characteristic elephant of upper Bed II and the whole of Bed IV.

Family: **DEINOTHERIDAE**
Genus: ***Deinotherium*** Kaup
Species: cf. ***bozasi*** Arambourg

The deinotherium at Olduvai is the largest known species of the genus, larger even than *Deinotherium gigantissimum*. Its remains have been found in Bed I and in lower Bed II, but not in any of the higher levels. It is possible that the Bed II examples may prove to be a more advanced species than that in Bed I, while perhaps neither is the same as *D. bozasi* from Omo.

DISCUSSION

On the basis of the available evidence there are four distinct forms of elephant in the Olduvai fauna, three of which belong to the genus *Elephas* and the fourth to *Deinotherium*. These elephants help us to assess the age of the Olduvai deposits, for the one which resembles *africanavus*, which is elsewhere characteristic of the Lower Villafranchian, occurs in Bed I. *Elephas* cf. *exoptatus* occurs in lower Bed II and is elsewhere characteristic of the Upper Villafranchian. Typical *E. recki* appears for the first time in Bed II and continues to the top of Bed IV. Dr Hopwood originally listed a mastodon as occurring in Bed II. He referred it to the genus and species *Anancus kenyensis*. This identification was based upon parts of two flat-sectioned tusks which resembled tusks of some of the mastodons of South America in their internal structure. Our better knowledge of the fauna of Olduvai shows that these large flat tusks belong to a gigantic extinct pig, *Afrochoerus nicoli*. We must, therefore, remove *Anancus* from the Olduvai faunal list.

Order: PERISSODACTYLA

Family: RHINOCEROTIDAE

Subfamily: *DICERORHININAE*

In the earlier reports dealing with the fossils from Olduvai, there were records of specimens representing the 'white rhinoceros'. These differed from the living species only in minor characters. They were only regarded as having subspecific value and the material was therefore allocated to *Ceratotherium simum germano-africanum*. Dr Dietrich subsequently reported on more complete fossil material, found at Vogelflüss (Laetolil) some twenty miles south of Olduvai. As a result, he described (1942*b*) a new genus of rhinoceros allied to *Ceratotherium*, but to which he gave the new generic name of *Serengeticeros*. The deposits in which this new genus and species was found have been considered by some authors as equivalent to Bed I, Olduvai.

Fossil material representing the rhinoceros group is represented by a considerable number of specimens in the British Museum of Natural History. This material includes four partial mandibles. Additional rhinoceros material, discovered since 1935, is in America where it was submitted to Dr Horace Elmer Wood, Jun. Recent additions to the material are in the Coryndon Museum Centre for Prehistory and Palaeontology, Nairobi.

Genus: ***Ceratotherium***

Species: ***efficax*** Dietrich

Some of the specimens in the British Museum of Natural History from Bed I and also from lower Bed II seem to be the same species as *Serengeticeros efficax*. I consider that there is insufficient evidence, at present, either in respect of the material in Dr Kohl-Larsen's collection, or in the material from Olduvai, to warrant generic rank. The material is, therefore, referred to the longstanding genus *Ceratotherium*, while retaining Dr Dietrich's specific name, *efficax*.

Species: ***simum germano-africanum***

In 1925 Dr Hilzheimer of Berlin described an incomplete skull and jaw, which had been obtained from Olduvai during Reck's 1913 expedition, as *Rhinoceros simus germano-africanus* (Hilzheimer, 1925). The specimen which he described certainly appears to be very similar to the living species and not to those referred to above as *efficax*. From the illustrations, however, it seems that it was barely fossilised and may have come from deposits much younger than Bed IV, such as Bed V or later.

A number of specimens which have been found in upper Bed II and in Bed IV are here referred to *Ceratotherium simum*. It is possible that they may have to be treated as a new species when they have been studied in more detail.

Genus: ***Diceros***

Species: ***bicornis***

There are a few specimens in the collections at the British Museum of Natural History from Olduvai which appear to be referable to the living species of black rhinoceros. They certainly belong to this genus but a study of more complete material may necessitate a revision in respect of species.

DISCUSSION

It would seem that the rhinoceros group is represented at Olduvai by not less than three distinct types. One of these, which occurs in Bed I and lower Bed II, is markedly different from the living white rhinoceros and is provisionally regarded as belonging to Dr Dietrich's species *efficax*. The second, which is at present only known with certainty from higher levels, seems to be similar to the living white rhinoceros. The third is the black rhinoceros, or some species closely related to it.

Family: CHALICOTHERIIDAE
Subfamily: *CHALICOTHERIINAE*
Tribe: **Schizotheriini**
Genus: ***Ancylotherium*** Gaudry
Species: cf. ***hennigi*** Dietrich

Amongst the fossils from Bed I at Olduvai are a few foot bones which indicate the presence of a large chalicothere. This material was provisionally placed with *Metaschizotherium hennigi*, which was described by Dr Dietrich (1942*b*) on the basis of material from Vogelflüss (Laetolil). Until more complete material from Olduvai becomes available it is not possible to be certain of the specific determination of the Olduvai representative of the group, but there is a reasonable probability that it does belong to Dr Dietrich's species.

Professor P. M. Butler has sent the following notes:

CHALICOTHERIIDAE. A number of foot bones from THC I may be tentatively referred to *Ancylotherium* (= *Metaschizotherium*) *hennigi* Dietrich (1942). The material consists of incomplete specimens of metacarpals II and III, a scaphoid, a lunate, a cuneiform, a basal phalanx and two middle phalanges, seemingly all from one individual. The carpal bones show a relationship to *Ancylotherium pentelicum* Gaudry & Lartet from Pikermi, but differ from this in certain respects. The basal phalanx is very similar to a specimen from Kaiso, Uganda, described by Andrews (1924), but it belongs to a different digit. It is not possible to compare this material with *Metaschizotherium transvaalensis* George (1950), which was based on a molar tooth from Makapansgat, but the distinction of the latter species from *Ancylotherium hennigi* is open to question.

Family: EQUIDAE
Subfamily: *EQUINAE*

Members of the subfamily Equinae are very commonly represented in the Olduvai fauna (pl. 19) and a great deal of material is now in the hands of Professor Stirton of Berkeley, University of California. Whereas specimens which clearly represented *Equus* and the hipparionids were sent to California, other material, which was clearly the same as Van Hoepen's *Eurygnathohippus* of South Africa, was retained in Nairobi (pl. 20).

The recent discovery of a skull of *Stylohipparion* in which the anterior dentition is preserved has clarified the problem of *Eurygnathohippus*. The specimens upon which the genus was based proved to have been the incisors and canines of *Stylohipparion*.

Order: ARTIODACTYLA
Family: SUIDAE
Subfamily: *SUINAE*

INTRODUCTORY NOTE

As I have indicated in some of my other publications on fossil Suidae, the teeth of members of this family, if considered by themselves out of the context of the mandibular morphology and skull structure, can be very misleading. I have, for example, published a note (1958*a*) illustrating a molar tooth taken from the skull of a modern *Sus cristatus*, which would certainly be classified as a tooth of the extinct genus *Pronotochoerus* if found alone. Similarly, Dr Broom published a note (1948) on a fossil pig which he called *Phacochoerus antiquus*, in which the second and third molars certainly suggested this living African genus. When a skull of this species was found, it became clear that in spite of dental similarity to *Phacochoerus* this extinct pig was much more closely allied to *Potamochoerus*.

Dr Dietrich published some Suidae teeth from Olduvai as representing the genus *Phacochoerus*, but we now know that these belong to the extinct genus *Afrochoerus*. If we had only the cheek teeth this would not be apparent.

There have been occasions in the past when I, too, have ascribed certain individual teeth to genera and species represented by more complete specimens in other levels. Thus, I recorded certain Suidae which are common in Bed II as also occurring in Bed I, basing this view on fragmentary material. We now know that this was incorrect.

The present study of much more complete material from Bed I has cleared up some of these errors. It has also strengthened my conviction that it is impossible to identify Suinae on molar and pre-molar teeth alone.

Before dealing with the new Suinae from Bed I, I propose to give brief notes on the already published pigs of Olduvai.

SUINAE ALREADY PUBLISHED

Tribe: **Suini**

Genus: ***Mesochoerus***

Species: cf. ***heseloni***

Mesochoerus heseloni is the species which is characteristic of the Omo fossil beds which Professor Arambourg (1943) described under the generic name of *Omochoerus*. He maintains that his genus is distinct from *Mesochoerus* on the grounds that its molars are lower-crowned and have no cement over the enamel. In the type species of *Mesochoerus* from South Africa and the species *olduvaiensis* the teeth are more hypsodont and carry some cement. I do not consider the absence of cement and the brachyodont nature of these teeth are sufficiently strong characters to justify a generic difference, when in *all* other morphological details *Omochoerus* is so similar to *Mesochoerus*.

It is now clear that in the very top of Bed I at Olduvai and also at the base of Bed II there are certain molar teeth of a *Mesochoerus* type which are low-crowned and without cement. These should perhaps be ascribed to *M. heseloni*. In other words, a *Mesochoerus* of the more primitive type does appear to be present at the top of Bed I and the base of Bed II. It is possible, however, that these teeth may belong to *Promesochoerus*.

Species: ***olduvaiensis*** Leakey

The type specimen comes from upper Bed II and represents the *Mesochoerus* common throughout East Africa in Middle Pleistocene times. It has been fully described in the monograph on *Some East African Pleistocene Suidae* (Leakey, 1958*b*) and need not be discussed further here.

Genus: ***Potamochoerus***

Species: ***majus*** Hopwood

= *Koiropotamus majus* Hopwood

Apart from the anomalous genus *Hylochoerus*—the forest hogs—the genus *Potamochoerus* is the only wild African representative of the tribe Suini living today south of the Sahara.[1] The extinct larger species *Potamochoerus majus* was abundant in upper Bed II and continued into Bed IV (pl. 21). In my monograph, I erroneously described certain suid teeth from Bed I as representing *Potamochoerus majus*. These were molar teeth from site FLK I. Further specimens have now been found at the same site under conditions which prove that they represent a wholly distinct genus. They belong to the new and much more primitive *Ectopotamochoerus dubius* (see p. 30).

Tribe: **Phacochoerini**

Genus: ***Pronotochoerus***

Species: cf. ***jacksoni***

In my monograph on the Pleistocene Suidae (Leakey, 1958*b*), I made no reference to *Pronotochoerus* at Olduvai since there was nothing to suggest its presence. However, during recent excavations, some teeth have been found which suggest that a *Pronotochoerus* may be present in Bed I.

Genus: ***Notochoerus***

Species: ***compactus*** Leakey

Notochoerus compactus is the characteristic member of this genus in upper Bed II at Olduvai, although it never seems to have been common. It has been described fully in my Suidae monograph and will not be described further here.

Species: ***hopwoodi*** Leakey

This is the *Notochoerus* of Bed IV at Olduvai, although it is rare. It has been dealt with in my monograph and need not be further discussed here.

[1] I do not consider the feral *Sus* on Pemba Island is a true wild pig since it is known to be descended from domestic pigs released there by the Portuguese two or three centuries ago.

Genus: ***Tapinochoerus***

Species: ***meadowsi*** (Broom)

Tapinochoerus meadowsi is the representative of this genus which occurs in considerable numbers in upper Bed II at Olduvai and continues into Bed IV. It is also recorded from Olorgesailie. It has been fully dealt with in my Suidae monograph.

Species: ***minutus*** Leakey

This is a small form of *Tapinochoerus* which appears in Bed IV. It therefore belongs to the close of the Middle Pleistocene. What little data we possess are given in the Suidae monograph.

Genus: ***Orthostonyx***

Species: ***brachyops*** Leakey

This is a very aberrant member of the Phacochoerini tribe in which the upper canines are directed upwards somewhat as in *Sus barbatus* of the Far East. It is, at present, only known in upper Bed II.

Genus: ***Afrochoerus***

Species: ***nicoli*** Leakey

This is a gigantic member of the Phacochoerini tribe which is now known to have extended over most of Africa in Middle Pleistocene times from Cape Town to Algeria. It is abundant in upper Bed II and in Bed IV. Among its characteristics are the very curious upper and lower canines, which are described in my monograph (pl. 22, top). The lower tusks of this giant pig were the basis of the supposed existence of a mastodon in Bed II.

Genus: ***Phacochoerus***

Species: ***altidens*** Shaw and Cooke

Subspecies: ***robustus*** Leakey

The Pleistocene deposits of East Africa have not so far yielded any remains of a *Phacochoerus* similar to the living species, but the genus is represented in the lower and middle parts of the Olduvai beds, that is to say in Bed I and in Bed II, by the species *altidens*. This species was first described by Dr Shaw and Dr Cooke (1941) from South Africa. The form in which it occurs in Beds I and II appears to differ from the type and has been given subspecific rank as *robustus*. It may be necessary to assign it full specific rank when further material is obtained.

Species: ***altidens altidens*** Shaw and Cooke

In Bed IV there is a pig which seems to be identical to *Phacochoerus altidens altidens*. Its teeth, at least, are indistinguishable from the South African specimens of this race.

THE NEW SUINAE FROM BED I

Tribe: **Suini**

Genus: ***Promesochoerus*** gen.nov.

DIAGNOSIS OF GENUS

An aberrant member of the tribe Suini resembling the genus *Mesochoerus* in the molar–pre-molar series but differing from that genus in the following very important characters. The upper and lower canine teeth are short-crowned with relatively long tapering *closed roots* which do not look like the canines of Suini. The lower canines are set in the mandible so that their tips project more or less at right angles to the long axis of the mandible. There are six lower incisors in the adult, whereas in all species of *Mesochoerus* the lateral incisors are lost very early in life.

GENOTYPE

The species *Promesochoerus mukiri* described below.

Species: ***mukiri*** sp.nov.

DIAGNOSIS OF SPECIES

A *Promesochoerus* of medium size in which the length of the molar–pre-molar series is a little less than in *Mesochoerus heseloni*. The lower canine exhibits wear facets both on the posterior face and transversely on the upper face. (It is not clear how the mandible can have been articulated in order to make it possible for the upper canines to wear both these facets on the lower teeth.) The third pre-molars are small in size compared with those of *Mesochoerus*, while the fourth pre-molars are more rectangular. The crown pattern of the lower third molars closely resembles that of a true *Mesochoerus*. The third molars have lower crowns than *M. olduvaiensis*, and resemble *M. heseloni*. The first lower molars were presumably erupted early in life, for they are the most worn teeth in the whole molar–pre-molar series. The wear extends down to the alveolar margin at the time when the pre-molars are scarcely worn and the talonid of the third molar is only just appearing.

The short-crowned upper canines are remarkable for the fact that the only wear facet is transverse on the anterior face. Before the transverse anterior facet becomes strongly marked, the upper canines exhibit a slightly rounded anterior face. As far as can be judged, there is a sex difference in the upper canines. There are some teeth in the collection with much less massive crowns and shorter roots, which appear to be female. The upper canines exhibit very faint traces of beading and ribbing which is presumably ancestral to the subsequent heavy beading and ribbing of the upper canines of *Mesochoerus* and *Potamochoerus*. The upper and lower canines carry a coating of cement on the roots as well as on the crowns.

TYPE

The type is a mandible no. G. 356, 1960, from FLK I (pl. 23).

FIRST PARATYPE

Four associated canines from FLK NN I, no. 701, with associated incisors and some molars and pre-molars (pl. 24).

SECOND PARATYPE

Two small and little-worn upper canines, no. 6084, presumed to be those of a female, from site FLK N I, layer 1.

HORIZON

Specimens representing *Promesochoerus mukiri* have been found at sites FLK NN I, FLK I and FLK N I. The horizon is thus the upper part of the Villafranchian.

DESCRIPTION OF TYPE

The type consists of the greater part of a mandible which has been crushed so that the *corpus* has been everted on either side. The symphysis has also been crushed but not everted, so that there is less distortion in this area. The complete dentition is preserved.

In this specimen the teeth are more worn on the left than on the right side, with the third and fourth pre-molars worn right down to the dentine. The first left molar has been worn almost to the alveolar margin and the anterior half of the second molar is also very worn. The right third and fourth pre-molars are less worn, but the whole crown of the first molar has been worn away. The anterior half of the second molar is a little worn, but not as much as on the left side. The posterior half of the second molars and the whole of the third molars are only slightly worn. They exhibit the detail of the enamel pattern extremely well. Both the second lower pre-molars are preserved; they are small and show a minimum of wear. The molar–pre-molar formula is P_3, M_3. The pre-molars are generally similar to those of the genus *Mesochoerus* but of different proportions. The fourth pre-molar is less elongated and more rectangular, while the third pre-molar is more elongated than in *Mesochoerus*. The structure of the fourth pre-molar is more like that seen in *Potamochoerus* than in *Sus*, and in this respect is close to *Mesochoerus*. The structure of the third pre-molar, in the type specimen, resembles that seen in *M. heseloni* rather than in *M. olduvaiensis*. The crowns of the second and third molars exhibit a pattern almost identical to that seen in all species of *Mesochoerus*. The third molars would undoubtedly be taken to represent a *Mesochoerus* and not a more archaic genus, if found isolated.

The length of the molar–pre-molar series is 134 mm., which is considerably greater than in the living African members of the Suini such as *Potamochoerus*. It is comparable to this measurement in *Mesochoerus olduvaiensis* and *heseloni*. In spite of the crushed state of the specimen it seems clear that the position of the lower canines and the incisors has not been greatly altered.

The lower canines exhibit the peculiar structure and wear which is characteristic of the genus and which can be seen in plate 24 on the paratypes. This degree of attrition is seen not only in the type but also in the first paratype and all the other known associated dentitions of the genus. It may, therefore, be regarded as characteristic. The six incisors exhibit a degree of wear far more extreme than might be expected from the wear on the cheek teeth.

The length of the diastema from the alveolar margin at the root of the second lower pre-molar to the alveolar margin of the canine can only be given approximately, as the bone is damaged. It is about 90 mm. The bilateral measurement at the external alveolar margins of the lateral incisors is 90 mm.

DESCRIPTION OF FIRST PARATYPE

A series of associated teeth from site FLK NN I, layer 1, is of importance because it shows us the nature of the upper canines of *Promesochoerus mukiri*, absent in the type. The incisors and lower canines of this series are very similar to those of the type except that the wear on the posterior face of the lower canines is more limited. The roots of the canines are short and completely closed, with a deep medial groove running down both faces of the root. The ends of the roots are rugose. The associated upper canines have somewhat longer and narrower roots and very short crowns. The wear on the crowns is transverse across the anterior face. No examples of upper canines have been found yet in a skull, but the nature of the wear facets in relation to the lower canines suggests that the upper canines were set vertically downwards in the maxilla. The roots of the upper canines also exhibit a deep lateral groove on either side.

DESCRIPTION OF SECOND PARATYPE

The two specimens which comprise the second paratype are upper canines with shorter roots than in the first paratype. They are assumed to represent a female. The wear on the anterior face has only just begun. The unworn upper canines of *Promesochoerus mukiri* were, therefore, similar to the lower canines but were not so sharply curved.

DISCUSSION

The genus *Promesochoerus* is an aberrant and entirely unexpected pig to find in Pleistocene deposits, even though these are of Villafranchian age. The canine teeth are most distinctive.

A search for comparable canine structure in other Suinae resulted in the discovery of teeth of somewhat similar type, but smaller size, in some Miocene genera. The Lower Miocene genus *Hyotherium*, which is represented in Europe by the species *typum* and in East Africa by a species which has not yet been described, is the most similar. An upper canine of the East African species of *Hyotherium*, from the Lower Miocene beds of Rusinga Island, is illustrated in plate 24 for comparison with those of the second paratype. The Middle and Upper Miocene Suinae from Europe revealed no comparable teeth. It is thus possible that Suinae with this peculiar canine structure did not survive in Europe after the Lower Miocene, although this stock persisted in East Africa.

The cheek teeth of *Promesochoerus* are so similar to those of *Mesochoerus* that it would be difficult to distinguish them without associated canines and incisors. The retention of the lateral incisors and the structure of the short-rooted, short-crowned canine teeth clearly distinguish *Promesochoerus* from *Mesochoerus*.

The above description of *Promesochoerus mukiri* is presented only as a *provisional statement*. More material of this new pig awaits study and will be dealt with in a more detailed report on the new East African Suidae.

Genus: ***Ectopotamochoerus*** gen.nov.

DIAGNOSIS OF GENUS

A genus of the Suini in which the molars and pre-molars are reminiscent of *Potamochoerus*, but differing from that genus in a number of morphological characters. The most important of these are the following: the roots of the canines taper rapidly and are either fully closed, or nearly so; they are relatively small and short, having no clear dividing line between the crown and the root, such as is seen in *Promesochoerus*; the pulp cavities are completely unlike those in *Potamochoerus*; four pre-molars are retained in both the upper and lower dentitions of adults, whereas in *Potamochoerus* the pre-molars are reduced to three by the time adult status is reached.

GENOTYPE

The species *Ectopotamochoerus dubius* as described below.

Species: ***dubius*** sp.nov.

HOLOTYPE

A crushed skull and mandible from site FLK N I, nos. N I 1235 and 1236 of 1961 (pls. 25 and 26).

HORIZON

The horizon of this genus and species is the upper part of the Villafranchian.

DIAGNOSIS OF SPECIES

A species of *Ectopotamochoerus* in which the diastema in the lower jaw is small and with a small diastema between the first and second upper pre-molars. The molars are of the *Potamochoerus* rather than the *Sus* type, and the talon and talonid of the third molars are longer than in most species of *Potamochoerus* but not quite as elongated as in *Mesochoerus*. The ascending ramus of the mandible is wide and set further forward than in *Potamochoerus*. There is a swelling on the external aspect of the mandible in the region of the molar teeth very similar to that which characterises *Mesochoerus*. The incisor teeth are massive and resemble those of *Potamochoerus* and are not of *Mesochoerus* type, but they exhibit certain features in the ribbing of the enamel on the posterior face which is reminiscent of *Mesochoerus*. There are six incisors, as in *Potamochoerus*.

DESCRIPTION OF TYPE

The type specimen of *Ectopotamochoerus dubius* consists of a crushed skull together with the left side and anterior region of an associated mandible, in which canine teeth are present but the right corpus and the right cheek teeth are missing.

The lower molars and pre-molars closely resemble those of *Potamochoerus majus*. The third molar is long and has a strongly developed talonid, more so than in *P. majus* or the living *Potamochoerus*, but not quite as long as in *Mesochoerus*. The second molar is more elongated than in *Mesochoerus* and the resemblance is with *Potamochoerus majus* rather than with the living species. The first molar is also more elongated than in any mesochoerine, although a little similar to *Mesochoerus heseloni*. The tooth is, however, relatively narrower than in any *Mesochoerus*.

The third and fourth pre-molars are closely comparable to the same teeth in *Potamochoerus* and not to those of *Mesochoerus*. The second pre-molar, as in *Potamochoerus*, is very reduced in size, whereas in *Mesochoerus* it is always a relatively large tooth. The diastema between the second pre-molar and the canine is short. In this character *Ectopotamochoerus dubius* resembles the living *Potamochoerus* more than *P. majus*. The ascending ramus rises steeply and nearly vertically from the region of the talonid, whereas in *Potamochoerus* the ascending ramus slopes backwards. The type specimen shows resemblances to *Mesochoerus heseloni* in the external lateral swelling of the corpus of the mandible in the region of the first and second molars. The length of the diastema is approximately 35 mm., while the length of the molar–pre-molar series is 148 mm., compared with 106 mm. in the modern species.

The lower canines are most unusual and distinguish this genus clearly from *Potamochoerus*. The lower canines are comparatively small and triangular in cross-section. They are set in the mandible in much the same plane as the pre-molars, rising nearly vertically upwards and scarcely everted. In the type specimen the roots of the lower canines are not completely closed. It is, however, a young individual and there are other specimens in which the canine roots are completely closed. The canines exhibit a deep groove on the posterior face and a shallow groove on each of the lateral faces. The only wear is on the posterior facet at the tip.

The lower incisor teeth are of the *Potamochoerus* type but are more massive than in the living *Potamochoerus* and a little like those of *P. majus*.

The skull of the type specimen is very severely crushed, and will not be described in any great detail.

The upper third molars are elongated but the talon is shorter than the talonid. These teeth are comparable to those of *Potamochoerus majus* and could be confused with those of this species. As in *Sus*, but not in *Potamochoerus*, there is a first upper pre-molar. In *Sus* this is situated close to the alveolar rim of the canine without a diastema separating it from the second pre-molar. In *Ectopotamochoerus*, however, it is set forwards so that there is a diastema of some 18–20 mm. separating the first and-second pre-molars. The sockets for the canines are set some 15–20 mm. in front of the first pre-molar, forming a second diastema between this tooth and the canine. There is also a diastema between the canines and the lateral incisors. The upper fourth pre-molar is of the *Potamochoerus* type, being more rectangular than in *Sus*. The third upper pre-molar is also molariform and is quite unlike those seen in *Sus* or *Mesochoerus* or *Potamochoerus*. The second upper pre-molar is small, relative to the third and fourth pre-molars. The first pre-molar was not very small, if one may judge by its sockets.

Turning to the anterior teeth, the sockets of the upper incisors are preserved on the right side. These indicate that the central incisors were the largest and the laterals the smallest. The positions of the root sockets are somewhat as in *Potamochoerus* and are more extended along the pre-maxilla than in *Sus* or *Mesochoerus*.

The upper canines resemble those of *Promesochoerus* but are worn in an entirely different manner. They are short-crowned, short-rooted teeth with a closed, or nearly closed, pulp cavity. They have thick cement over the enamel of the roots, which extends up on to the crowns. These upper canines display clear traces of beading of the enamel such as is found in all *Potamochoerus* species. On the labial face the canines carry a deep groove with a lesser one on each lateral face.

The length of the upper molar–pre-molar series on the right side is 164 mm., compared with 109 mm. in *Phacochoerus*.

DISCUSSION

Ectopotamochoerus is another aberrant member of the Suini. Like *Promesochoerus*, it is notable for the small size of the roots of the canines, which taper rapidly. In certain individuals the pulp cavity is completely closed and in others nearly so. While the cheek teeth in this genus closely resemble those of *Potamochoerus*, rather than *Mesochoerus* or *Sus*, some characters are present which stand nearer to *Mesochoerus*.

Genus: ***Potamochoerus***

Species: ***intermedius*** sp.nov.

HOLOTYPE

The holotype is a damaged and crushed skull from site FLK NN I, no. 177, 1960.

HORIZON

Upper Villafranchian.

DIAGNOSIS OF SPECIES

A *Potamochoerus* with molars and pre-molars larger than in any known example of *P. majus*; the canines resemble those of *Hylochoerus* rather than *Potamochoerus*; the arrangement of the pillars and intermediate valleys of the second and third upper molars is also more like *Hylochoerus* than *Potamochoerus*; there is a bony buttress of *Potamochoerus* type over the root area of the upper canines; a third upper pre-molar is present when the third molars are fully erupted; the incisors are of the *Potamochoerus* type.

DESCRIPTION OF HOLOTYPE

The holotype (pl. 28) consists of a very crushed skull in which the molars and pre-molars are preserved on the left side. Both canines are present. The damaged left pre-maxilla carries the central incisor together with the root sockets of the median and lateral incisors. The right median incisor is also preserved. The upper third molars are much larger than in the largest *Potamochoerus majus*, having an anterior–posterior length of 56 mm., and a maximum width of 28 mm., compared with 43·5 mm. and 22 mm., in the largest known upper third molar of *P. majus*. In a random specimen of the living *Potamochoerus* the figures are 35 mm. and 21 mm. Each transverse pair of pillars is separated from the next pair by a wide valley. This is occupied by a cusp of *Hylochoerus* type but higher in relation to the main pillars. There is a similar hylochoerine structure in the second upper molar. This has an anterior–posterior length of 31 mm., and a width at the alveolar margin of 21 mm. The first molar is a large tooth measuring 23 mm. and 21 mm. Its general structure is, however, more potamochoerine. The fourth upper pre-molar is of *Potamochoerus* type, being quite unlike those seen in either *Sus* or *Hylochoerus*. The third upper pre-molar is again characteristic of *Potamochoerus* and quite unlike those in *Sus*, *Hylochoerus* or *Mesochoerus*.

Although the skull is damaged and crushed, a potamochoerine type of buttress is present over the right upper canine. Half of the first pre-molar is present as well as its root socket. It is set about 5 mm. in front of the second pre-molar. In the living *Potamochoerus* as well as in *P. majus* there are only three upper and three lower pre-molars in the adult, although up to the age of the eruption of the third upper molar, a small first pre-molar is sometimes retained, separated from the second by a small diastema. In this respect *P. intermedius* retains a juvenile potamochoerine character when adult.

The left pre-maxilla is preserved with the central incisor in position. This is markedly of potamochoerine type, although very much larger. The root sockets of the median and lateral incisors have the same position relative to the canines and pre-molars as in *Potamochoerus*. The upper canines are not of the *Potamochoerus* type and are more reminiscent of *Hylochoerus*. The cross-section of these teeth as well as the ribbing and beading is of hylochoerine type.

DISCUSSION

Unfortunately, our knowledge of *P. intermedius* is based on a very crushed skull in which, however, the greater part of the dentition is preserved, so that many important characters can be studied.

In the upper canines, as well as in the structure of the second and third molars, this pig is definitely reminiscent of the genus *Hylochoerus*. On the external margin of the maxilla, behind and supporting the very large canine, there is a bony buttress which is characteristic of the genus *Potamochoerus* and which is not known to occur in any other member of the Suini tribe. Although the third upper molars have paired pillars set in a *Hylochoerus* pattern, the talon is longer than in any known *Hylochoerus*. In *Hylochoerus*, moreover, the dentition tends to be reduced, with only a fourth upper pre-molar which is an elongated tooth unlike those of *Potamochoerus*. In the present species the third and fourth pre-molars are retained in the adult and are characteristic of *Potamochoerus*.

It should be noted that the two tusks which were designated the paratype of Dr Hopwood's original description of *Potamochoerus majus* had been found as isolated specimens in Bed I. I have previously suggested that they probably did not belong to *Potamochoerus majus*. The discovery of *P. intermedius* makes it clear that Dr Hopwood's original paratype belongs to this new species and must be assigned to it.

Tribe: **Phacochoerini**

Genus: ***Notochoerus***

Species: cf. ***euilus*** (Hopwood)

Remains of a *Notochoerus* occur in the material excavated from site FLK N I where it is represented by specimens no. 335 (pl. 29) and no. 441 of 1960.

The specimens consist of the right half of the mandible, lacking the incisors, with part of the ascending ramus preserved and the greater part of the dentition. This is represented by the canine, fourth pre-molar and the molars. In the right half of the mandible the pre-molar is displaced away from its normal position. Both the canines are essentially notochoerine in character, having convex lateral faces without any marked groove. They have longitudinal striations as well as fine cross-lines, producing the characteristic *Notochoerus* lattice pattern of the enamel. The pulp cavity is open. The posterior face is concave, covered with enamel and without cement. There is a faint trace of a groove, which is not median, but lies nearer the lingual face.

The molars exhibit an enamel pattern similar to that in the *Notochoerus* from Kaiso which Dr Hopwood named *Hylochoerus euilus*. There is no trace of any pre-molar other than the fourth, although the third molars on both sides are not fully erupted. The diastema between the fourth pre-molar and the canine is short. The right canine is set vertically in the socket and is only slightly everted.

For the time being, these specimens are referred to *Notochoerus euilus*.

DISCUSSION

The Olduvai *Notochoerus* from Bed I, described above, has been provisionally referred to *N. euilus*. We must, therefore, consider its relationship to *Notochoerus* material from other Lower Pleistocene sites in East Africa. In 1958 I suggested that *Gerontochoerus scotti* from Shungura, *N. serengetensis* from Laetolil and the suid that Professor Arambourg referred to as *N.* cf. *capensis* from Omo, should all be regarded as representing *N. euilus*. The new material from Olduvai suggests the possibility that this view will require revision. It may be necessary to separate the original Kaiso material and the Bed I material from the more robust forms of the other sites, which will have to be termed *N. scotti* since this has precedence over *serengetensis*.

The *Notochoerus* dealt with above from site FLK N I, is not the same as that from Laetolil and Omo. Nor is it the same as *N. compactus* nor *hopwoodi* from Beds II and IV at Olduvai. The new find supports the suggestion, drawn from other lines of evidence, that the Kaiso fauna is more archaic than that of Omo or Laetolil.

Genus: ***Tapinochoerus***

Species: indet.

Amongst the fossil material from site FLK I there are some specimens resembling a *Tapinochoerus*. These consist of parts of a mandible found in 1959 and part of a second mandible found in 1960 (pls. 30, 31 and 32). In each specimen the third molar is present, while in the 1959 mandible the cheek teeth are complete. The lower dentition consists of three molars and a fourth pre-molar, which is very reduced in size. The talonids of the third molars are damaged. In the 1960 specimen only the third molar is present, but the talonid is intact.

DISCUSSION

It is not possible, with such limited material, to assign this *Tapinochoerus* from Bed I to any definite species. It is clearly not the same as *T. meadowsi* from Bed II and Bed IV, nor is it the same as *T. minutus*. There is a possibility that it might represent the lower dentition of a primitive *Orthostonyx*, since we know that, without the canine teeth, this genus is not easy to distinguish from *Tapinochoerus*.

The genus *Tapinochoerus* had not, hitherto, been recorded from any Villafranchian deposit. It is clear, however, from the presence of an evolved *Tapinochoerus* and its close relative, *Orthostonyx*, in the lower part of the Middle Pleistocene, that there must have been an ancestor in the Villafranchian. The new material from Bed I, is, unfortunately, too fragmentary to solve this problem.

Family: **HIPPOPOTAMIDAE**

Genus: ***Hippopotamus***

Species: ***gorgops*** Dietrich

Hippopotamus gorgops is the characteristic hippopotamus of upper Bed II. It has been recorded in former Olduvai faunal lists as also being present in both Beds I and IV. The species was described by Dr Dietrich (1926) and, although there is now more complete material than was then available, I do not propose to give a revised diagnosis at the present time. There is no doubt that *H. gorgops* persists into Bed IV, but its presence in Bed I is less certain.

Genus: ***Hippopotamus***

Species: Indet.

A nearly complete skull of a *Hippopotamus* from the lower part of Bed I is now available. It has not yet been studied in detail but it appears to differ from *H. gorgops* in certain important characters.

Family: **GIRAFFIDAE**

Subfamily: ***PALAEOTRAGINAE***

Genus: ***Okapia***

Species: ***stillei*** Dietrich

DIAGNOSIS

The diagnosis given by Dr Dietrich in his publication in 1941 on the specimens from Laetolil is as follows: 'An Okapi-like creature with palaeotragine milk molars.' The diagnosis was accompanied by an illustration of the upper right deciduous molar and of specimens of the upper and lower dentitions and also one horn core. A description of the more important material was given. Dr and Mrs Leakey in 1935 also obtained *Okapia* material from this site.

NEW MATERIAL FROM OLDUVAI

Amongst fossils found in Bed I during the excavations in 1960 are specimens which are provisionally referable to *Okapia stillei.* They include material from sites FLK I and FLK N I. Specimen no. 637 appears to be a horn core of an adult giraffid of the *Okapi* type.

Subfamily: ***GIRAFFINAE***

Genus: ***Giraffa***

Species: ***gracilis*** Arambourg

DIAGNOSIS

Professor Arambourg (1947) described this species from the Omo fossil beds in the following terms (translation): 'A species of size comparable in the length of the limb bones and of its neck with *Giraffa camelopardalis* but with a skeleton more lightly constructed and with finer proportions in all its parts. Dentition smaller and mandible weaker than in *G. camelopardalis.* Third lower pre-molar with metaconid developed and internal wall continuous. Metapodials remarkable for the width of their articulation extremities.'

SYNTYPES

The three syntypes are specimens in the Paris Museum of Palaeontology.

ADDITIONAL MATERIAL

Material representing *Giraffa gracilis* occurs in lower Bed II and possibly in other beds, but it has not, as yet, been identified in Bed I. There is a nearly complete skeleton from site HWK at the base of Bed II and some other isolated specimens (pls. 33, 34, 35).

Species: ***jumae*** sp.nov.

DIAGNOSIS

A species of *Giraffa* which is generally more massive than the largest recorded specimen of *G. camelopardalis,* from which it also differs in the following characters: the ascending ramus of the mandible is wider and stouter, the corpus of the mandible is deeper and also longer, so that the overall picture is of different proportions; the teeth approximate to those of the largest known specimens in the living species; the anterior portion of the corpus of the mandible, from the premolars to the posterior region of the symphysis is inclined upwards towards the front and is then bent downwards to the incisor region (see pl. 36); in males the 'third horn', set on the mid-line near the upper limit of the nasal bones, is much less strongly developed than in adult male giraffe of the living species.

HOLOTYPE

A nearly complete skull and mandible as well as a large part of the post-cranial skeleton from Rawe. This specimen is in the British Museum of Natural History, catalogue no. M. 21466.

Note. The holotype is taken from Rawe because the most complete material of this new species is from that site. The same species is certainly represented at Olduvai Gorge in Beds II and IV.

HORIZON

The Rawe deposits as well as Beds II and IV are of Middle Pleistocene age.

DESCRIPTION OF TYPE

No detailed description of the type specimen will be given here, but the mandible is figured (pls. 36 and 37) with the corresponding mandible of a present-day male giraffe. The principal diagnostic difference between the species is clearly shown. A full description of the type will be given in the detailed study of the fossil giraffids of East Africa, in due course.

DISCUSSION

Various authors have listed the presence of a giraffe, comparable in size to the living species, in the Pleistocene deposits of East Africa. Dr Hopwood listed *Giraffa* cf. *capensis* as being present in Beds I, II and IV, but the specimen which was supposed to have come from Bed I was found only on the surface of Bed I. The whole of this material is probably referable to *G. jumae*.[1] Professor Arambourg listed *G.* cf. *camelopardalis* among the fossils of Omo, but the material was only fragmentary and was placed with *camelopardalis* because it was clearly different from the more common *gracilis* of the Omo beds. Dr Dietrich also listed *G. camelopardalis* from Laetolil, but he had only very fragmentary material. It now seems reasonably clear that the Lower and Middle Pleistocene species is not the same as the living form.

Subfamily: ***SIVATHERINAE***
Genus: ***Libytherium***
Species: ***olduvaiensis*** (Hopwood)

Dr Hopwood originally described the presence of a short-legged giraffid which he called *Helladotherium olduvaiensis*, but he subsequently revised his identification and called it *Sivatherium olduvaiensis*. The sivatherines of Africa have been discussed by Dr Dietrich, Professor Arambourg and Dr Singer and others. As a result, there was general agreement that the African representatives of the antlered giraffes all belonged to the genus *Sivatherium*, with the possible exception of a supposedly distinct genus, *Libytherium*.

Libytherium was described by Gaudry, and his type specimen was a mandibular fragment with only two pre-molars, which set it apart from all other giraffids. A recent re-examination of the original fossil has been made by Professor Arambourg who has now shown that the type specimen had been wrongly pieced together from fragments and that, in fact, it had a normal complement of pre-molars. Consequently, it is now considered necessary to treat all the African Pleistocene antlered giraffes as members of the genus *Libytherium*.

Fossil remains of a large-sized sivatherine, which must now be called *Libytherium olduvaiensis*, are certainly present in Beds I, II and IV of Olduvai. Much of this material has been described by Dr Singer and Dr Boné in a South African publication (1960).

[1] A new specimen of a very well preserved skull of a fossil giraffe from site EF–HR in upper Bed II fully confirms the setting up of the new species *jumae* for the Middle Pleistocene giraffes of East Africa. In the Rawe skull certain characters were thought to be possibly due to the crushing of that specimen. It is now apparent that they were not. The horn cores of the new skull extend back from the external rim of the orbits in a straight line instead of from a point on the frontals some distance behind the orbits.

CHAPTER IV

MAMMALIAN FAUNA: BOVIDAE

Fossil remains of Bovidae are exceedingly plentiful in the Olduvai deposits. There are two reasons for this. In the first place, Bovidae normally represent a high proportion of the total animal population of most African habitats other than the dense forest. Consequently, it is only to be expected that fossil remains of this group should outnumber those of any other. Secondly, the vast majority of the fossils which we find in the excavations come from living-floors or camp-sites of prehistoric man and represent the remains of his meals. Man apparently preferred the flesh of the Bovidae to that of many other groups, as he does throughout the world today.

Some of the Bovidae described in this chapter have been named previously; others are described here for the first time. In one or two cases, species have been included which are not represented in the collections to which I have had access in London, or in Nairobi. They were specifically listed as present in the Olduvai fauna by Dr Schwarz on the basis of Professor Reck's material.

The following is a list of the subfamilies, tribes, genera and species, which are dealt with in the pages which follow.

Subfamily: *BOVINAE*

Tribe: STREPSICEROTINI

1. *Strepsiceros grandis* sp.nov.
2. *S. maryanus* sp.nov.
3. *S. stromeri* (Schwarz)
4. *Tragelaphus* cf. *scriptus*

Tribe: TAUROTRAGINI

5. *Taurotragus arkelli* sp.nov.

Tribe: BOVINI

6. *Bularchus arok* Hopwood
7. *Gorgon olduvaiensis* sp.nov.
8. *G. semiticus* (Reck)

Subfamily: *CEPHALOPHINAE*

Tribe: CEPHALOPHINI

9. *Philantomba* cf. *monticola*

Subfamily: *HIPPOTRAGINAE*

Tribe: REDUNCINI

10. *Kobus* sp.indet.
11. *Kobus* sp.indet.
12. *Redunca* sp.indet.
13. *Reduncini* gen. et sp.indet.

Tribe: HIPPOTRAGINI

14. *Hippotragus niro* Hopwood
15. *H. gigas* sp.nov.
16. *H.* cf. *niger*
17. *H.* cf. *equinus*
18. *Oryx* sp.indet.

Tribe: ALCELAPHINI

19. *Damaliscus angusticornis* Schwarz
20. *D. antiquus* sp.nov.
21. *Parmularius altidens* Hopwood
22. *Parmularius* sp.indet.
23. *P. rugosus* sp.nov.
24. *Alcelaphus kattwinkeli* Schwarz
25. *A. howardi* sp.nov.
26. *Beatragus antiquus* sp.nov.
27. *Xenocephalus robustus* gen. et sp.nov.

Subfamily: *ANTELOPINAE*

Tribe: NEOTRAGINI

28. Cf. *Nesotragus* sp.indet.

Tribe: ANTELOPINI

29. *Gazella* cf. *praecursor* Schwarz
30. *Gazella* sp.indet.
31. *Gazella* sp.indet.
32. *Gazella* sp.indet.
33. *Gazella* sp.indet.
34. *Gazella* sp.indet.
35. *G.* cf. *granti*
36. *Gazella* sp.indet.
37. *Gazella* sp.indet.
38. *G.* cf. *wellsi*
39. *Gazella* sp.indet.
40. *Phenacotragus recki* Schwarz

Tribe: AEPYCEROTINI

41. *Aepyceros* sp.indet.

BOVIDAE incertae sedis

42. *Thaleroceros radiciformis* Reck
43. *Alcelaphini*? sp.indet. (*a*)

44. *Alcelaphini*? sp.indet. (*b*)
45. *Alcelaphini*? sp.indet. (*c*)
46. *Bovinae*?
47. *Strepsicerotini* sp.indet.?

Subfamily: *CAPRINAE*

Tribe: CAPRINI

48. *Pultiphagonides africanus* Hopwood
49. *Pelorovis oldowayensis* Reck
50. *Caprini* incertae sedis
51. *Caprini* gen. et sp.indet.
52. *Caprini* gen. et sp.indet.

Family: **BOVIDAE**

Subfamily: ***BOVINAE***

Tribe: **Strepsicerotini**

Genus: ***Strepsiceros***

INTRODUCTORY NOTE

There is some disagreement as to where the kudu, bongo, eland, nyala, sitatunga, bushbuck and harnessed antelope stand in the Bovidae, and of their relationship to each other. Professor Simpson, in his 'Classification of Mammals' (1945) places them all in the subfamily Bovinae. He allows only two living genera, *Strepsiceros* and *Taurotragus*, for this large group of animals. In the first of these he places the kudu, together with the nyala, sitatunga, bushbuck and harnessed antelope (in other words, those in which the females of the species are hornless), while he includes in the second group the eland and the bongo. I am not able to concur with him in this. Apart from the fact that the bongo females have horns, this species does not resemble the eland sufficiently to be placed in the same genus. I agree that the tribe Strepsicerotini should be divided into two genera which I would call *Strepsiceros* and *Tragelaphus*. If the bongo had to be retained within this tribe I would place it with *Tragelaphus* in spite of its horned females. It seems to be better, however, to remove eland and bongo from the Strepsicerotini and place them in a distinct tribe, Taurotragini. This would leave *Strepsiceros* and *Tragelaphus* in the Strepsicerotini; the first for the kudu, sitatunga and nyala, the second for the bushbuck and harnessed antelope.

This suggestion agrees to some extent with the views of both Dr Dietrich and Dr Schwarz, except that they would both make a distinct subfamily, Tragelaphinae, for this group. Dr Dietrich would also separate the sitatunga from the kudu and give it full generic status. Dr Schwarz would include the kudu with the bushbuck in the genus *Tragelaphus*. It is clearly time that a complete revision of the whole of this group was undertaken, paying particular attention to cranial characters, rather than to horns and teeth. In this study, we shall treat the kudu, the nyala and sitatunga as *Strepsiceros*, while the eland and bongo will be regarded as distinct genera in another tribe.

In the various published lists dealing with the fauna of Olduvai, the Strepsicerotini (in Simpson's meaning) are reputed to be represented by several genera and species, namely, *Strepsiceros strepsiceros* (the greater kudu), *Strepsiceros imberbis* (the lesser kudu), *Tragelaphus buxtoni* (the mountain nyala), *Tragelaphus scriptus* (the bushbuck), *Taurotragus oryx* (the eland), and *Tragelaphus spekii* (the sitatunga). This list cannot be fully justified, for some of the identifications were wrong, and others were based on inadequate fragments.

Species: ***grandis*** sp.nov.

DIAGNOSIS

A *Strepsiceros* of about the same size as the largest members of the living species, *Strepsiceros strepsiceros*, and considerably larger than the average. It is characterised by the following points which distinguish it from the living species. Between the base of the horn cores there is a transverse saddle-shaped depression some 60–70 mm. wide, and 50–60 mm. from back to front; the pedicles of the horn cores lie at the extreme edges of the frontal bones, above the orbits and, in this region, the cores have an angular and almost keeled cross-section instead of the rounded one of the greater kudu. In this character the new species resembles *Strepsiceros* (*Tragelaphus*) *buxtoni* to some extent. The cross-section of the horn cores near their base is roughly triangular instead of markedly oval as it is in *Strepsiceros strepsiceros*; posteriorly some 50 mm. from the base, the surface of the horn core, where it faces backwards towards the occipital condyles, is more or less flat, instead of markedly rounded. In this respect, too, the horn cores recall the mountain nyala. Behind the orbit, at the base of the horn cores, there are deep grooves which somewhat resemble those in the living species, *S. strepsiceros*, but which are much deeper and more strongly marked. The angle between the face and the top of the brain case is much less obtuse than in either the greater or lesser kudu or the nyala, and in this character it recalls the condition seen in the sitatunga. The vault of the skull, between the horn cores and the nuchal crest, is relatively shorter than in any of the living Strepsicerotini. The occipital region is relatively lower and wider than in the living kudu and nyala and is set more or less at right angles to the top of the cranium. The cranium has a constriction behind the horn cores and widens markedly towards the back in a manner not seen in any living species of the genus. The supra-orbital foramina resemble those of the kudu and differ greatly from those of the mountain nyala. The basi-occipital region resembles the condition to be seen in both *S. strepsiceros* and *S. imberbis* but differs in points of detail. It is quite unlike the condition of this region in the nyala, the sitatunga and the bongo. The upper molars are much more massive than the largest recorded in the living greater kudu. They are somewhat cracked and will not be described in detail.

HOLOTYPE

A large part of the skull with both horn cores nearly complete. The right upper dentition is preserved but is separate from the face. The specimen is registered in the British Museum of Natural History collection as no. M. 21461.

HORIZON

The specimen was excavated in 1931 from the upper part of Bed II at the site known as 'Reck's Man site'. The age of the specimen is, therefore, Middle Pleistocene.

REMARKS

The specimen was figured in *Stone Age Africa* as a fossil kudu. It was subsequently listed by Dr Hopwood as *Tragelaphus buxtoni* and exhibited under this name in the British Museum of Natural History. It is clearly a very large extinct kudu and not a mountain nyala.

DESCRIPTION OF HOLOTYPE

The type specimen (pls. 38 and 39) consists of the greater part of a brain case with both horn cores nearly complete, except for the tips. The whole of the facial region below the orbits is broken away, but parts are preserved separately. The forehead, below the supra-orbital foramina, is somewhat crushed, and the left orbital region, except for the lower rim, is missing. This region has been repaired with plaster of paris without any attempt to reconstruct the orbital margin and in such a way that the distance between the base of the horn core and the orbital region is too long. The horn cores project upwards and outwards and slightly forwards relative to the plane of the forehead. They then twist backwards, upwards and forwards, then outwards again and finally backwards once more. They thus have the form of an open cork-screw spiral. The horn cores exhibit a rugose crest at their base on the anterior face. This starts just above the supra-orbital foramina and develops into a well-defined keel, continuing right round the horn cores almost to their tips. The spiral twist of the horn cores results in this keel once more facing forwards at a point some 500 mm. in a straight line from the base of the horn cores. The surface of the horn cores, at the base where they face each other, is relatively flat; so is the posterior face and also the surface above the orbits. The horn cores thus have a roughly triangular cross-section, instead of the oval form to be seen in the greater kudu.

There is no true second keel, such as may be seen in the nyala, in which species the keel starts above and behind the orbits. There is, however, a slight angulation of this region which is reminiscent of the mountain nyala and this may perhaps be the reason why Dr Hopwood attributed the specimen to the mountain nyala rather than the kudu.

The basi-occipital region is very long and massive, with a wide shallow valley between the two elements. In general structure this region is similar to that of the greater and lesser kudu, although not in points of detail. It differs completely from the form to be seen in nyala, sitatunga and bongo.

The horn cores had been very considerably gnawed by rodents before fossilisation.

Measurements

	mm.
External width at the base of the horn cores	178
Minimum width on frontal below horn cores and above the orbits	161
Width at the orbits (estimated by doubling the right side)	186
Bilateral diameter of the horn cores at the base	67·5
Anterior–posterior diameter of the horn cores at the base	80
Distance from the mid-point between the horn cores and the *foramen magnum*	152
Height of the occipital region	80
Width of the occipital region	203
Distance from the base of the horn cores to the rim of the orbit (right side)	51
Width between the horn cores	80

ADDITIONAL MATERIAL

There are a few other specimens of horn cores in the British Museum of Natural History collection which may, provisionally, be regarded as representing *Strepsiceros grandis*. They are: a fragmentary horn core M. 14549, collected on the surface of Bed III in 1932, which was catalogued as *Tragelaphus buxtoni*; specimen M. 21469 collected from the surface of Bed IV at site CK in 1935, and another specimen M. 21470. These specimens are all more massive than the type specimen, suggesting that the species *grandis* may have been much larger than the present-day greater kudu.

DISCUSSION

Strepsiceros grandis is more closely allied to the kudu in the strict sense than to the nyala and the sitatunga, although exhibiting some characteristics which can be found in these species. This confirms the view that the nyala, sitatunga and kudu are fairly closely related. They certainly stand closer to each other than they do to the bongo and eland, both of which I consider to belong to distinct and separate genera of a different tribe.

Species: ***maryanus*** sp.nov.[1]

DIAGNOSIS

A *Strepsiceros* differing from all known living and extinct species, in the following characters.

The horn cores are set somewhat forward, so that when the skull is orientated with the top of the vault horizontal they rise vertically above the posterior half of the orbit and its back wall. The angle between the forehead and the vault of the skull is approximately 90 degrees. In all the other Strepsicerotini this angle is much more obtuse, but the living sitatunga and *Strepsiceros grandis* have certain resemblances to *maryanus* in this character. There is a deep depression lying behind the posterior rim of the orbits in the region of contact of the frontal, temporal, parietal and sphenoidal bones. This character is even more strongly developed in *S. grandis* but is only known, to a limited degree, in the living Strepsicerotini, although it is a little more pronounced in *S. strepsiceros* than in the others. In the occipital region the angle between the sagittal line and the nuchal region of the occiput is more obtuse than in any of the living Strepsicerotini, slightly resembling the sitatunga and lesser kudu. In this respect *S. maryanus* differs markedly from *S. grandis*. The vault of the skull extends backwards behind the base of the horn core relatively further than in the greater kudu or in *S. grandis*. This condition is similar to that seen in the lesser kudu and the nyala, although the whole relationship of the horn cores to the cranial vault is completely different in these species. There is a well-marked saddle-shaped depression between the horn cores. In this character *maryanus* resembles the nyala and *S. grandis*. The horn cores exhibit two keels in their upper half. The principal keel originates at the base of the horn cores above the supra-orbital foramina, as it does in the greater kudu and in *S. grandis*, but in no other member of the genus. The second keel can be seen in the upper region of the horn cores but it becomes very faint as it approaches their base. There are slight indications that it rises from a position above the posterior rim of the orbits. In the living greater kudu this second keel is poorly defined, whilst in *maryanus* it is more like that seen in the lesser kudu.

(It should be noted that in the Strepsicerotini, as a whole, there is frequently a *major* difference between the external appearance of the horn itself and the bony structure of the horn core and, in consequence, it is essential to compare a fossil core with the cores, i.e. with the horns removed.)

[1] The specific name is after my wife who has done so much work with me at Olduvai.

In *S. maryanus* the bilateral width at the base of the horn cores is considerably less than the maximum occipital width. The figures in the type specimen are 115 mm. and 154 mm. respectively. In this feature *S. maryanus* resembles the greater kudu and is markedly different from *S. grandis*. The cross-section at the base of the horn cores is narrow from side to side and is elongated anterior–posteriorly. In two specimens of *S. maryanus* the bilateral measurements are 55 mm. and 59 mm. respectively, while the anterior–posterior measurements are 75 mm. and 77 mm. In two random samples of living greater kudu the corresponding figures are 62 mm. and 58 mm.; 70 mm. and 66 mm. respectively. The indices, therefore, in the greater kudu are 88·8 and 87·5, while in *S. maryanus* they are 73·3 and 76. *S. grandis* is intermediate between the two, with an index of 84·3. The horn cores rise more vertically from their base in *S. maryanus* than in any living or extinct *Strepsiceros*. They then curve backwards, upwards and backwards again, with a well-defined corkscrew spiral twist.

The basi-occipital region is very similar indeed to that seen in *S. grandis* with the anterior element leading to the basi-sphenoid much less bent in relation to the main axis of the basi-cranium than in the greater kudu, but somewhat similar to the lesser kudu and sitatunga. There is a wide shallow valley on the posterior element, very much as in *S. grandis*, but quite different from that seen in any of the modern Strepsicerotini.

On the mid-line of the upper surface of the cranium there is a shallow depression behind the horn cores. A less marked but comparable depression can be seen in some, but not all, of the sitatunga.

HOLOTYPE

The holotype of *S. maryanus* is a well preserved cranium lacking the bullae and the face, but with both horn cores complete (pls. 40, 41 and 42). It was collected in 1959 at site HWK East at the base of Bed II.

HORIZON

The type specimen came from the base of Bed II but other specimens come from Bed I. The horizon is, therefore, the Upper Villafranchian.

DESCRIPTION OF TYPE

Since the lengthy diagnosis has been based on the type specimen, it is not necessary to describe it again in any detail, but the following notes are given.

The cranium, including the occipital condyles, the mastoid processes, the para-occipital processes and the basi-cranial region and the horn cores, are well preserved. On the left side, below the horn core, part of the upper rim of the orbit is preserved, but on the right side it is broken away. The left supra-orbital foramen is preserved. The bones of the forehead, between the supra-orbital foramina, are nearly at right angles to the top of the cranium. The brain case itself, as far as the parietals and temporals are concerned, is narrower immediately behind the horn cores than it is in the region of the mastoid process. It expands gradually backwards to the mastoid region without any constriction. The crests on the temporal and squamosal regions are strongly marked, but those on the anterior part of the parietal are weak. Apart from a very small depression (mentioned in the diagnosis) the top of the skull is relatively broad and flat. The nuchal crest is not nearly so rugged as in the greater kudu but is more like that in the lesser kudu. The whole occipital region is much wider and lower than the form seen in the lesser kudu, sitatunga or nyala. This feature has not been included in the diagnostic characters of the species, since it is shared by greater kudu and *S. grandis*.

The occipital condyles are more everted upwards than in either of the living kudu, and in this character too the fossil comes closer to some of the sitatunga. Although the bullae are missing, it is possible to infer from the shape of the basi-cranium that they were morphologically more like those of the lesser kudu and sitatunga than the greater kudu. In the latter, the bullae are set very low in relation to the occipital condyle and at a different angle from the rest of the Strepsicerotini. The measurements are set out in the table below.

Measurements

	mm.
Bilateral width of horn cores at base	138
Anterior–posterior diameter of the right horn core	75
Bilateral measurement of the right horn core	55

	mm.
Distance from the mid-point between the horn cores to the upper margin of the *foramen magnum*	132
Maximum width of the occipital region (on mastoid processes)	153
Height from the base of the *foramen magnum* to the highest point on the nuchal crest	90
Length of the horn cores in a straight line from the anterior point to the tip:	
As preserved	582
Estimated (approximate)	597
Length of the horn core on spiral as preserved (approximate)	748

DISCUSSION

In addition to the type specimen there are two other well-preserved crania of *S. maryanus*, both of which come from Bed I, together with many other fragments. One specimen, from site FLK NN I has associated bones and teeth, but a description of these must wait for the much more detailed report on the Bovidae. The other is a well-preserved cranium with horn cores, which are broken about half-way. Both these conform well with the type.

The first point to be noted is that if only fragments of horn core had been available *S. maryanus* might have been regarded as similar to the greater kudu. This emphasises that horn cores alone cannot be used to determine genera and species in the Bovidae, since *S. maryanus* clearly differs from *S. grandis* found in upper Bed II. There are, however, characters in *S. maryanus* which suggest that it may have been ancestral to *S. grandis* and to the living members of the genus. The striking differences between the two extinct species can be seen in pls. 38–42.

The living representatives of the genus *Strepsiceros* (in the sense in which I have used the term, confining it to the kudu, sitatunga and nyala), seem to represent different evolutionary developments. *S. maryanus*, from the Upper Villafranchian deposits, reveals some characters which are like those of the living kudu, others recalling nyala and many like the sitatunga.

The sitatunga appear to be more variable in their general characters than any other members of the Strepsicerotini. This may be due to the fact that they are now confined to limited swampy zones. Such isolation is accompanied by a great degree of inbreeding and may be the cause of the great differences seen in some major characters. It may be that a fuller study of the sitatunga will necessitate dividing them into two or more full species instead of the present subspecies. We know that the mountain nyala became so isolated from the bush nyala that the differences between these two justify full specific rank. It is possible that *S. maryanus*, with a combination of characters now seen in the nyala, sitatunga and the kudu, may represent a stage close to an ancestral stock of this group. *S. grandis*, higher up in Bed II, is already a true kudu.

Species: ***stromeri*** (Schwarz)

= *Tragelaphus spekii stromeri* Schwarz

= *Limnotragus stromeri* Dietrich

Dr Schwarz recorded the presence of a sitatunga in the Olduvai fauna, in the material collected by Professor Reck in 1913. He did not illustrate it, and made it only a subspecies of the living sitatunga.

This specimen probably came from Bed IV and it is possible that it represents a form allied to the living species; it is therefore included in the Olduvai lists for the time being.

Genus: ***Tragelaphus***

Species: cf. ***scriptus***

Dr Schwarz recorded the presence of *T. scriptus* in Reck's collection on the basis of fragmentary material. There are some other specimens which also suggest this genus and perhaps this species, but exact identification is not possible.

Tribe: **Taurotragini** new tribe

Genus: ***Taurotragus*** Wagner

DIAGNOSIS OF GENUS[1]

Wagner's original diagnosis of the genus *Taurotragus* has not been seen, but revised diagnoses have been given from time to time by various authors, including Dr Pilgrim.

[1] As we have already seen, Professor Gaylord Simpson, in his 'Classification of Mammals' (1945), treats *Boocercus*, the bongo, as a synonym of *Taurotragus*. This view is not accepted in this report. The genus *Taurotragus* is used here in its original definition, for the eland only.

Species: ***arkelli*** sp.nov.

The trivial name is after Dr Anthony Arkell, who found the type specimen whilst visiting Olduvai in 1941. I am greatly indebted to Mrs S. C. Coryndon for much valuable initial work on this species.

DIAGNOSIS

A species of *Taurotragus* differing from *T. derbianus* and *T. oryx* in the following principal characters.

The cranium is much more elongate behind the base of the horn cores than in either of the living species. The upper surface, between the temporal crests, is very wide and relatively flat, somewhat as in the bongo. The nuchal region of the occipital is set at an angle of more than 90 degrees to the top of the skull, whereas in both the living species of eland it is a right angle or less. In this, too, the fossil shows similarities to the bongo. The bases of the horn cores are very wide apart, so that the width across the frontals is very nearly as great as that across the occipital. In the living species of eland the occipital width far exceeds the frontal width. In the region behind the posterior rim of the orbit on the frontal bone, in front of the anterior squamosal suture, there is a rugose and deeply incised valley. This is unlike anything seen in the modern eland, but is present to a limited degree in the bongo. There is no raised boss between the horn cores separating the face from the top of the cranium, such as is present in both the modern eland and the bongo. The angle between the forehead and the top of the brain case is less obtuse than in the living eland. It is somewhat similar to that seen in bongo. The whole skull is generally massive and muscular although the horn cores are noticeably shorter than those on the skulls of *T. oryx* or *T. derbianus* of comparable size. There are two well-marked bony keels on the horn cores. The first of these starts near the mid-line of the forehead and is very strongly marked. The second begins as a definite keel behind the posterior rim of the orbit, just above the deep valley referred to above. In this, the new fossil species *arkelli* resembles *T. derbianus*. The second keel remains strongly marked throughout the entire length of the horn cores to the extreme tip, whereas in *T. derbianus* the upper part of the horn cores have no trace of a second keel.

HOLOTYPE

A cranium with one horn core complete and the other broken off half-way. The face and maxillae are missing from a point near the upper margin of the nasal bones (pls. 43 and 44). No. F. 3665.

HORIZON

Bed IV, Olduvai, upper part of the Middle Pleistocene.

DESCRIPTION OF TYPE

Although only the posterior part of the skull is preserved, a considerable amount of detail can be seen and there is no apparent distortion. Comparisons have been made with skulls of *T. oryx* in Nairobi, as well as with skulls of *T. oryx* and *T. derbianus* in the British Museum of Natural History.

The skull is generally similar in size to that of *T. oryx* (pl. 45), but the proportions are quite different and the skull is markedly longer than in any members of the two living species that have been examined. The greatest difference in general appearance is in the angles between the frontal, parietal and occipital bones. The supra-occipital ridge is comparatively wider in the fossil than in modern eland, with the 'waist' at the squamosal occipital suture not so sharply indented. The para-occipital processes are smaller and more slender. Although the left process is missing and the right one damaged, they appear to be set closer to the skull than in the modern species. The occipital condyles are set wider apart at their superior margin, and the posterior edge projects further back. There is a marked wide V-shaped area of rugosity on the mid-line of the supra-occipital, and a slender ridge, with a shallow groove on either side, running from the supra-occipital to the *foramen magnum*. These conditions are found in some modern eland skulls.

In profile, the horn cores are set in a line with the face, much as in the modern eland, but there is a very marked difference in the proportions and angles of the bones posterior to the horn cores. It is in this region, more than any other, that the differences between the living and fossil species become clear. In the fossil the length of the cranium posterior to the horn cores is nearly twice that seen in the modern skulls. The angle between the horn cores and the vault of the parietal is about 45 degrees, whereas in modern species this is about 20–25 degrees.

The orbital rims are incomplete, as are those parts of the temporals which form the posterior portions of the zygomatic arches. The squamous bone itself shows several points of difference from that seen in modern skulls. The bullae, though damaged, appear not to have been very inflated. The glenoid fossae are broken laterally, but appear to be comparatively longer anterior–posteriorly than in modern eland. They are situated more on the main body of the squamous bone than on its lateral projection. The post-glenoid foramen does not follow the posterior border of the whole of the glenoid fossa, but crosses it on its lateral edge, giving the fossa an extra flat projection on the posterior–lateral border. Only the superior borders of the orbits are preserved, and the region of the frontal bone lying between them rises more vertically towards the horn cores than in the living species.

The main and secondary keels of the horn cores arise from a similar position in both the fossil and the modern skulls, but the latter starts as a sharp keel, as it does in *T. derbianus*. In *T. arkelli*, these keels ascend more sharply and are distinct right to the tip of the core. There is a well-marked valley below and behind the horn cores, somewhat as seen in *Boocercus*. The ridge between the frontal and squamosal bones is not so well marked in *Taurotragus arkelli* as in the living species.

The facial aspect shows flattened frontal bones with a slight concavity between the supra-orbital foramina. A slight ridge in the mid-line extends from the level of the base of the horn cores to the nasals, as in *T. oryx*. There is no raised ridge between the horn cores such as is seen in the living species, but the horn cores are spaced at roughly the same distance and rise at the same angle as in modern eland skulls (when viewed from the front).

Only the upper portion of the nasal bones is preserved, where they are in contact with the frontals. The lachrymals are not clearly defined and only a small portion is preserved. They appear to form a smaller and less important part of the orbital rim than in the living species. The surface of the bone on the anterior part of the frontals, on either side of the lower part of the median ridge, is slightly roughened and striated. The parietals are elongated and are not so domed as in the modern skulls.

The basi-occipital is comparatively wider and thicker and does not project below the body of the occipital as far as in *T. oryx*. The anterior elements are not so vertical as in the modern skulls, and indeed are nearly horizontal.

Measurements of the type specimen	mm.
External width at base of the horn cores	161
Lateral diameter of the left horn core at base	62
Anterior–posterior diameter of the left horn core at base	75
Length of the cranium from the highest point between the horn cores to the upper margin of the *foramen magnum*	158
Distance from the base of the left horn core to the nuchal crest	69
Width of the occipital region on the para-occipitals	178
Height from the base of the *foramen magnum* to the top of the nuchal crest	111

COMMENTS

We have already seen that Professor Simpson, in his 'Classification of Mammals' (1945), placed the eland and bongo with the kudu, nyala, sitatunga, bushbuck and harnessed antelope, calling them all Strepsicerotini.

Professor Simpson's reasoning seems to be that the eland and bongo should be included in the Strepsicerotini because of certain cranial characters, as well as their spirally twisted, keeled horns and the fact that they have a pelage which is to some extent striped. His reason for separating them generically from his first group within this tribe is because in both bongo and eland the females carry horns, whereas in the other group the females are hornless. I feel that Professor Simpson was unduly influenced by the spirally twisted horns and the striped coats of these two groups, and I am unable to agree with him.

Professor Arambourg's *Tragelaphus nakuae*, which I would prefer to call *Boocercus nakuae*, shares many characters with the bongo, but differs markedly from it in the conformation of the cranium. The top of the cranium, the structure of the nuchal region and the relationship of the occipitals to the parietals are strikingly different. *T. nakuae* is wholly different from the eland in the structure of its horn cores and in certain other characters. It has a cranial vault which is in many respects much more like that of the eland, particularly the conformation on top of the brain case and of the occipital region. *T. arkelli* is essentially like the eland in the structure of its horn cores and in certain other features, but it also shares a large number of characters with the bongo.

The differences between the group which includes the bongo, the eland and *T. nakuae*, and that which includes the kudu, nyala, sitatunga, bushbuck and harnessed antelope, are so great that I feel it is essential to separate them into two distinct tribes of the subfamily Bovinae.

The living eland and bongo seem to me to be distinct and separate genera, linked to each other through the two fossil forms.

Tribe: **Bovini**

Genus: ***Bularchus*** Hopwood

Species: ***arok*** Hopwood

In 1936 Dr Hopwood described a new genus of the Bovini, which he called *Bularchus*. His generic diagnosis reads as follows: 'Bovidae of large size with massive horn cores, compressed from front to back, oval in cross-section, closely approximated at their bases, curving crescentwise upwards, outwards and downwards. So far as is known the horns are in the same plane as the face and are not spirally twisted.'

He took as type specimen for the genus and species a frontlet with two incomplete horn cores and a distorted brain case (pl. 46), no. M. 14947 in the British Museum of Natural History. He designated a fragment of left horn core and base of a skull (pl. 47), no. M. 14948, as paratype. The type specimen is in bad condition, having been cracked and crushed prior to fossilisation. The second specimen, though less complete, is in a slightly better state of preservation. Both specimens came from Bed II at Olduvai and were collected in 1935.

The excavations at site BK II during the 1952–8 seasons yielded a considerable quantity of material, including horn cores, skulls, maxillae and limb bones, which seem to represent *Bularchus*. One example of what is provisionally identified as *Bularchus* from BK II is seen in pl. 48 where it is contrasted with a modern ox skull.

Genus: ***Gorgon***[1]

Species: ***olduvaiensis*** sp.nov.

HOLOTYPE

An incomplete skull, with the frontals, parietals and part of the occipital preserved. No. M. 21451 British Museum of Natural History. The greater part of the right horn core is intact; the left horn core is broken off near the base (pls. 49 and 50).

HORIZON

This specimen was collected in 1935 and came from the junction of Beds III and IV, the age being Middle Pleistocene.

DIAGNOSIS

A species of the genus *Gorgon*, differing markedly from *Gorgon taurinus* in the following characters. The width between the horn cores is relatively much greater than the maximum width of the brain case; the width of the skull on the frontals, in front of the horn cores, is very great relative to the frontal width; the forehead between the orbits is domed and convex instead of flat or concave; the bone at the base of the pedicle projects laterally beyond the external rim of the orbits.

DESCRIPTION OF TYPE

The type specimen exhibits the basic characters of the new species. The convex frontal region between the orbits is bounded, on either side, by complex multiple supra-orbital foramina similar to those of *G. taurinus*. In the fossil, however, they are set further away from the orbital rim than in the living species (pl. 51). The bony pedicle of the horn core is smaller than in the living wildebeest; the anterior rim of the horn core rises from the forehead without a marked step. The distance between the inner rims of the horn cores has a minimum width of 100 mm. This is similar to the width on

[1] This genus is regarded as a member of the Bovini and not of the Alcelaphini.

a female of *Gorgon taurinus*. On a male, the measurement is 70 mm. It seems probable from this and from the general conformation, that it represents a female. This is confirmed by a second specimen which is larger and more massive. In the female of the living species, the greater width across the forehead is linked with less massive horn cores which are set upon more developed pedicles than in the males. No such pedicles exist in the fossil, although it is presumably a female. The ratio of the distance between the horn cores to that across the narrowest part of the forehead in front of the horn cores is 100 mm. to 171 mm., whereas in *G. taurinus* females it averages 100 mm. to 143 mm.

While certain characters in the frontal region of the skull and in the horn cores suggest that the type specimen may be a female, the relationship of the parietals and frontals to the nuchal crest is of the form only normally to be seen in adult males of the living species. The area of attachment for the temporal muscles lies in a deep valley beneath the base of the horn cores, in a manner typical of members of the Bovini tribe. In the new species, however, this narrow valley is deeper than in any male or female individuals of the living species.

A sufficiently large part of the brain case is preserved to make it possible to measure the maximum width behind the horn cores. It is 108 mm. compared to 112 mm. on both male and female examples of *G. taurinus*. In other words, although the fossil has a longer face and wider forehead, it has a smaller brain case than the living species.

The right horn core is preserved for about half its length and shows that the horn was everted sideways and slightly backwards and then forwards, as in present-day *G. taurinus*. The bilateral measurement of the base is 76 mm. and the anterior–posterior diameter is 60 mm. The measurements at the point at which the horn core begins to turn upwards are 40 mm. and 30 mm. respectively. There is no trace of ribbing.

ADDITIONAL MATERIAL

In the collections at the British Museum of Natural History there is a large fragment of a right frontal, with part of the right parietal and the right occipital preserved (pl. 52). No. M. 21452. It probably represents a male of *G. olduvaiensis*. The horn core is broken off near the tip and the preserved part is more massive and more oval in section than in the type specimen. On the basis of comparison with the living species this confirms that the holotype is a female and the second specimen a male. Apart from its larger size, this second specimen agrees closely with the type. At the base of the horn core the bilateral measurement is 94 mm. and the anterior–posterior is 65 mm. The corresponding measurements, where the horn begins to turn upwards, are 41 mm. and 37 mm.

There is one other specimen in the British Museum of Natural History collections which can be attributed to *G. olduvaiensis*. It is a left frontal with part of the horn core preserved, collected by Dr Hopwood in 1931. It was found on the surface of Bed I. The frontal suture at the base of the horn core is still filled with matrix which resembles material from the upper level of Bed I.[1]

G. olduvaiensis is represented in the Nairobi collections by a well-preserved horn core from site BK II from upper Bed II.

Species: ***semiticus*** (Reck)

In 1935 Professor Reck described a fossil from Olduvai under the name of *Rhynotragus semiticus*. This was subsequently transferred to the genus *Gorgon* by Dr Schwarz (1937) and regarded as a race of *G. taurinus*. No comparable specimen is known in the collections either at the British Museum of Natural History or at Nairobi. *G. semiticus* is provisionally included in the list of Olduvai Bovidae on Schwarz's identification.

DISCUSSION OF GENUS

In 1932 Professor Reck wrote a popular book on Olduvai. He included in it a description of two new members of the family Bovidae, one of which he called *Rhynotragus semiticus* and the other *Thaleroceros radiciformis*. The scientific validity of these names was questioned because they were published in a popular book. He therefore re-described both genera in 1935, giving a new diagnosis and illustrations of each. Subsequently,

[1] Since this was written, a skull which seems to represent *G. olduvaiensis* has been found *in situ* in the top of Bed I.

various authors treated the genus *Rhynotragus* as synonymous with *Gorgon*, although Professor Reck's illustration would not seem to support this. In 1937 Dr Schwarz listed Professor Reck's new genus and species as nothing more than a subspecies of *G. taurinus*.

Dr Dietrich has described a new genus *Parestigorgon*, which Dr Hopwood subsequently treated as a synonym of *Gorgon*. This is difficult to accept in view of the illustration of the horn cores of *Parestigorgon* which resemble those of *Damaliscus angusticornis*. *Parestigorgon* is completely different from the species *G. olduvaiensis* described above. Dr Dietrich has also referred to '*Connochaetes taurinus major*', but this identification was based entirely upon a few teeth. It is possible that this subspecies (which, in any event, should be *Gorgon taurinus major*) may prove to be *G. olduvaiensis*.

Subfamily: ***CEPHALOPHINAE***

Tribe: **Cephalophini**

Genus: ***Philantomba***

Species: cf. ***monticola***

Although no material representing this genus has been identified in the collections at the British Museum of Natural History, or those in Nairobi, Dr Schwarz (1937) recorded the presence of fossil remains of this genus in the collections from Professor Reck's 1913 expedition. It is therefore included in the list of Olduvai Bovidae.

Subfamily: ***HIPPOTRAGINAE***

Tribe: **Reduncini**

INTRODUCTION

The scarcity of fossils representing the Reduncini in the collections from Middle and Lower Pleistocene deposits in East Africa is remarkable. Dr Schwarz originally described (1937) the specimen which is now known as *Phenacotragus recki* as *Adenota recki*. It has since been taken out of the Reduncini and placed with the gazelles. Dr Hopwood and Dr Schwarz both listed the presence of reduncines of indeterminate affinities at Olduvai, but I was unable to find any such specimens in the collections at the British Museum of Natural History. Professor Arambourg has recorded a new species of *Redunca*, as well as a *Kobus*, at Omo. His *Kobus sigmoidalis* appears to be founded on inadequate material, but is probably a large waterbuck, resembling *defassa* of the present day. His *Redunca*, which he made into a new species, *ancystocera*, is based upon two horn cores. He lists the supposed differences between the fossil species and the living, but I am not convinced of the validity of the fossil species, since the Reduncini are notable for the variation in the size, shape and conformation of the horns. Dr Dietrich has listed material of a supposed *Redunca* from Laetolil, but the material is too fragmentary for a positive identification.

Genus: ***Kobus***

Species: indet. (*a*)

There is a frontlet with parts of both horn cores, no. P.P.R. 1, which was excavated from site BK II East in 1953. It has a number of characters which strongly suggest that it belongs with the genus *Kobus*, but it does not closely resemble any of the living species.

Species: indet. (*b*)

Another specimen, no. 579, 1957, found on the surface of SHK II is certainly not of the same species as that from BK II East. It, too, probably represents the genus *Kobus*. It consists of a right horn core with the upper part missing, but with part of the frontal and parietal bone preserved.

Genus: ***Redunca***

Species: indet.

In 1950, in a review of the fauna from the Lake Eyasi sites, Dr Dietrich listed a *Redunca* sp.indet. as being present and stated that it was also present in the Olduvai fauna collected by Professor Reck. It is provisionally included in the Olduvai fauna as there are some other broken horn cores recalling this genus which have been found in Bed IV.

Genus: indet.

Species: indet.

A frontlet (pl. 53) with the base of both horn cores and the upper rims of both orbits preserved is unquestionably a member of the Reduncini. There are, however, clear differences from *Redunca*, *Adenota* and *Kobus*. I have not been able to compare it with *Pelea* and it is not possible to assign it to any genus or species at this stage. Among the characters to be noted are: the pedicles of the horn cores are set close together; the base of the horn core is flattened laterally with a greater anterior–posterior than bilateral measurement; there is a ridge following the line of the inter-frontal suture between the horn cores, of the type frequently seen in the Reduncini. The orbits are set very high on the frontals and their superior rims are everted nearly at right angles to the horn cores. Behind the base of the horn cores, and posterior to the rim of the orbits, there is no typical reduncine pit. The structure in this area has no parallel in any member of the tribe.

Tribe: **Hippotragini**

Genus: ***Hippotragus***

Species: ***niro*** Hopwood

DIAGNOSIS

Dr Hopwood's diagnosis, published in 1936, reads as follows:

A *Hippotragus* with horn cores of the same type as those of *Hippotragus niger* but with well-developed knots on the anterior surface at intervals of about 70 mm. Knots continued on the lateral and medial surfaces as indistinct rings.

The holotype is a broken right horn core from an immature animal. British Museum of Natural History no. M. 14561.

HORIZON

The specimen was collected on the surface of Bed IV and cannot be older, and may, therefore, be placed in the upper part of the Middle Pleistocene.

NOTES

The specimen (pl. 54) was found by Dr Hopwood in 1931 and no site number is given. In his remarks he states, 'This species agrees with *Hippotragus niger* and differs from *Hippotragus equinus* in that the medial surface of the horn core is flattened so that the cross section is more nearly D-shaped than regularly oval'.

Examination of specimens of the living species, *Hippotragus niger*, shows that in that species, although the horn core is more flattened on one side than on the other, it is the lateral surface and not the medial face which is most flattened, while in the fossil *H. niro* it is emphatically the medial face which is most flattened while the lateral one remains curved. This character, together with the clear marks of the cross-ribbing (what Dr Hopwood called 'knots') serves to distinguish his species, *niro*, from the living species *niger*. It also differs from *H. equinus*, the roan antelope. This species is distinguished both from *niro* and *niger* in having horn cores which are more or less oval in cross-section throughout their length.

Dr Hopwood's type specimen exhibits a shallow, ill-defined valley rather like that seen in the korrigum hartebeest, but less clearly marked, lying between the rim of the orbit and the base of the horn core. This character alone would not justify taking Dr Hopwood's specimen out of the genus *Hippotragus* and transferring it to *Damaliscus*, although this may become necessary in view of the ribbing of the cores. This never seems to occur in living members of the genus *Hippotragus* but is often seen in the alcelaphines. It is possible that when *Hippotragus niro* is better known it may be found to stand as an intermediate form between *Damaliscus* and *Hippotragus*.

Dr Hopwood regarded his type specimen as representing a juvenile, but there is little evidence to support this view. The horn core, although incomplete, is clearly much shorter relative to its thickness than in any modern *Hippotragus* species, but this does not imply that the specimen is juvenile. In this character as well the fossil resembles *Damaliscus*.

The horn core rises from a point very close to the frontal suture, a character reminiscent of some of the fossil Alcelaphini such as *Damaliscus angusticornis* and *Parmularius altidens*. The type specimen has a bilateral width at the base of 36·5 mm., and an anterior–posterior diameter of the horn core of 58 mm. The length from the base (as preserved) is 233 mm. An additional 30 mm. or 40 mm. may have been broken off. The spacing of the ribbing is far wider than in the horn cores of any known living African antelope. This character seemed, at first, to be suggestive of the Ethiopian ibex and consequently the horn cores of this genus were examined. It was found, however, that although the horns themselves are markedly ribbed at wide intervals, the ribbing is never present on the cores of this genus.

ADDITIONAL MATERIAL

In the collections in Nairobi there are five horn cores which unquestionably represent *Hippotragus niro* (pl. 55). These horn cores exhibit the same traces of transverse cross-ribbing and there can be no doubt that this is characteristic of the species. They also exhibit a D-shaped cross-section in the upper half of the horn core, with the flat surface on the medial face and not, as in *H. niger*, on the lateral face. The new material makes it clear that, although the horn cores of *H. niro* are fully as massive, at their bases, as those of a very large male *H. niger*, the total length was considerably shorter than in any living race of the sable antelope.

In the discussion of the type specimen, I expressed doubt as to whether it should be placed in the genus *Hippotragus*. One of the new specimens, no. 53/282, fortunately has a large part of the frontal bone preserved, which is lacking in the type specimen. This includes the inter-frontal suture, the right upper rim of the orbit, the anterior part of the frontal–parietal suture and the supra-orbital foramen. This specimen leaves little doubt that we are dealing with a genuine member of the genus *Hippotragus*, although a very aberrant one.

In this specimen the horn cores are set much closer to the inter-frontal suture than in *H. niger* or *H. equinus*, so that there can have been little space between them for the horn integuments. In the region behind the orbit there is a groove on the frontal bone at the base of the horn core. This is more pronounced posteriorly than anteriorly and is set much in the same direction as in *H. niger*. It is quite unlike the corresponding groove in *H. equinus*. The position and shape of the supra-orbital foramina are characteristic of the genus *Hippotragus*. The ribbing on the horn cores, as in the type specimen, is widely spaced. There is the characteristic hippotragine L-shaped groove for a blood vessel, leading from a point on the lateral side, beneath the horn core, to the post-orbital groove.

All the new specimens were found *in situ* in upper Bed II, while the original type specimen was from Bed IV.

Species: ***gigas*** sp.nov.

DIAGNOSIS

A *Hippotragus* of gigantic proportions, with characters recalling both *H. equinus* and *H. niger*. In addition to its greater size, it differs from both in the following points.

The nuchal region of the occipital is set at a more obtuse angle to the top of the cranium than in either of the living species, in which this angle is nearer to a right angle; the para-occipital processes are much shorter and wider in proportion to the size of the skull than in either of the living species; the basi-occipital, in its general morphology, is similar to that of *H. niger*, as far as the posterior and middle part of the element is concerned, but is relatively wide and shallow near the *foramen magnum*; the bending of the basi-occipital region takes place further back than in any member of the Bovidae that has been examined. The bones of the cranium are very thick: in this character the fossil resembles *H. equinus* rather than *H. niger*. As in the other Hippotragini, the horn cores of the female are shorter and less curved than the male. They are as massive as in a large male of *H. equinus* (pl. 57). The horn cores of the male carry a well-marked external, lateral, shallow groove on the upper middle third and another deeper groove on the internal surface towards the tip of the core.

TYPE

An incomplete male cranium (pls. 56 and 58) with the left horn core preserved intact and the right horn core nearly complete. No. P.P.T. 2. Part of the vault, most of the basal region and the face are missing.

PARATYPE

A female skull (pls. 59 and 60) with a well-preserved cranium which is damaged in the region of the bullae. The basi-occipital and the face are missing. No. P.P.T. 3.

HORIZON

The type specimen comes from Bed II. The paratype is from site VEK at the top of Bed I. The horizon therefore includes the upper part of the Villafranchian.

DESCRIPTION OF TYPE

The type is a large and apparently male skull, very considerably damaged. Unfortunately, the upper margins of the orbits are missing, but it is clear from the preserved parts that the orbits were set vertically beneath the horn core when the skull is so orientated that the upper margin of the vault is horizontal. The form of the supra-orbital foramen preserved on the left side is much as in *H. equinus*. The cranium is very elongate. This feature is characteristic of the whole *Hippotragus* genus, but it is more exaggerated in *gigas* than in either of the living species. The measurements of the type are given below with those of the paratype.

DESCRIPTION OF PARATYPE

A cranium, believed to be female, in which both horn cores are preserved intact. It has a rather larger part of the cranium preserved than in the type. The horn cores are shorter and set more vertically on the cranium and are less curved. This is a normal sexual difference in members of the genus *Hippotragus*.

Measurements

	Type male (mm.)	Paratype female (mm.)
Length of the cranium from point on mid-line between the horn cores to the upper margin of the *foramen magnum*	228	185
Length from mid-point between the horn cores to posterior margin of the nuchal crest	163	134
Lateral width at the base of the horn cores	155	156
Diameter of the left horn core at the base:		
Lateral	72	52
Anterior–posterior	86	66
Occipital width on mastoids	147	—
Height from the base of the *foramen magnum* to the nuchal crest	107	—
Length of the left horn core:		
Straight	578	360
Curve	640	368

These two specimens represent a very large extinct species of *Hippotragus* exhibiting certain characters which are found in one or other of the living species.

ADDITIONAL MATERIAL

There is another specimen (pl. 61) undoubtedly representing a male *H. gigas*, in the collection in the British Museum of Natural History. It was collected in 1931 by Dr Hopwood from the surface of Bed II. The specimen consists of a large part of a very massive right horn core, with the tip missing. Most of frontal region, including the pedicle is preserved. The core fragment is 367 mm. long in a straight line. At its base the bilateral diameter is 70 mm., and the anterior–posterior diameter 84 mm. At the point of fracture the bilateral diameter is 46 mm. and the anterior–posterior diameter 64 mm. The specimen is catalogued as no. M. 21448 at the British Museum of Natural History.

In the Nairobi collections there is a part of a very large horn core, no. F. 992, 1941, from the surface of Bed IV; it probably represents a male of this species. Two other horn cores, each with a part of the frontal bone preserved, are presumably females of the same species. They are only two-thirds the size of the male specimen and closely resemble the female paratype. These two specimens are registered as no. 281, 1953, from site SHK II, and no. P.P.R. 5, from site BK II East, also collected in 1953.

Species: cf. ***niger***

Dr Hopwood listed *H. niger* as present in Bed II. The collections in the British Museum of Natural History have been examined for material representing this living species. There are two fragments of horn cores—one, no. M. 14564, from Bed IV, and the other, no. M. 21450, from the surface of Bed II —which suggest *H. niger*, but the material is too incomplete to warrant positive identification. It is equally possible that they represent an extinct species whose horn cores are similar to those of *niger*.

Two fragments of horn cores in the Nairobi collections represent a similar animal. One, no. F. 971, was found in 1941 on the surface of Bed IV. It consists of the upper third of the horn core with the tip missing. The other, no. P.P.R. 2, is from site BK II and was found in 1953. It is slightly larger and represents the same part of the horn core as the previous specimen. It had been extensively gnawed by a large rodent prior to fossilisation.

Species: cf. ***equinus***

Dr Hopwood listed *H. equinus* as present in Bed II. An examination of the material available at the British Museum of Natural History reveals two incomplete frontals, each with broken horn cores, which are apparently close to *equinus*. Neither of the specimens is sufficiently diagnostic to state that *equinus* is present at Olduvai. The first specimen is no. M. 14530 from the surface of Bed I. The matrix adhering to it suggests it may have come from Bed II. The second specimen is no. M. 21449 from the surface of Bed II. Both were collected in 1932. This second specimen may represent a female of *H. gigas* rather than a male of *equinus*.

In the Nairobi collections there is a specimen, no. 365, excavated in 1957 from site BK II. It consists of the right frontlet and the greater part of a horn core of a *Hippotragus* which is very similar to the living *H. equinus*. This specimen and those in the British Museum of Natural History confirm the view that the fauna of Bed II includes an antelope closely allied to the present-day roan. This was contemporary with *H.* cf. *niger*, *H. niro* and *H. gigas*.

DISCUSSION

The available material indicates that members of the genus *Hippotragus* were strongly represented at Olduvai by four well-defined species which were clearly contemporary and co-existent. Two of these, *H. gigas* and *H. niro*, are clearly extinct. The other two may be the same as the living species. The genus *Hippotragus* has long been regarded as standing close to the Alcelaphini and some of the characters of *H. niro* seem to present an important connecting link between the two groups.

Genus: ***Oryx***

Species: Indet.

There is a specimen, no. G. 390 (pl. 62), from site FLK I, which appears to be referable to the genus *Oryx*, but it is more massive than any living species of *Oryx*. The specimen consists of a left frontlet with a complete horn core. It exhibits features in the frontal bone which are characteristic of *Oryx*. The horn core tapers very rapidly, with a bilateral measurement at the base of 43 mm. and an anterior–posterior measurement of 48 mm. Approximately 200 mm. from the base, the measurements are 21 mm. bilaterally and 25 mm. anterior–posteriorly. The total length is 380 mm. This is very short compared with the figures for the much less massive horn cores of *Oryx beisa*.

Tribe: **Alcelaphini**

NOTE

In all the living species of the Alcelaphini the skulls exhibit an elongate groove behind the orbits. In form and in detail it differs in each genus and species of the tribe, but it is clearly a constant character in both sexes of all the known living alcelaphines. The difference in the size and shape of the groove in the various living species will often serve to distinguish one from another without reference even to the shape of the horn cores. For example, in *D. korrigum* the groove is of about the same size and shape as the bowl of a narrow teaspoon whereas in *D. albifrons* it is shallower, narrower and more elongate.

Grooves of this type do not occur in any living African bovidae except the alcelaphines. They are therefore regarded as an important character in deciding whether or not fossil bovidae should be placed in this tribe. In the descriptions which follow, this feature will be referred to as the alcelaphine groove.

Genus: ***Damaliscus***

Species: ***angusticornis*** Schwarz 1937

INTRODUCTION

In 1937 Dr Schwarz published a study of the fossil antelopes from Olduvai. He described a number of specimens, mostly from the material in Berlin and Munich collected by Professor Reck in 1913. He also made occasional references to the British Museum of Natural History material collected from 1931 to 1935.

In this report he described a new species of the genus *Damaliscus* under the name of *angusticornis*. He took as his type a very crushed specimen in the Berlin Museum. His paratype was part of a frontal with a fragmentary horn core, no. M. 14553, which was in the British Museum of Natural History (pl. 63).

DIAGNOSIS

Dr Schwarz's diagnosis, broadly translated, reads as follows:

Differs from *Damaliscus korrigum* essentially in having the horn cores set much closer together at the base. The distance separating the cores at the base in the type specimen is barely 20 mm. compared with 30–35 mm. in recent species. The type specimen is very crushed and the occipital region bent upwards. It is therefore not possible to be certain of the vertical position of the horn cores or whether this position is natural or due to distortion. A second, but unnumbered, specimen lacks the horn cores but has only half the cranium preserved and so provides no contribution (to the problem of the occipital). The angle between the parietal and frontal of this second specimen seems to be greater than in the living species so one gets the impression that the normal position would obtain. The horn cores are very massive, much more so than in recent examples of *D. korrigum* which lives in the same region today. This is a special character of the new species and is seen in the specimen of the Hopwood collection, B.M. 14553 which is designated the paratype.

The above diagnosis was published without illustration. It is not at all satisfactory and has made it very difficult to determine the essential characters of *D. angusticornis*. We must treat the uncrushed paratype as being typical, for it exhibits some of the characters upon which Dr Schwarz laid great stress. The horn core is certainly very massive, and the distance from the base of the horn core to the mid-line suture is about 10 mm.

HORIZON

There is no evidence to show which bed at Olduvai yielded the type specimen, but the paratype was on the surface of Bed II, while other comparable material in the British Museum of Natural History also comes from Bed II. There is no evidence of the presence of this species in Bed I. The horizon is, therefore, Middle Pleistocene.

DESCRIPTION OF PARATYPE

Although Dr Schwarz designated specimen no. M. 14553 in the British Museum of Natural History as paratype, he neither described it nor figured it.

The specimen consists of a fragment of the brain case and the frontal region, with part of the right frontal bone and an incomplete horn core. The inter-frontal suture is preserved, as well as a small part of the left frontal bone with the base of the pedicle. From this specimen it can be seen that the horn cores were set very close together, a character which is said to be diagnostic. The horn cores are also very massive. The base of the right horn core measures 52 mm. bilaterally and 58 mm. anteriorly–posteriorly, compared with 54 mm. and 62 mm. in a large adult male of *D. korrigum*.

Dr Schwarz stated in his diagnosis that the horn cores were more massive than in the living species of *D. korrigum*. This is not entirely correct, as may be seen from the figures quoted above. In the living species there is a considerable sexual dimorphism and the measurements on a horn core of a random female specimen are 45 mm. and 53 mm. It is, therefore, possible that the paratype represents a female and that Dr Schwarz's type in Berlin was larger; unfortunately, he gave no measurements.

In the British Museum of Natural History, amongst material collected in 1935, there is a relatively complete cranium of this species, with horn cores but with the facial region missing. There are also a number of other specimens which compare closely with the paratype. These specimens, together with some in the Nairobi collections, represent *D. angusticornis* and are described. They have been used as the basis for the revised diagnosis of the species.

In this new material there are some specimens considerably more massive than in the living species. The paratype exhibits one marked character not mentioned in Dr Schwarz's diagnosis but which was presumably the reason for choosing the name *angusticornis* for the species. A short distance from the base, the horn core tapers very rapidly and becomes much more slender. Thus, the measurements at the base are 52 and 58 mm. but have dropped to 30 and 34 mm., only 110 mm. higher up.

REVISED DIAGNOSIS

A species of the genus *Damaliscus* distinguished from the living species by the following characters. The angle between the face and the vault of the cranium is nearly a right angle; the angle between the occipital region and the vault is very obtuse, somewhat as in *Alcelaphus*; the orbits are set very low relative to the brain case; the horn cores rise vertically in the same plane as the face, they are very massive at their base and taper rapidly; there is a keel on the upper part of the horn cores; the alcelaphine groove is intermediate in type between that of the living species of *Damaliscus* and *Alcelaphus*.

DISCUSSION OF 'DAMALISCUS ANGUSTICORNIS'

The additional material which is available has made it possible to extend our knowledge of the diagnostic characters of this species, and to understand its position in relation to the alcelaphine group.

It is certain that *angusticornis* was correctly placed within the genus *Damaliscus*, while its position as a full species with highly specialised characters has become apparent. There is no evidence to suggest that it is in any way ancestral to the living members of the genus. It has some characters that are nearer to *Alcelaphus* than to *Damaliscus*. Certain characters which Dr Schwarz thought were due to deformation in his specimen are now known to be characteristic and normal.

DESCRIPTION OF FIRST ADDITIONAL SPECIMEN

The most complete specimen is a brain case with nearly complete horn cores but lacking the face and the orbital region. Part of the basi-cranial region is also missing. Its number is M. 21425 in the British Museum of Natural History collection (pls. 64 and 65). It was found *in situ* at SHK II. This specimen conforms very well with both the type and the paratype in respect of those characters which are diagnostic of the species *angusticornis*. In particular, the horn cores are set close together; they are relatively massive at the base; and they taper rapidly.

Dr Schwarz suggested that there was some doubt whether the vertical position of the horn cores, relative to the frontals, was genuine or due to post-mortem distortion of his type. The specimen now being described makes it abundantly clear that the position of the horn cores on the frontal bone, relative to the top of the skull, is normal for the species. This is clearly seen in plate 64. This character, together with the tapering of the horn cores, is most typical for the species. It was also suggested that the occipital region of the type might have been displaced upwards. There may have been a little upward displacement of the occipital region in the type specimen, but the additional material makes it clear that the relationship of the nuchal region of the occipital to the top of the brain case is most unusual. The occipital region is at a very obtuse angle to the vault and the position is very similar to that in *Alcelaphus* (pl. 64). The bone which forms the vault of the skull is unusually thick. There is a slight lateral depression on the parietals on either side of the mid-line with a corresponding bulge in the middle. The rugosity of the nuchal crest coincides with the occipital suture. The bone in the region of the orbits is broken away, exposing the frontal sinuses, which extend upwards to include the whole area of the pedicle of the horn core and its base.

The following measurements have been taken on this specimen.

Measurements

	mm.
External width at the base of the horn cores	120
Width between the horn cores on the pedicle	27·5
Distance from the bregma to *foramen magnum*	122
Distance from bregma to upper limit of rugosity on the lambda point	47
Distance from the upper margin of the *foramen magnum* to the mid-point on the nuchal crest	69
Distance from the basi-occipital to the mid-line on the nuchal crest	82
Estimated width of occipital region (based on right side)	146
Lateral diameter of horn cores	52
Anterior–posterior diameter of horn cores	62
Height at which horn core rapidly shows compression or thinning	60
Total length of right horn core as preserved (it is estimated that perhaps 20 mm. more may be missing)	197
Diameter of horn cores 110 mm. from the base:	
Laterally	28
Anterior–posteriorly	34

As in all other specimens of *D. angusticornis* the lateral compression half-way up the horn cores corresponds to the formation of a slight anterior keel. The horn cores diverge slightly at the base with an angle of about 20 degrees from the mid-line.

DESCRIPTION OF SECOND ADDITIONAL SPECIMEN

The second additional specimen (pl. 66) was collected in 1935 at site VEK. Although it was found on the surface it has matrix adhering to it which suggests that it came from Bed II. Its number is M. 21422 of the British Museum of Natural History. This specimen is important because certain parts of the skull are preserved which are missing in the other specimens.

In its basic characters it entirely fits the diagnosis of the species. It is a very massive example, probably a male. On the left side the horn core is preserved to a length of about 160 mm. Part of the face is also preserved, including the upper rim of the left orbit. In the mid-line, the face is broken. A very large sinus extending into the base of both horn cores is exposed and also extends under the frontal bone between the orbits. On the right side the orbital region is broken away. The frontal–parietal suture is preserved but the parietals have broken off a short distance behind it. On the right side an alcelaphine groove is preserved. It is long and narrow and set between two low keels extending from the external wall of the back of the orbit upwards towards the horn core. On the left side a part of this groove is also preserved.

On the right side the frontal curves downwards and gently outwards to the external upper margin of the orbit. This region is much longer than in a living *D. korrigum*, and there is a considerable resemblance to the condition in some of the lelwel type of hartebeest. In that group the great height between the base of the horn cores and the outer upper rim of the orbits is associated with an exaggerated development of the bony pedicles which are fused into a single unit well above the level of the brain case. In *D. angusticornis*, however, the distance between the base of the horn cores and the rim of the orbit is due to a quite different factor. The orbit is set in a much lower position relative to the brain case than in any living alcelaphines examined. The position of the orbits may be regarded as another of the basic characters of *angusticornis*. The whole of the frontal bone in the region behind the orbit is preserved on the left, together with a part on the right side.

The following are the measurements for this specimen:

Measurements

	mm.
External width across the base of the horn cores	133·5
Width between the horn cores at the base	26
Width at the orbit (twice width on left side to centre line)	174
Lateral diameter of horn core, at base	57
Anterior–posterior diameter of horn core at base	61
Height on horn cores at which the lateral compression starts	64
Measurement of the horn cores 110 mm. from base:	
Lateral diameter	34
Anterior–posterior diameter	28
Length of right horn core as preserved	202
Estimated length of right horn core (approximate)	250
Length from base of horn core on left side of the external rim of orbit	85

DESCRIPTION OF THIRD ADDITIONAL SPECIMEN

The third additional specimen, no. M. 21423 in the British Museum of Natural History, consists of a frontlet and parts of broken-off horn cores. The upper margins of the external rims of both orbits are preserved. It was collected on the surface of site SHK II in 1935. Since it was found on the surface it is not certain that the specimen belongs to Bed II, and some of the adhering matrix suggests that it may belong to Bed IV. It differs slightly from the other specimens but is regarded as an individual variation of *D. angusticornis*. This variant may prove to be constant in Bed IV and to represent a further evolutionary stage of the damaliscine antelopes.

The principal differences between this specimen and the others described are as follows:

(1) There is a bulging ridge on either side of the frontal, from the base of the pedicles towards the orbital margin, lateral to the supra-orbital foramina.

(2) There is a distinct flattening of the frontal bone between these ridges.

(3) The frontal bone, from the base of the horn core to the upper external rim of the orbits, flares outwards more than in other specimens.

(4) The alcelaphine groove is elongate, somewhat shallower and less well defined than in the specimens described above.

Since similar variability occurs between the sexes in several living species of *Damaliscus*, it would be unwise to regard this specimen as anything but a minor variation within the species *angusticornis*, possibly a variation due to sex.

The following measurements may be given for this specimen:

Measurements

	mm.
External width at base of horn cores	118·5
Internal width between the base of the horn cores	31
Distance from the base of the horn core to the external orbital angle on the right side	81
Width across the orbits (approximate, the left side is slightly damaged)	172

DESCRIPTION OF FOURTH ADDITIONAL SPECIMEN

In the Nairobi collection of material excavated from the site BK II East, there is a brain case, with horn cores, of *D. angusticornis*. The chief importance of this specimen lies in the fact that it was found at the identical site and the identical level as the type specimen of *D. antiquus*, which is described below, showing that the two distinct species were co-existent. It exhibits the basic characters of *angusticornis*, especially the size and shape of the horn cores, as well as the nature of the alcelaphine grooves behind the orbits.

D. angusticornis is also represented in the Nairobi collection by a considerable number of isolated horn cores, in addition to the specimens which have been described.

Species: ***antiquus*** sp.nov.

DIAGNOSIS

A species of *Damaliscus* with certain resemblances to *D. angusticornis*, from which it differs in the following characters.

The horn cores are inclined slightly backwards from the forehead. They curve slightly outwards and backwards and then forwards (pls. 68 and 69). They do not taper as rapidly as in *angusticornis*, nor does the cross-section change in the same manner; on some specimens traces of cross-ribbing are visible. The alcelaphine groove extends from behind the back rim of the orbits towards the base of the horn cores, but is less deeply indented and less well defined than in *angusticornis*, although in general of the same shape. The groove is slightly everted backwards, with a sharp keel defining its anterior margin. On the top of the cranium there is a low, flat-topped protuberance extending across the parietals and backwards to the rim of the nuchal crest. It is similar to that seen in *angusticornis*, but much more strongly developed. In general the skull is of larger size and more rugose.

COMMENT

The specimen now treated as the type of *D. antiquus* was first thought to be a subspecies of *angusticornis*. Since the two were co-existent this is unlikely. The possibility was also considered that the differences in the horn cores might be no more than sex characters. This is not the explanation, since the differences extend to many other parts of the skull.

TYPE

A cranium with incomplete horn cores and with the greater part of the face missing, no. P.P.T. 3 excavated in 1955 from site BK II (pl. 67).

PARATYPE

A cranium with incomplete horn cores damaged in the region of the occipital condyle, no. M. 21428 in the British Museum of Natural History. It was collected in 1935 from site VEK, eroding from Bed I (pls. 68 and 69).

HORIZON

Since specimens of *Damaliscus antiquus* have been recorded from Bed I as well as from upper Bed II, the horizon extends from the Upper Villafranchian to the Middle Pleistocene.

DESCRIPTION OF TYPE

The type specimen is a large cranium from which both the horn cores have been broken off about 135 mm. from the base. The occipital condyles and basi-occipital region are intact, as well as the whole of the brain case. The bullae are missing, as well as part of the right temporal and the glenoid fossa region on the right side. The whole region of the face below the supra-orbital foramina is missing. On both sides the upper rims of the orbits are incomplete, and the rest of the orbital rim is missing. On the frontals, roughly from the mid-point of the anterior region of each horn core, a slight swelling extends towards a point external to the supra-orbital foramina. This slight bulging results in a flat depression on the forehead between the supra-orbital foramina and the top of the skull. The type specimen exhibits all the characters described in the diagnosis (except the conformation and bending forward of the horn cores—since this part of the horn cores is missing). The relationship of the nuchal region of the occipital to the top of the skull is almost exactly the same as in *angusticornis*. The greater rugosity of the protuberance masks the angle between the two regions.

DESCRIPTION OF PARATYPE

The paratype consists of the greater part of a cranium, with both horn cores preserved but broken off at the tips. The left horn core is the better preserved and exhibits the curving backwards and outwards and forwards of the upper part, described in the diagnosis (pls. 68 and 69). The contrast between the horn cores of *D. angusticornis* and *D. antiquus* is clearly shown in the plates. The occipital condyles and the basi-occipital region are damaged. As in the type, the face is broken away just in front of the supra-orbital foramina. The paratype exhibits the general characters listed in the diagnosis and there is no need to describe it in detail.

Only a limited number of measurements can be given for these two specimens.

Table of measurements of type and paratype

	Type (mm.)	Paratype (mm.)
External width at the base of the horn cores	121	121
Internal width between the base of the horn cores	30	34
Width at the external orbital angle (based on doubling the measurement on one side)	—	188
Lateral diameter of the horn cores at base	51	55
Anterior–posterior diameter of the horn cores at base	60	62·5
Length of left horn core as preserved	134	177
Estimated full length of the horn cores	150	—
Estimated width across the occipital region	150	142
Distance from the base of the *foramen magnum* to the top of the nuchal crest	82	84
Base of the horn core to external rim of the orbit (estimated)	74	78

ADDITIONAL MATERIAL

D. antiquus is represented by three further specimens in Nairobi. These are no. 1284/57 found *in situ* at site SHK II, no. P.P.R. 4 and no. F. 948, both collected in 1941. The first of these was *in situ* in Bed II and the second on the surface. These specimens confirm the presence of *D. antiquus* in upper Bed II, but it was less common than the contemporary *D. angusticornis*.

Genus: ***Parmularius*** Hopwood

Species: ***altidens*** Hopwood

Among the fossils which were excavated from Bed I during the 1931–2 season was a well-preserved skull, with mandible attached, of an unusual bovid. It was found projecting from a tuffaceous deposit some 15 ft. below the upper level of Bed I, and therefore belongs to the Upper Villafranchian.

Dr. Hopwood made this specimen into the type of a new genus and species of alcelaphine.

DIAGNOSIS

His generic diagnosis reads as follows:

Bovidae with relatively short sub-parallel horn cores and a long narrow face. Horn cores on a short pedicle curved gently backwards and outwards, neither keeled nor twisted. Lachrymal fossa of medium depth; supra-orbital pit small, lying at the upper end of a shallow groove, not sunken. Nasals long and narrow. Face bent down on basi-cranial axis at about 65°. Brain case with a median conical parietal boss. Genotype—*Parmularius altidens* sp.n.

Dr Hopwood's specific diagnosis was as follows: 'A *Parmularius* with very hypsodont teeth. Height of first molar 42 mm.' His description of the holotype was as follows: 'An associated skull and mandible with the cervical vertebrae and other parts of the skeleton.' After listing certain measurements he concluded that it was apparently an aberrant member of the hartebeest group.

COMMENTS

The specimen is very well preserved (pl. 70). The state of wear of its third molars, as well as the closure of many of the sutures, indicates that we are dealing with an adult. If we had only this specimen it would be difficult to decide whether the very slender horn cores were characteristic or not. Additional specimens found in 1961 make it reasonably certain that the type specimen represents a female and that the horn cores in males are more robust than in the holotype. Dr Hopwood probably underestimated the length of the horn cores, and his statement that they are 'sub-parallel' is not borne out by the additional material.

In view of the fact that a large number of specimens of *Parmularius altidens* are now known from Bed I, and that there are two more evolved species in later deposits, revised diagnoses are necessary.

REVISED GENERIC DIAGNOSIS

Alcelaphini characterised by the presence of a conical boss on the parietals on the mid-line behind the horn cores. These rise from relatively low, clearly differentiated pedicles; lachrymal fossa present but not strongly defined; face long and narrow. Genotype, *Parmularius altidens*.

REVISED SPECIFIC DIAGNOSIS FOR 'PARMULARIUS ALTIDENS'

A *Parmularius* with strongly hypsodont molar teeth; nasal bones long, narrow, and flatter than in most alcelaphines; lachrymal fossa shallow; inter-orbital foramina set high, nearly at the level of the upper rim of the orbits, with well-defined shallow grooves about two inches long running from the foramina towards the lachrymal fossa and parallel to the frontal suture; horn cores set very close together on top of the head, so that the lateral margins of the frontal bones turn sharply outwards from the base of the horn core towards the upper margin of the orbits; distance from base of horn core to upper margin of orbit much longer than in present-day blesbok (*Damaliscus albifrons*) but similar in morphology; gap between the pedicles supporting the horn cores narrow, deep and U-shaped; horn cores curved slightly forward, then backwards and outwards; cone on the mid-line of the parietals very strongly marked; facial element not so strongly bent in relation to the basi-cranial axis as in present-day alcelaphines; upper molar teeth with well-defined small medial pillars in the first and second molars and the crowns relatively long anterior–posteriorly and narrow bucco-lingually; the teeth have a well-marked coat of cement on the external face.

DESCRIPTION OF TYPE SPECIMEN

The original description of the type specimen may be amplified as follows.

The skull is almost complete except for a small region around the right orbit, the ends of the two horn cores and a part of the pre-maxillary region. The sutures are closed but not fully fused, and the specimen probably represents an adult, but not an aged, individual. The nasal bones are remarkably long, flat and narrow, differing noticeably in this respect from all the living alcelaphines. The orbits are somewhat oval from back to front as in all the alcelaphine tribe, being noticeably longer than wide. The occipital region is morphologically similar to the living alcelaphines. The angle between the top of the parietals and the nuchal crest of the occipital is very wide. There is a well-defined sharp keel running from the lateral margin of the orbit towards the base of the horn cores forming the anterior limit for the attachment of the temporal muscles. The forehead between the orbits is flat and not depressed.

The mandible is relatively short compared with that of present-day alcelaphines. It is massive posteriorly, being deep in the region of the second and third molars. It narrows rapidly from the region of the first molar forwards to the diastema, where the bone is very slender. There are three pre-molars and three molars; the second pre-molar is small and peg-shaped. The anterior region of the mandible, carrying the incisors, is very reduced.

The following table sets out the most important measurements of the skull and mandible.

Measurements

	mm.
From bregma to anterior point of pre-maxilla (approximate)	339
External width across base of horn cores	72
Maximum width at angles of orbits (based on doubling left measurement from mid-line to orbital angle)	142
Length of nasal bones (measured on right side)	149
Maximum width of nasal bones	18
Minimum width of nasal bones	16
Maximum length of orbits from back to front (measured on left side)	55
Maximum height of orbits (approximate)	46
Diameter of horn core measured at base (left side)	28
Anterior–posterior measurement of horn core at base (left side)	31
From bregma to upper limit of *foramen magnum* (approximate)	82
Width of occipital region	108
Height from base of *foramen magnum* to uppermost point of nuchal crest	65
Distance from base of horn core, left side, to the external angle of the upper margin of the orbit	64
Length of upper molar–pre-molar series (at alveolar margin, approximate)	94
External palatal width at second molars	82
Internal palatal width at second molars	54
Mandibular length (approximate)	285
Mandibular height from base of angle to top of condyle	111
Mandibular height from base of angle to top of coronoid	153
Height of mandible at mid-point of third molar	50
Height of mandible at mid-point of second pre-molar	32
Minimum height of mandible	20
Length of diastema	71·5
Minimum width of mandible behind symphysis	25

DESCRIPTION OF FIRST ADDITIONAL SPECIMEN

A horn core (pl. 71) which is attributable to *P. altidens* was found by Dr Hopwood in Bed I in 1931 and is registered in the British Museum of Natural History as no. M. 14514. Its identification as *P. altidens* is based on a number of distinctive characters, including the very close position of the base of the horn core to the frontal suture, the keel running from the external angle of the orbit to the base of the horn core, as well as the cross-section of the horn core itself. The specimen is 240 mm. long from the point where the horn core joins the pedicle on the internal aspect, to the tip, which is slightly damaged. The transverse measurement at the base is 37 mm. and the anterior–posterior measurement 43 mm. The horn core curves backwards and slightly outwards at first and then inwards at the tip. The massive nature of this horn core suggests that it belonged to a male.

DESCRIPTION OF SECOND ADDITIONAL SPECIMEN

Specimen Ba. 126, 1960, from site FLK I consists of a frontlet with the right horn core almost complete and the left horn core broken off about half-way (pl. 72). It represents a large *Parmularius altidens*, presumably male. The horn cores are considerably more massive than in Dr Hopwood's original type or the second specimen in the British Museum of Natural History, which was presumed to be male. The specimen has the characteristic keel on the anterior side of the 'alcelaphine groove', as in *P. altidens*, but the parietals with the cone are missing. The gap between the horn cores is very narrow, and wider on the pedicle than on the base of the horn core itself. There can have been only just room for the horn integument.

DESCRIPTION OF THIRD ADDITIONAL SPECIMEN

This is a frontlet, FLK N I, no. 1315, 1960, with the lower part of two horn cores preserved. It is presumably a male, although not quite so massive as the previous specimen. The typical *Parmularius altidens* cone is present in the mid-line of the parietals. The cone is broken through its posterior edge and its structure can be seen clearly. It consists of solid bone and is considerably larger and higher than in the original type (pl. 73).

NOTES OF FURTHER ADDITIONAL SPECIMENS

(*a*) A horn core from FLK I, no. F. 206, 1960 (pl. 74), appears to represent another adult male. On it some faint ribbing[1] is visible. The anterior keel of the alcelaphine groove is typical of *P. al-*

[1] On the horn cores of many *Parmularius* specimens, faint indications of transverse ribbing can be seen, as found in the horn cores of old males in the lelwel group of hartebeest.

tidens, although not so clearly defined as in some specimens.

(*b*) A right horn core, FLK I, no. G. 230, 1960, with faint ribbing. It is provisionally assigned to *P. altidens* in view of the curvature and the bulging inwards, just above the pedicle.

(*c*) Two unfused frontal bodies, FLK N I, nos. 1166 and 1392, 1960, of a juvenile *P. altidens* with the characteristics of the genus and species clearly visible.

(*d*) A complete left horn core, FLK N I, no. 1410, differing from the others in having a very tapered tip and less pronounced backward curvature.

(*e*) Part of a cranial vault, FLK N I, no. 178, 1960, of what is probably a juvenile, or a female, showing the characteristic cone on the parietals as well as the keel in front of the alcelaphine groove. The para-occipital processes are well preserved. They are long, slender and curved inwards.

DISCUSSION

Parmularius altidens is certainly a member of the Alcelaphini. Its affinities seem to be more with the blesbok and the korrigums (*Damaliscus*) than with the lelwel and Lichenstein's hartebeest (*Alcelaphus*). The evidence for this can be seen in the pedicles of the horn cores, the form of the alcelaphine groove and the relatively short face. In *Damaliscus*, moreover, there is a small bulge on the parietals, in the mid-line, which seems to be a relic of the cone of *Parmularius*. Professor Simpson, in his 'Classification of Mammals', places *Parmularius* between *Damaliscus* and *Alcelaphus*. The new evidence fully supports this view and is contrary to Dr Dietrich's suggestion that *Parmularius* belongs to the Hippotragini.

Species: ***rugosus*** sp.nov.

DIAGNOSIS

A species of the genus *Parmularius* differing from *P. altidens* in the following characters. The parietal cone is much more reduced and is replaced by a slight boss; the whole skull is more rugged and muscular: the rugosity is particularly strong on the bony wall of the orbits, on the keel of the alcelaphine groove, at the base of the skull, on the forehead and in the occipital region. The palate is more markedly oval and the horn cores curve more outwards. In this respect it resembles the living blesbok (*Damaliscus albifrons*) from which, however, it differs clearly in its other characters and resembles the genus *Parmularius*.

HOLOTYPE

A nearly complete but somewhat crushed skull lacking the mandible. British Museum of Natural History no. M. 21430. The left horn core, parts of the pre-maxillary region and the posterior rim of the orbits, as well as the tip of the right horn core, are missing (pls. 75 and 76).

HORIZON

The type specimen comes from the base of Bed IV at site HWK. It belongs, therefore, to the upper part of the Middle Pleistocene. It was associated with an industry of an Acheulean stage of the Chelles-Acheul culture.

DESCRIPTION OF HOLOTYPE

The skull is relatively complete, with the right horn core broken off about 80 mm. from the base, while the left is broken off at the pedicle. From about the mid-points of the anterior faces of the pedicles, marked ridges extend down the forehead, curving slightly outwards, to disappear just above the supra-orbital foramina. The upper rims of the orbits are thick and rugose when compared with *Parmularius altidens* and with modern male blesbok of similar dimensions. As a result of this rugosity there is a flat depressed region on the anterior part of the frontal between the orbits. The supra-orbital foramina are not set as high as in *P. altidens*, nor as in modern blesbok. There is a marked notch on the rim of the orbits, half-way down the frontal region, but this is not as marked as in *P. altidens*, although much more so than in modern blesbok. The nasal bones are long and narrow but much less flat than in *P. altidens*. They conform more closely in shape and size to those of modern blesbok. The maxillary region above the roots of the molar teeth is flared outwards, a feature linked with the oval palate which is a diagnostic character of the species. The lachrymal fossae are more marked than in *P. altidens* but not so strongly developed as in the blesbok. The general morphology of the brain case is similar to that of *P. altidens*, except that the mid-parietal cone is reduced to a slight boss which is more marked than the residuary cone which can be seen on some *Damaliscus* species. In the basi-occipital region the bullae are far less prominent that in any of the modern or fossil *Damaliscus* examined. In this respect *rugosus* corresponds closely to *altidens*. The posterior margins of the maxillae, behind the third molars, as well as the vomer region are thick and rugose. The palate is both more oval and deeper than in modern *Damaliscus* species.

Only a few dimensions of the type specimen will be given, in view of its crushed and damaged state.

Measurements

	mm.
External width at the base of the horn cores (estimated by doubling figure on right side)	78
Internal width between the horn cores (estimated by doubling figure on right side)	22
Width at the upper margin of the orbits (approximate)	138
Length of the nasal bones (approximate)	142
Orbital length (based on right side)	56
Diameter of the horn core at base	34
Anterior–posterior width at the base of the horn core	37·5
Distance from bregma to base of *foramen magnum*	84
Width of occipital (approximate)	117
Height of occipital (approximate)	63
Distance from the base of the horn core to the rim of the orbit (right side)	62
Length of molar–pre-molar series (right side)	86·5
External palatal width at second molars (approximate)	91
Internal palatal width at second molars	57·5

Species: indet.

Among the specimens in Nairobi, collected in 1941, is the left side of a frontal, no. F. 3001. It includes the upper rim of the orbit, the greater part of a horn core and a part of the left parietal. It seems to represent a large *Parmularius* (pl. 77). This specimen came from upper Bed II. It is of special interest because it seems to represent a species standing between *Parmularius altidens* of Bed I, and *P. rugosus* of Bed IV. Among the characters which indicate the specimen is a *Parmularius* is the nature of the alcelaphine groove, which is of the *Parmularius* and not the *Damaliscus* type. There is also a characteristic *Parmularius* type of cone behind the horn cores on the mid-line of the parietals. The horn core is much more massive than the largest known *P. altidens* or *P. rugosus*.

DISCUSSION OF THE GENUS 'PARMULARIUS'

The genus *Parmularius* was first found at Olduvai in 1931, where it is now known to be represented by three different species. Of these, the original *P. altidens* is found in Bed I. In upper Bed II there is a much larger and little-known species, suggesting that this genus was affected by the current giantism. In Bed IV there is a smaller species—*rugosus*—which differs from both the earlier ones and has a number of resemblances to *Damaliscus albifrons*. Indeed, it seems probable that the blesbok are the direct living descendants of *Parmularius* and they may have to be removed from the genus *Damaliscus* and placed with *Parmularius*.

Genus: ***Alcelaphus***

Species: ***kattwinkeli*** Schwarz

Dr Schwarz described this species from material in the collections made by Professor Reck in 1913.

There are no specimens which can positively be identified as *Alcelaphus kattwinkeli* in the British Museum of Natural History, or the Nairobi collections. The species is included in the Olduvai list on the basis of Dr Schwarz's identification. The specimen illustrated in pl. 78 may perhaps represent *A. kattwinkeli*.

Species: ***howardi*** sp.nov. Leakey

The trivial name is after Mr S. Howard, who found the type specimen at the site SC when working at Olduvai in 1935.

DIAGNOSIS

A species of *Alcelaphus* resembling Lichenstein's hartebeest, from which it differs in the following characters: the horn cores are less massive at their base; their development upwards is more even; they exhibit marked ribbing on the lower part; there is no bulge on the mid-line of the forehead. At their base, the horn cores are markedly rugose on the anterior face and are somewhat flattened from front to back. Initially, they are inclined slightly backwards as well as outwards and upwards; there is an inward twist of the whole core at the point where it turns more strongly upwards and finally inwards.

HOLOTYPE

A frontlet with horn cores, British Museum of Natural History no. M. 14950 (pl. 79).

HORIZON

The specimen comes from site SC in upper Bed II and is of Middle Pleistocene age.

DESCRIPTION OF TYPE

The type consists of a frontlet with the left horn core intact and the right one broken off short half-way.

COMMENTS

This specimen was catalogued in the British Museum of Natural History as *Alcelaphus kattwinkeli*, but it does not agree with the description or the illustrations of that species. It has certain resemblances to *Xenocephalus*, to which it may possibly be ancestral.

Genus: *Beatragus*

Species: *antiquus* sp.nov.

DIAGNOSIS

A species of *Beatragus* larger and more massive than *hunteri*, from which it differs in the following characters. The horn cores rise from the corners of the frontal bones more forwards and sideways, they then twist inwards and upwards, then backwards and upwards again, to terminate in very long tapering points; on the lower third of each horn core there is distinct ribbing.

TYPE

A right frontal bone with the lower part of a massive horn core, registered in the British Museum of Natural History as no. M. 21445.

PARATYPE

An almost complete left horn core with a part of the external rim of the orbital angle (pl. 80), registered in the British Museum of Natural History as no. M. 21446.

HORIZON

These two specimens come from Bed I and were collected in 1935. The horizon is Upper Villafranchian.

DESCRIPTION OF TYPE

The type consists of a right frontal bone. Almost the whole of the frontal suture is preserved, from the bregma to a point nearly in contact with the maxillae. The horn core is preserved for a length of 225 mm. measured in a straight line. The base of this horn core is very massive, having a bilateral measurement of 68·5 mm. compared with only 53 mm. in a very large modern specimen of Hunter's antelope. The anterior–posterior measurement at the base of the horn core is 67 mm. compared with only 52 mm. on the modern specimen mentioned above. The supra-orbital foramen is large and set high in relation to the top of the orbit. The horn core rises from the frontal forwards and outwards, and the angle of its outward projection is much lower than in the living species. The distance from the base of the horn core to the mid-line is about 22 mm. The distance from the mid-line to the external rim of orbit is 78 mm. At the point where the horn core begins to turn sharply upwards, the bilateral diameter is only 41 mm., and the anterior–posterior diameter only 36 mm. There is marked ribbing on the lower part of the horn core.

DESCRIPTION OF PARATYPE

The paratype consists of an almost complete and slightly less massive horn core, which in all essential respects agrees with the first specimen. It has been partially reconstructed, which makes accurate measurement impossible in the basal region. As in the type, there is evidence of ribbing on the lower part of the horn core. At the point where the horn turns upwards the measurements are: bilateral diameter 35 mm., anterior–posterior diameter 31 mm. The length in a straight line from the anterior part of the base of the horn core to the tip, as preserved (about a centimetre or more is missing), is 555 mm. On the curve it measures 670 mm.

ADDITIONAL MATERIAL

Among the specimens excavated from site FLK N I is a horn core, no. 5123, 1960, which is unquestionably *Beatragus antiquus*. Like the type of *B. antiquus*, it is much more massive, although relatively short, compared with the modern *Beatragus*. This specimen had been extensively gnawed by rodents prior to fossilisation.

COMMENTS

There can be no doubt that these specimens represent an early extinct ancestor of the living species of *Beatragus*. It should also be noted that the upper third of the horn core of the paratype has a cross-section and form which are indistinguishable from the top of the horn core of an *oryx*. In consequence, any attempt to identify *oryx* in the fossil fauna of this period in East Africa on the basis of broken horn cores would be unwise. A horn core in the collection at the British Museum of Natural History has been attributed to the living oryx in Dr Hopwood's lists, but this is doubtful.

Genus: ***Xenocephalus*** gen.nov.

DIAGNOSIS OF GENUS

A genus of the Alcelaphini exhibiting some characters of *Damaliscus*, some of *Alcelaphus*, but differing from both.

The cranial vault and the horn cores resemble those of *Alcelaphus*, but the horn cores rise from small, separate pedicles on the corners of the frontals, as in *Damaliscus*. The molar teeth differ from those of both genera but are a little closer to *Damaliscus*. The face is very long, as in *Alcelaphus*, but not so narrow. The alcelaphine groove is more like that in *Alcelaphus*. The genus exhibits superficial resemblances to *Gorgon*.

GENOTYPE

The genotype is the species *Xenocephalus robustus* described below.

Species: ***robustus*** sp.nov.

DIAGNOSIS OF SPECIES

A species of *Xenocephalus* in which the horn cores are very massive and are directed backwards, outwards and upwards, then upwards and slightly forwards, then inwards and backwards again. There is a shallow depression on the vault of the cranium between the rim of the nuchal crest and the back of the horn cores (see pl. 81). The face is exceedingly long; the third upper molar has a marked heel.

HOLOTYPE

The greater part of a skull. British Museum of Natural History, no. M. 21447 (pls. 81 and 82). It was found in 1935 at site TK IV, in Bed IV.

DESCRIPTION OF TYPE

In the type specimen the brain case and the upper part of the skull are relatively well preserved, as well as the palate. The whole of the facial region is broken and distorted laterally, while the middle part of the face is crushed. The nasal bones are missing. On the right side a large part of the maxillae and the lachrymal region is also missing. The occipital condyles have been broken off. On the right side the horn core is broken off at the base of the pedicle to expose the sinuses, but on the left side the horn core is preserved from base to tip. The frontal region, between the orbits, is wide and flat and superficially recalls the condition seen in *Gorgon*. On the right side the whole of the orbital rim is missing. On the left side the upper orbital rim is intact; anteriorly it is crushed and broken away at the back. The anterior parts of the maxillae are nearly complete but the pre-maxillae are missing. The upper face, from the level of the orbits to the diastema, is very long. The vault of the palate is partly crushed. On both sides there are three large molar teeth, and the sockets for the fourth pre-molars are also present. There are no root sockets for the second and third pre-molars; these had been shed and the alveolar margins had become fused. Since the wear of the molar teeth does not suggest an advanced age, the loss of the second and third pre-molars is unexpected and may prove to be diagnostic of the species. The third molars carry a well-defined medial pillar. Such a pillar is not present in the second and third molars. The third molar has a backward prolongation, or heel, quite unlike the condition seen in any living or fossil alcelaphine examined. Since the pre-maxillae are missing, it is not possible to estimate the total facial length, but it must have been very great. On the temporal bone behind the jugal projection, there is a relatively shallow, wide pit with rugose edges, at about the level of the external auditory meatus. This resembles the pit to be seen in Lichenstein's hartebeest. It is unlike the condition seen in *Damaliscus*, but *Beatragus* has a similar pit. The region around the external auditory meatus is partly damaged and not wholly clear of matrix. It is, however, more like *Alcelaphus* than *Damaliscus*.

The whole of the occipital region, in particular the nuchal crest, is very massive and rugged. The distance from the anterior rim of the crest to the mid-line between the horn cores on the frontal bone is about 46 mm., resembling Lichenstein's hartebeest in this respect. There is a depression between the nuchal crest and the horn cores. This slight depression very clearly distinguishes this new genus and species from *Alcelaphus* and other living hartebeest. The external basal rims of the bullae have a contour which are roughly semi-circular and resemble the structure seen in Lichenstein's hartebeest.

On the left side the bony wall behind the orbit has not been fully cleared of matrix so that the morphology is difficult to interpret. On the right side, below the broken-off horn core, a part of an alcelaphine groove is visible. The form of the groove resembles that in *Alcelaphus lichensteinii*.

The left horn core is very massive and long, even when compared with a very large specimen of *Gorgon taurinus*. It is not as massive, at its base, as in Lichenstein's hartebeest, although it is longer over-all. The horn cores are directed backwards, upwards and outwards, then upwards and slightly forwards, then curve inwards and backwards. There are traces of cross-ribbing on the lower and middle part of the horn core, such as are to be seen on the horn cores of various Alcelaphini.

Measurements

The following measurements of the type specimen can be given:

	mm.
Length of face from the base of the horn cores to tip of pre-maxilla (approximate)	452
External width at the base of the horn cores	152
Width at orbits (calculated from the left side)	198
Lateral diameter of the horn cores at base	79
Maximum width of the horn cores	81
Anterior–posterior diameter of the horn cores at base	71
Anterior–posterior diameter at point of maximum width	47
Length from mid-point between the horn cores to the upper margin of the *foramen magnum*	122
Length from base of *foramen magnum* to the top of the nuchal crest	92
Maximum occipital breadth	172
Length from the base of the horn core to the rim of the orbit on the left side	85·5
Length of molar–pre-molar series at alveolar margin (only the fourth pre-molar is preserved)	109
External palatal width at the second molar	108
Internal palatal width at the second molar	68

DISCUSSION

Xenocephalus robustus is only represented with certainty, in the fossil collection from Olduvai, by the type specimen. Some fragmentary horn cores may relate to this genus and species, but in view of the fact that other genera and species have somewhat similar horn cores positive identification is not possible.

The type specimen was regarded by Dr Hopwood as representing the living genus *Gorgon* species *taurinus*. He consequently listed the living wildebeest in the Olduvai fauna.

Subfamily: *ANTELOPINAE*

Tribe: **Neotragini**

Genus: cf. ***Nesotragus***

Dr Schwarz (1937) identified a species of *Nesotragus* among the fossils collected by Professor Reck from Olduvai. Although nothing comparable has been seen in the collections at the British Museum of Natural History, or in Nairobi, *Nesotragus* sp.indet. is included in the list of Olduvai Bovidae on the basis of this record.

Tribe: **Antelopini**

Palaeontologists and zoologists differ greatly on the question of the classification of the gazelles and the impala, as well as various gazelle-like antelopes, such as living gerenuk, dibatag and springbok and the fossil *Phenacotragus*.

Professor Simpson treats all of these genera as members of the Antelopini, within the subfamily Antelopinae, in which he also includes klipspringer, pygmy antelope, royal antelope, steinbok and dik-dik. Dr Dietrich, however, places dibatag, gerenuk, gazelles and springbok in a subfamily Gazellinae, but separates impala into a completely distinct subfamily containing only this genus. This subfamily he calls Aepycerotinae. Like Professor Simpson he places *Phenacotragus* in the larger group of Gazellinae.

Dr Schwarz has a different concept. He makes a single subfamily, Antelopinae, but includes in it not only the gazelles, gerenuk, dibatag, impala and springbok, but also all of the hartebeest.

Re-examination of the living and fossil material of these groups strongly supports Professor Simpson's classification, with the possible exception of the impala. This I suggest ought to be placed in a different tribe (of the same subfamily) called Aepycerotini. The cranial characters of the impala appear to be sufficiently distinct from all the others in this group to justify such a distinction. There would thus be three tribes within this subfamily: Neotragini, Antelopini and Aepycerotini.

Dr Hopwood listed only two members of Professor Simpson's Antelopini in the Olduvai fauna. These are Dr Schwarz's *Phenacotragus recki* and his *Gazella gazella praecursor.* Dr Schwarz regarded his race *praecursor* as merely a subspecies of *G. gazella*, but he places a very large number of quite distinct species into *G. gazella*, allowing them only subspecific rank. It seems desirable to treat Dr Schwarz's *G. g. praecursor* as representing a full species, noting that he regarded it as close to *Gazella thomsoni.* Unfortunately, Dr Schwarz did not figure his type.

Dr Dietrich created a number of full species of gazelle on the basis of measurements of upper and lower dentitions. He listed no less than three from the Lower Pleistocene of Laetolil. These he named *hennigi, janenschi* and *kohllarseni.* It is most unlikely that all three are valid. There was a great range of size in his material, but we know that this varies greatly with age, race and sex.

Genus: ***Gazella***

Species: ***praecursor*** (Schwarz)

There are two horn cores in the British Museum of Natural History, nos. M. 14507 and M. 14508, collected in 1931, which represent a type of gazelle similar to *G. thomsoni.* Since Dr Schwarz stated that his *G. praecursor* was similar to, and perhaps ancestral to, Thomson's gazelle, these two specimens are provisionally regarded as representing his species.

OTHER GAZELLES

(*a*) In 1935 my wife discovered a very rich deposit containing bones of many individuals of a small gazelle-like antelope at site SHK in upper Bed II. Both immature and adult specimens were present. They have not yet been described and will be the subject of a special report by Mr Alan Gentry. A few illustrations are given here (pls. 83 and 84). A brief examination suggests that they stand closer to Dr Schwarz's *Phenacotragus* in both cranial and horn-core characters, than they do to the living gazelles of East Africa. They may, however, be distinct from *P. recki.*

(*b*) A horn core and part of a frontal collected from site VEK II in 1935, British Museum of Natural History no. M. 21457, superficially resembles *Phenacotragus*, but the horn core is flatter from side to side and has certain resemblances to the horn cores of the so-called 'Mongolian gazelle' of Asia. It appears to be quite distinct from either of those mentioned above.

(*c*) There are two other horn cores which are clearly distinct from any of the types mentioned above but which are almost certainly those of a gazelle. These two specimens are catalogued as no. M. 14513 (which was collected on the surface of Bed I in 1931) and no. M. 22360 (which was collected in Bed I at site DK in 1935). The horn cores are very short, and are narrow from side to side and very wide from back to front at the base. They bend backwards and taper rapidly to a thin oval-sectioned tip.

(*d*) Another important specimen (pl. 85) in the collection of the British Museum of Natural History is registered as no. M. 21462. It was collected in 1931 from Bed I at site VEK. It consists of an almost complete cranium with the upper part of the face and the bases of both horn cores. The horn cores are shorter and not so strongly curved as those listed above. This seems to represent yet another different gazelle.

(*e*) There are two further horn cores at the British Museum of Natural History, which seem to represent a different and larger member of the genus *Gazella*.

(*f*) *Gazella* cf. *granti*. A specimen from site BK II East includes the frontal bones with the mid-frontal suture and the bases of both horn cores. It is comparable to *G. granti* in size and morphology and is provisionally referred to that species.

A number of other gazelle-like antelopes are present in the material from the excavations in Bed I. Three, if not more, distinct species are present.

(*g*) The first of these species is represented by a pair of horn cores, nos. 229 and 230, 1960, found close together in layer 13 at site FLK I. They carry indications of cross-ribbing and there are very deep anterior grooves. These two specimens have some similarities to 'group (*b*)' above. There are also some resemblances to 'group (*a*)'. They may well have been ancestral to both groups. A specimen, no. 1139, 1960, from site FLK N I, is probably of the same species.

(*h*) The second gazelle in the material excavated from Bed I is represented by specimen no. 1307, 1960, from site FLK N I. It has some resemblance to 'group (*c*)' above.

(*i*) *Gazella* cf. *wellsi*. A third gazelle from Bed I is represented by the two horn cores and a crushed palate, no. 6334, 1960, from site FLK N I. This is a small gazelle and its horn cores are much less curved than in the other types, projecting upwards for a short way, then curving slightly and finally becoming almost straight. Dr Basil Cooke, who saw this specimen, expressed the opinion that it was probably identical to *G. wellsi*.

(*j*) Among the fossil material excavated from upper Bed II at site BK II are a large number of horn cores indicating the presence of a very large antelope with affinities to the gazelles (pl. 86). The specimens preserved in Nairobi are: no. 876, 1957, BK II; no. 65, 1953, BK II East; no. 139, 1953, BK II East; no. 64, 1953, BK II East; no. 49, 1957, BK II; no. 1495, 1957, BK II; no. F. 946, 1941, Olduvai surface; no. 473, 1959, HWK II; as well as two unnumbered specimens from BK II.

(*k*) There is a specimen in the Nairobi collections, obtained in 1941 from the surface of Bed IV, which is similar in certain respects to the specimens listed above, but which differs in other characters. It possibly represents a more evolved species or subspecies.

Genus: ***Phenacotragus***

Species: ***recki*** Schwarz

= (*Adenota recki*)

This genus was described by Dr Schwarz on the basis of a well-preserved skull with one complete horn core.

HOLOTYPE

The holotype is in the Munich Museum, where it is registered as OR-NR VIII 343. The specimen was collected by Professor Reck in 1913, but there is no record of what horizon it came from. An excellent cast (pl. 87) of the type specimen is in the British Museum of Natural History, where it is registered as no. M. 21460.

OTHER MATERIAL

Two other specimens in the British Museum of Natural History may be regarded as representing *Phenacotragus recki*. They are a right frontal with horn core, no. M. 14563, collected in 1931 from site VEK; and a specimen collected from site HWK at the base of Bed II in 1935.

Tribe: **Aepycerotini**

Genus: ***Aepyceros***

Species: indet.

Among the fossils excavated from BK II in 1957 is a specimen, no. 662, which represents a left frontal and part of the horn core of a member of the Aepycerotini, probably the genus *Aepyceros*. It suggests the presence of impala in upper Bed II.

BOVIDAE

Incertae sedis

Genus: ***Thaleroceros***

Species: ***radiciformis***

Professor Reck, in 1935, published a new genus and species, which he called *Thaleroceros radiciformis*. It was based upon a single specimen of a horn core and frontal bone. Only one other specimen has since come to light which could possibly be regarded as *Thaleroceros*. It was found in 1955 at site FLK (pl. 88) and is in the Nairobi Collection. It approximates sufficiently closely to Professor Reck's type to confirm that the latter was not a freak. It seems certain that this genus represents one of the rarer Bovidae of Olduvai.

Cf. **Alcelaphini**

(*a*) In the collections in Nairobi there is a frontlet and part of a brain case (pls. 89 and 90) of a large member of the Bovidae, with part of the right horn core preserved. This specimen is recorded as it undoubtedly represents an unknown member of the Bovinae, displaying some unusual characters. When orientated with the vault of the cranium horizontal, the horn core rises vertically above the orbit, as in *D. angusticornis*. The frontal bone is flat and in the same plane as the horn core. The brain cavity itself extends very far forward, so that the frontal lobe reaches almost as far as the front part of the orbits. This character can be clearly seen by the internal structure of the brain case. The horn core is very massive at the base and tapers very rapidly. It has the same measurements laterally and anterior–posteriorly at the base, 93 mm.

(*b*) In the collections of the British Museum of Natural History there is a specimen, no. M. 21429 (pl. 91), collected in 1932 on the surface of Bed II at site VEK. It is badly crushed and was heavily weathered before fossilisation and is therefore very difficult to interpret. It consists of the greater part of a brain case and horn cores, the tips of which are missing. Viewed from the front the specimen has superficial resemblances to *Damaliscus angusticornis*; but the nature of the occipital region suggests that it is not a *Damaliscus*.

The only measurements that can be given for this specimen are:

	mm.
External width of the skull at the base of the horn cores	127
Width between the horn cores	12
Distance from the base of the *foramen magnum* to the top of the nuchal crest	97
Distance from the bregma to the upper margin of the *foramen magnum*	93

(*c*) During excavations at site FLK NN I in 1960, a brain case with horn cores was uncovered in deposits overlying the main living-floor. This specimen recalls the one described above. It was also very badly cracked and weathered prior to fossilisation.

Both specimens exhibit affinities with certain alcelaphines, for example in the basi-cranial region, but both differ markedly from the known genera of this tribe.

(*d*) Among the specimens excavated in 1960 there is one which may belong to the Alcelaphini but which is quite distinct from any listed above. It is specimen no. 231 from site FLK I. The horn core shows strong transverse ribbing.

(*e*) Among the specimens excavated from site BK II are a number of examples of another alcelaphine, as far as may be judged by the presence of an alcelaphine groove at the base of the horn cores. These horn cores, one of which is illustrated (pl. 92), are certainly not of the same species as any of those mentioned earlier in this report.

Subfamily: ***CAPRINAE***

Tribe: **Caprini**

Genus: ***Pultiphagonides*** Hopwood 1934

Species: ***africanus*** Hopwood 1934

INTRODUCTION

Dr Hopwood's type specimen for this genus and species consisted of the greater part of a skull (pls. 93 and 94), registered in the British Museum of Natural History as no. M. 14688. The locality of this specimen is given as 'Bed II Olduvai Gorge'. It was found during the 1931–2 season. It exhibits no trace of its original field identification number; but a good deal of somewhat pebbly and consolidated sandy matrix, of dark colour, still adheres to it.

ORIGINAL DIAGNOSIS

Dr Hopwood's original diagnosis of the genus placed it in the family Bovidae but made no attempt to indicate its probable subfamily or tribal status. His generic diagnosis reads as follows: 'Bovidae, with widely divergent (120°) horn cores, gently twisted, spiral of the left one clockwise; amount of twist from half to three fourths of one revolution: horn cores in plane of face, with a definite keel above and below, the upper one apparently the stronger. Genotype—*Pultiphagonides africanus*, sp.n.'

His specific diagnosis was as follows: 'A *Pultiphagonides* with long narrow nasal bones: no lachrymal fossa?; supra-orbital pits fairly large and somewhat sunken; face bent down on the basi-cranial axis at about 45°; teeth moderately hypsodont, with well-marked ribs but no basal pillar.' The description of the holotype reads as follows: 'An adult skull lacking the right horn core, much of the basi-cranial region, and the premaxillae. Registered M. 14688.' After giving a certain number of measurements, the brief note concluded with the following remarks: 'This species bears no very close resemblance to any of the living African Bovidae, and its affinities are uncertain. The generic name is taken from Plautus.' No illustration of the type specimen accompanied the original description, nor can I find any record of one having been published since.

NEW EXAMINATION

The type specimen has been carefully re-examined. As a result it seems clear that while Dr Hopwood's statement that *Pultiphagonides* has no close link with any living African member of the Bovidae is correct, it does have certain resemblances to another extinct genus from Olduvai, namely, *Pelorovis*. It is not suggested, however, that the link between these two genera is close. Both *Pultiphagonides* and *Pelorovis* may be aberrant representatives of the subfamily Caprinae. Both were placed in the Caprini by Professor Gaylord Simpson in his 'Classification of Mammals' published in 1945.

REVISED DIAGNOSIS OF THE GENUS 'PULTIPHAGONIDES'

The following revised diagnosis of the genus is suggested. It differs in only a few small characters from that originally given by Dr Hopwood.

Bovidae of the subfamily Caprinae and tribe Caprini, with broad forehead and widely divergent horn cores which are set at the lateral angles of the wide frontals with the brain case projecting posteriorly to a marked degree; the horn cores are gently twisted spirally in such a manner that the left one is clockwise and the right one counter clockwise, as in sheep and goats; the horn cores are set more or less in the same plane as the face; lachrymal fossae are apparently absent; maxillae with lateral bulges above the roots of the second molars.

REVISED SPECIFIC DIAGNOSIS FOR THE SPECIES 'AFRICANUS'

A species of the genus *Pultiphagonides* with relatively broad and arched nasal bones and a marked 'waist' on the maxilla immediately in front of the orbits (pl. 93); the face is relatively short and there is an inter-orbital bulge on the forehead in the mid-line between the orbits; the anterior part of the face is set at an angle of about 45° downwards from the basi-cranial axis; the post-orbital foramina are simple and relatively deep, placed rather forward, without anterior grooves; teeth are relatively, but not extremely, hypsodont.

DESCRIPTION OF HOLOTYPE

The following is a revised description of the type specimen. It consists of a fairly complete skull from which, however, the right horn core, the greater part of the pre-maxillary region and part of the base of the skull, including the right occipital condyle, are missing. The whole specimen is cracked and weathered, so that the finer detail cannot be interpreted. The left horn core has been broken off some 130 mm. from the base and it is not possible to estimate its original length. It exhibits a slight trace of an upper keel starting posteriorly. It is somewhat compressed from back to front, with a slight spiral twist. The horn core rises from a short pedicle on the frontal bone close to the upper margin of the orbit. The simple post-orbital foramina are set low on the face, approximately half-way down the orbit. They differ greatly from the form seen in most of the larger African Bovinae. There is no sign of any lachrymal fossa. On the forehead, between the orbits, slightly above the level of the orbital foramina there is a slight bulge or 'boss' on either side of the mid-line. The brain case projects backwards from the forehead for a distance of some 75 mm. and there is a strongly rugose nuchal crest at the upper limit of the area of attachment of the neck muscles. The occipital region is low and wide. It is also relatively flat, as in the Bovinae and not at all as seen in the Hippotraginae or the Antelopinae. In this respect it is different from the more normal Caprinae but is like *Pelorovis*. The palate is relatively wide (pl. 94). All three molars are preserved on both sides. The left fourth pre-molar is also intact. It seems probable that there were three pre-molars. The teeth are mediumly hypsodont and the anterior segment of each molar tooth is set in slight echelon to the posterior segment. Owing to the deep weathering of the specimen it is not possible to be certain whether the outer surface of the molars originally carried much cement, but slight traces are visible. None of the molar teeth show any sign of median or other accessory pillars, while the valleys between the two segments of each molar, on the palatal side, are very deep and extend down to the alveolar margin. The nasal bones are broken anteriorly so that their exact length cannot be measured; but they are noticeably arched and not flat as stated by Dr Hopwood. They rise well above the pre-maxillary level.

Owing to the damaged state of the type specimen very few exact measurements can be given, but the following approximate measurements will give some idea of the relative dimensions of different parts of the skull.

Measurements

	mm.
Width at lower margin of the external aspect of the horn core (calculated by doubling the width to the mid-line on the left side)	144
Width at the upper margin of the orbital processes	154
Width between the bases of the horn cores at the top of the skull (calculated by doubling the measurement on the left side to the mid-line)	96
Width between the margin of the supra-orbital foramina	70
Maximum bi-nasal width	41
Minimum bi-nasal width	30
Width of the maxillary 'waist' in front of the orbits	78
External palatal width at second molars	89
Internal palatal width at second molars	50
Length of the molar series at the alveolar margin	66
Approximate width of the occipital region (calculated from the left side)	140
Approximate height on the basi-occipital mid-line from the basi-occipital region of the *foramen magnum* to the nuchal crest	57

Genus: ***Pelorovis***

Species: ***oldowayensis*** Reck

In 1928 Professor Reck published a description of a new genus and species which he called *Pelorovis oldowayensis*. This description was well illustrated by photographs and line-drawings.

The material which was excavated from site BK II during the 1952–7 seasons includes a considerable number of specimens, amongst which are skulls, horn cores, mandibles and an almost complete skeleton which belong to *P. oldowayensis* (pl. 95).

A preliminary study throws some doubt upon the true position of the genus *Pelorovis*. It is possible that it is not really a member of the Caprini. Moreover, some of the specimens which have been provisionally included in this genus and species may be quite distinct. A full study and revision of the material is therefore required.

Incertae sedis

(*a*) A fragment of frontlet and the greater part of a small horn core, no. 5196, 1960, from site FLK N I, clearly represents a caprine. The horn core has the characteristic reverse spiral twist of the sheep and goats. It resembles a miniature example of *Pelorovis*.

It should be noted that amongst the other material excavated from this site there is a metacarpal of Caprini affinities but more like that of sheep than of goats. It may well belong with the horn core mentioned above.

(*b*) Specimen no. 749, 1960, from FLK NN I, is also probably an example of the Caprini. It is not of the same type as (*a*) above.

(*c*) There are four specimens which undoubtedly represent a very large member of the Caprini. One, no. F. 3000, was collected from the surface of Bed II during 1941. Two specimens, nos. 234 and 280, were found *in situ* at site SHK II during 1952. The last, no. 92, was collected from the surface of SHK II in 1957. They all represent the same genus and species.

Specimen no. F. 3000 is more massive than the others from SHK II. It consists of a right frontlet with the lower part of the horn core in position.

The horn core has an anterior–posterior diameter of 66 mm. and a lateral diameter of 44·5 mm. at the base. The frontal lobe of the brain case extends further forward than the most anterior part of the horn core. The horn core rises almost vertically and curves gently outwards so that the twist is counter-clockwise. Seen in profile, the angle between the top of the brain case and the frontal bone is nearly a right angle, as in the living Caprini.

Specimen no. 280 has characters similar in all respects to the previous specimen. It is a left frontlet with rather more of the horn core preserved.

Specimen no. 92 of 1957 is a left horn core, slightly more massive than no. 280 and with a greater length preserved. Specimen no. 234 of 1952 is the upper part of a horn core broken about a third of the way up, with only the tip missing. This specimen shows clearly that in this species the horn cores were long and slender and curved gently upwards, outwards and slightly forwards.

(*d*) There are two specimens in the Nairobi collections—one, no. 596, from site GC at the base of Bed IV, and one, no. F. 1003, from the surface of Bed II (site not known)—which represent very unusual members of the Caprinae, whose horn cores have a very flat cross-section. Similar horn cores are represented by fragments in the British Museum of Natural History collections.

SUMMARY

In this chapter some indication has been given of the wealth and diversity of the fossils representing members of the Bovidae in the Olduvai collections. They range from forms such as *Pelorovis oldowayensis*, *Bularchus arok*, *Hippotragus gigas* and *Strepsiceros grandis* to rare examples of the small antelopes such as *Philantomba* cf. *monticola* and a very small caprine in Bed I.

In a few cases a relatively detailed account of new genera and species has been attempted, but for the most part the detailed study of the material is not complete. In any event, excavations are still continuing and already, since the bulk of this chapter was written on the material available in 1961, a number of new species of Bovidae have been found, together with much better material representing some of the species discussed here.

Perhaps the most remarkable fact that emerges so far is the scarcity of fossils representing the smaller members of the Bovidae, for example, the duikers, the dik-dik, the oribi, the steinbok and the klipspringer. This cannot be due to the bones of these species being small and possibly escaping notice since thousands of bones of much smaller animals are in the collections. There must therefore be some other and at present unexplained reason for the scarcity of small antelope remains on the living-floors of the early hominids.

CHAPTER V

NON-MAMMALIAN FAUNA

Ever since 1931, when we found remains of crocodiles and fish at several sites, we have known that a non-mammalian fauna was present in the Olduvai deposits, but it is only during the past three years that any considerable quantity of fossils of this natuıe has been found *in situ.*

It has not been possible to obtain detailed reports upon any of the non-mammalian fossils in time for inclusion in this volume. The following summary, however, indicates the range of material. The following categories have been found:

Mollusca
Pisces
Amphibia
Reptilia
Aves

MOLLUSCA

Fossil mollusca occur in Beds I, II and IV. In Bed IV the commonest are those representing *Unio* sp. In Beds I and II there is a small bivalve akin to *Corbicula*, as well as specimens allied to *Melanoides tuberculata.*

PISCES

Fossil fish remains occur in Beds I, II and IV. Some are found in sands and silts, where they became imbedded in lake deposits at death. The majority of the specimens, however, come from the living-floors of early man and were clearly part of his diet.

The presence of such fish as *Tilapia* and Cichlidae in certain deposits suggests that at these periods the lake contained a considerable amount of water.

So far, only a brief preliminary report on the fish remains is available, based on material from two living-sites in Bed I, two in Bed II, and various sites in Bed IV. The identifications have been made by Dr Humphrey Greenwood of the British Museum of Natural History and are as follows:

Bed I. At site FLK NN I the genera *Clarias* and *Tilapia* are present in levels 2 and 3, while level 3 also has Cichlidae.

At site FLK I, on the *Zinjanthropus* living-floor, remains of *Clarias* and of Pecomorphi (probably cichlids) were recorded.

Bed II. The sites at SHK II and BK II yielded remains of *Clarias* and possibly *Tilapia.*

Bed IV. *Clarias* remains are plentiful at a number of different levels in Bed IV.

At all sites, *Clarias* specimens predominate. The skeletal elements represented are fairly comprehensive. The *Tilapia* and Percomorph fragments, on the other hand, are mainly vertebral.

It is impossible to give more than a generic identification since specific characters are not present. The ornamentation of the *Clarias* skull fragments agrees with that of the extant *C. mossambica–C. lazera* group.

AMPHIBIA

A considerable number of fossil bones representing Amphibia were found in 1959–61 during excavations of living-floors in Bed I.

REPTILIA

Reptiles are represented in the collections by remains of Chellonia, Squamata (including Serpentes) and Crocodilia. Most of the specimens in the first three groups come from excavations of living-floors of the Oldowan culture in Bed I and were apparently part of the diet of early man. Remains of crocodiles also sometimes occur on the living-floors in Beds I and II, but they are most common in deposits without occupational material, especially in Bed I, where they are very plentiful in the lowest levels. They also occur in Beds III and IV.

It has not been possible to obtain detailed reports on this material, but the following preliminary notes have been received.

(*a*) *Chellonia*

Tortoises are very well represented on the early Oldowan living-floors and seem to have been a regular article of diet. This is particularly true of the living-floor at site FLK NN I, where the remains of the juvenile hominid were found. Other sites in Bed I and in Beds II and IV have also yielded tortoise remains, but none have yet been identified. There are a few specimens belonging to the genus *Trionyx*.

Specimens from site FLK NN I have been submitted to the British Museum of Natural History, and the following report has been received from Miss A. G. Grandison:

The tortoise bones that you left with the Palaeontology Department have been identified by Dr E. E. Williams of the Museum of Comparative Zoology, Harvard, as belonging to *Pelusios castaneus*, Scweigger 1812. This name was resurrected by Laurent in 1956 to apply to a population that is known to occur in the L. Edward, Ruanda Urundi, Katanga, Kasai and Lower Congo areas; specimens from adjacent areas also probably belong to this species but have been erroneously identified as *P. subniger* although in the shape of the plastron and in the digital scalation there are marked differences in the two species and there seems no doubt that *castaneus* is distinct.

Wermuth and Mertens in their 1961 checklist of the Testudines include *castaneus* in the synonymy of *P. subniger* but apparently they did not examine the *castaneus* material.

(*b*) *Squamata*

During the excavation of various living-floors in Bed I a number of bones representing small lizards and chamaeleons were found, as well as some bones of a species of *Varanus* lizard.

(*c*) *Serpentes*

Snakes appear to have been included in the diet of the makers of the Oldowan culture. Vertebrae and teeth are represented in the collections from the living-floors in Beds I and II and are particularly plentiful at sites FLK NN I and FLK I. Elsewhere remains of snakes are rare, but a nearly complete skeleton has been found in Bed IV at site FC.

(*d*) *Crocodilia*

This material has not yet been studied. There is a large collection including parts of skulls and mandibles, limb bones, vertebrae, scutes and very numerous isolated teeth. None of these specimens suggest the presence of the narrow-snouted crocodiles *Euthecodon brumpti* and *Crocodilus cataphractus* which are so common in the Omo deposits. All the material seems to represent the living Nile crocodile *Crocodilus niloticus*. The nearest occurrence of this reptile today is in Lake Victoria and it has not survived in the Rift Valley lakes such as Manyara and Eyasi.

I have received the following preliminary note on the Reptilia and Amphibia from Dr Hofstetter of the Paris Museum of Natural History.

Quoi qu'il en soit, la faune étudiée comprend essentiellement:

Pipidae:	Un *Xenopus* assez abondant.
Bufonidae:	Un *Bufo* assez abondant et probablement d'autres Anoures non encore identifiés.
Agamidae:	Un *Agama*, ou peut-être 2 espèces.
Chamaeleontidae:	Un *Chamaeleon* très voisin de l'actuel *Ch. jacksoni*.
Scincidae:	Un *Riopa*.
Boidae:	Un grand *Python* proche de *P. sebae*. Des vertèbres d'un petit Boïdé, qui pourrait être le jeune du précédent.
Colubridae:	Au moins trois genres.
Elapidae:	Quelques vertèbres, apparemment de *Naja*.
Viperidae:	Une grande *Bitis*, distincte, par son maxillaire, de toutes les espèces actuelles.

AVES

Amongst the fossil material collected by Professor Reck in 1913 were remains of an eagle, *Aquila rapax*. Bones of an ostrich were also found during the 1913 expedition. Since 1951, fossil avian remains have been found in increasing numbers at sites in Beds I and II. In particular, the excavation of the Oldowan living-floors at sites FLK NN I, FLK I and FLK N I yielded an exceptionally large and varied collection of bird bones. These and the limb bones of a very large ostrich, also from Bed I, are now being studied by Dr Wetmore of the Smithsonian Institution, Washington, D.C. In view of the quantity of material a full report cannot be expected for some years, but Dr Wetmore has sent the following preliminary note.

The bird material from Olduvai is the richest find of avian fossils known to date from the whole of Africa. Preliminary sorting and examination, prior to the cleaning and hardening that is required before detailed study is possible, indicate a considerable variety in species. These show a marked difference in ecological conditions since numerous individuals are of kinds that live in aquatic habitats in contrast to the arid conditions that hold at present near the site.

Specimens in fair state of preservation from the level (FLK I) at which the *Zinjanthropus* remains were found include fragments from an ostrich, numbers of bones of flamingos and avocets, and parts from a large goose. There is one especially important find, in which a number of skeletal parts have been associated in such a manner as to indicate that they come from one individual.

The site (FLK NN I) where the hominid find was made adds ducks and a hawk, in addition to further bones of flamingos and avocets.

The abundant specimens from the higher level (FLK N I) include a greater diversity of kinds from a variety of medium-sized passeriform species to such aquatic forms as cormorants and ducks, accompanied by a spur-winged plover, and other shorebirds. Bones of francolins, doves, a grebe and an owl, and two species of swift are further indication of the variety present. The collection from this level includes many of the smaller parts of the skeleton, such as toe bones, vertebrae, quadrates, and wing phalanges, unusual finds in avian fossil material.

The finds *in toto* are of great importance as a record of the early avifauna of the region.

CHAPTER VI

REVIEW OF THE FAUNAL EVIDENCE

INTRODUCTION

In chapter II the faunal evidence which was available in 1951 was reviewed. It was pointed out that many of the conclusions drawn at that time were no longer tenable. This was due, in part, to the relative lack of *in situ* specimens and, in part, to faulty identifications of material that was often very fragmentary.

In chapters III, IV and V the present state of our knowledge has been set out. Very many new specimens have been collected, and the faunal list has, in consequence, been considerably extended. But the position is still far from satisfactory. Only a small part of the available fauna has been studied in any detail. There is now so much material in many of the groups that it will take many years for the various specialists to complete their reports. The position is further complicated by the fact that field work is still continuing, so that every year species new to the collections are found, together with further and often more complete specimens of species already known. Since so many of the reports on the faunal material are only of a preliminary character, this review of the fossil evidence must, itself, be regarded as wholly tentative.

In 1951 it was possible to list only 51 mammals as probably present in the Olduvai deposits. Another 20 were known to occur in similar formations at Laetolil, some twenty miles to the south. At the time, these were regarded as likely to be found at Olduvai in due course. The number of mammals listed in the present volume has risen to 152. Moreover, since the chapters dealing with the fauna were written, at least ten other new species have been found. These have not been described here.

The fauna of Bed I is now considerable. It has been recovered, in the main, from living-floors of the makers of the Oldowan culture. As a result, much of the material representing the large mammals is very fragmentary, since early man most effectively broke open skulls and marrow bones to get the maximum food value from them. There are also the remains of a very much larger number of small mammals such as rodents. Although much remains to be done we now have a reasonably good idea of the fauna of Bed I times.

When we turn to Bed II the position is complicated by the fact that the fauna must now be divided into two parts. The lower part of Bed II, which precedes a major geological and cultural break, had been only partly explored when this volume was being prepared, so that our knowledge of the fauna of that period was necessarily very limited. There is, however, a very large collection of fossil material from the upper part of Bed II, from sites such as BK II, SHK II, FC II, MNK II and FLK II.

The fauna of Bed III will not be discussed in this chapter. Very few of the specimens that have been found in Bed III are of diagnostic value and some are derived. In other cases material that has been described as coming from Bed III was not really from that deposit but from 'red beds' in Bed IV or Bed II which, in the earlier days of exploration, were sometimes misidentified as Bed III.

Bed IV is represented in our collections by relatively few specimens from excavated sites. There is no doubt that when extensive work on Bed IV is undertaken the faunal list for that deposit will be considerably enlarged.

DISCUSSION OF THE VALUE OF FAUNA AS AN INDICATION OF ECOLOGICAL SETTING

Before turning to a consideration of the fauna of the different deposits and its value in assessing the age of the beds, it is necessary to discuss, briefly, the significance of mammalian fauna as a guide to ecology. In particular, we must consider whether our knowledge of ecological conditions under which modern genera and species of mammals live, justifies us in drawing conclusions as to the existence of similar conditions from the presence of related fossil forms.

It is all too common for conclusions to be drawn about the past on the basis of a very limited knowledge of zoo-geographical conditions and problems. It is not generally recognised that many of the larger mammals are remarkably adaptable. Although they may today prefer certain types of habitat, they are by no means confined solely to these areas. A few examples from the present-day fauna of Africa will serve to illustrate this point. The giant forest hog, *Hylochoerus meinertzhageni*, was first found in Kenya forests at a high altitude and low temperature. It was therefore widely regarded as characteristic of this type of ecological setting. In fact, this animal ranges from the moorland zone at 9000 ft. down through the bamboo and juniper forests. It can be found both in dense forest and in open bush country. Moreover, it also occurs in the lowland tropical forests of the Congo. Because of the very variable nature of its habitat this species could clearly never be used as an indication of ecological conditions. Leopards in Africa range from the forests and bush country at the sea coast up to the forests high on Mounts Kenya and Kilimanjaro, occasionally even to the snow line. They also occur in the open plains of the semi-desert regions of the Kenya Northern Frontier Province. While zebra and giraffe are most commonly found in savannah and open plains with scattered thorn bush, they can also be found well within tropical forest zones such as those bordering Lake Manyara. These few examples serve to show how unwise it is to regard the usual habitat of large mammals as necessarily constant. If the habitat of large living mammals, belonging to a single species, varies so widely it is clear that the presence of extinct fossil species—even if related to the living forms—cannot be used as a basis for deducing ecological or climatic conditions.

It is just possible that some of the smaller mammals, such as the rodents and insectivores, may provide a surer indication of ecological conditions. *Heterocephalus*, the naked mole rat is today found only in semi-desert, arid areas. The gerbils and jerboas also seem to be essentially creatures of the drier regions. But it is not possible to be sure that their ancestors were similarly restricted in habitat. During Lower Pleistocene times a rodent with the dental and cranial characters of *Heterocephalus* may have been covered with fur and may have lived under colder and wetter conditions than would be suitable for the normal present-day representatives of the group.

While we may justifiably conclude that the presence of fish, hippopotamus, crocodiles and aquatic birds indicates the presence of a relatively stable body of water, it is necessary to use great caution when attempting to assess climatic conditions on the basis of most mammals.

THE FAUNA OF BED I

A large part of the fauna available from Bed I consists of small mammals, birds and reptiles which have not yet been studied. These cannot, therefore, be used for an appreciation of the faunal age of Bed I except in very general terms. There is available, however, a great deal more evidence about the large mammals of Bed I than there was in 1951. This makes it clear that the earlier view that Bed I should be grouped with Beds II, III and IV as representing the Middle Pleistocene must be modified. The evidence provided by the elephants, pigs, antelopes and carnivores indicates a Lower Pleistocene age for Bed I with Upper rather than Lower Villafranchian affinities. It seems to be legitimate, now, to refer in general terms to the fauna of Bed I as a distinct stage, provided it is clearly understood that progressive changes were taking place and that the material from the upper part of the bed differs, in certain elements, from that of the middle and lower parts.

Let us consider the elephants. Bed I has *Deinotherium bozasi* throughout. In the lower part there is a species of *Elephas* with molars that are low-crowned, with wide plates, which recalls an advanced form of *E. africanavus*, but which M. Coppens considers may be ancestral to *E. recki*. In the uppermost levels of Bed I the *Elephas* is more highly evolved and closely resembles the form of *E. recki* recorded from the Omo deposits. There is no trace of a *Stegalophodon* nor of *Anancus*, both of which occur in the Lower Villafranchian of East Africa at such sites as Kanam and Kaiso. The typical form of *Elephas recki* does not seem to appear until the upper part of Bed II. The presence of an elephant recalling *E. africanavus* might, by itself, be regarded as evidence of Lower Villafranchian age but the context indicated above suggests that an Upper Villafranchian age is more probable.

Turning next to the pigs. Bed I contains some very archaic forms such as *Ectopotamochoerus* and *Promesochoerus* as well as a primitive *Notochoerus*, which is similar to the form found at Kaiso. There are also very primitive members of the *Potamochoerus* and *Tapinochoerus* genera. Four of these pigs have not previously been recorded, whereas the fifth has affinities with a species from the Lower Villafranchian of Kaiso. The *Promesochoerus* may also perhaps be represented at Kaiso, since there are molar teeth from that site, previously attributed to *Mesochoerus heseloni* and *Sus limnites*, which would equally well belong to *Promesochoerus*.

It is clear that the pigs of Bed I are more primitive than those of Omo and have affinities with the Kaiso forms, while the absence of such pigs as *Nyanzachoerus* suggests that the horizon is younger than that of Kanam. The evidence of the pigs, therefore, together with that of the elephants, indicates an Upper Villafranchian age for Bed I.

Too little is known at present about the Bovidae of the Lower Villafranchian of East Africa for comparisons to be made with the material from Bed I. Species such as *Parmularius altidens*, *Damaliscus antiquus* and *Beatragus antiquus* are all more primitive and quite distinct from the published Bovidae from Omo and also from the corresponding species found in upper Bed II.

In addition to the genera and species mentioned above, an *Okapia*, a *Libytherium* and a giraffe occur in Bed I. The giraffe has not yet been identified, but since the *Libytherium* continues into Bed II and even into Bed IV, whilst the *Okapia* occurs at Laetolil, Omo and in Bed II, none of these animals are of value for comparative dating.

The general picture presented by the fossil mammals from the uppermost levels of Bed I suggests that a change of climate (and consequently of fauna present in the area) may have been in progress. At this level there are gerbils and naked mole rats, an antelope of affinities with the Hunter's antelope, numerous gazelles and a marked increase in equids. The elephant resembling *E. africanavus* seems to have disappeared but the pigs do not differ from those of the lower and middle parts of the bed, and are all notably more primitive than those from Omo.

Thus, the whole fauna of Bed I seems to fit best into the Upper Villafranchian. It may be regarded as later than that of Kaiso and Kanam but older than that of Omo. This is a new concept. In the past Omo was regarded as contemporary with Kaiso and Kanam, but the reappraisal of the Omo forms given in chapter III makes it clear that this view is no longer correct.

THE FAUNA OF BED II

In 1931 Bed II was regarded as representing a continuous series of deposits, and no attempt was made to check the fauna from the different levels. The geological evidence now available makes it clear that there was a major break in sedimentation within Bed II. The faunal and cultural evidence both support this view. In particular, the fauna which precedes the break differs in a number of important respects from that which follows it. In many ways the fauna and culture from the lower part of Bed II is closer to that of Bed I than to the upper part of Bed II, but there seem to be sufficient differences to warrant it being treated separately for the present.

Deinotherium bozasi continues in the lower part of Bed II but it is now associated with a more advanced elephant of the *E. recki* type. A slightly more primitive elephant also seems to occur. This association is very similar to that found at Omo and suggests a somewhat comparable age.

The Suidae of the lower part of Bed II, so far as we know them at present, include *Mesochoerus heseloni* (which is also found at Omo) but there is no trace of either *Promesochoerus* or *Ectopotamochoerus*. The pigs therefore also support the view that the base of Bed II may be of the same age as the Omo deposits.

The giraffe which is characteristic of the lower part of Bed II is *Giraffa gracilis*. This species was first found at Omo. It is not known to occur in Bed I and does not seem to have continued to the upper part of Bed II, where *Giraffa jumae* is found. *Damaliscus angusticornis* appears in the lower part of Bed II and also continues into upper Bed II and even later. The available evidence shows that the culture in the lower part of Bed II is an advanced Oldowan rather than an early Chellean.

In view of the combined faunal and cultural evidence as well as the known existence of a major climatic break separating the lower and upper parts of Bed II, it is best, for the present, to regard the basal part of Bed II as belonging to the closing stages of the Upper Villafranchian rather than to the start of the Middle Pleistocene.

A very extensive fauna has been obtained from deposits in the upper part of Bed II at sites BK II and SHK II.[1] The prevalence of many giant herbivores in upper Bed II times suggests optimum feeding conditions. The most important of these animals which so far as we know appear for the first time at this level are *Hippopotamus gorgops*, *Afrochoerus nicoli*, *Mesochoerus olduvaiensis*, *Potamochoerus majus*, *Orthostonyx brachyops*, *Pelorovis oldowayensis*, *Bularchus arok*, *Strepsiceros grandis* and *Tapinochoerus meadowsi*. *Libytherium olduvaiensis* has now become very common. The more primitive elephants have disappeared, together with *Deinotherium bozasi*, *Mesochoerus heseloni* and *Giraffa gracilis*. The Equidae are particularly well represented in upper Bed II times by two genera, *Equus* and *Stylohipparion*, but the study of this group is not yet available. The carnivores now include a very large sabre-tooth feline, which is quite distinct from that of Bed I and a large felid which shows affinities to the modern tiger. The fauna of this period is remarkable for the number of genera and species which appear in the Olduvai deposits for the first time, whilst most of the archaic forms, which were present in Bed I and the base of Bed II, have disappeared.

The overall picture presented by the fauna of the upper part of Bed II indicates that it is of Middle Pleistocene age.

THE FAUNA OF BED III

It is not possible to discuss the fauna of Bed III at present. A few sites containing some cultural and faunal remains are now known but have not yet been excavated.

THE FAUNA OF BED IV

The fauna of Bed IV is now somewhat better known than it was in 1951. Three sites have been excavated recently by Dr J. Waechter and Dr M. Kleindienst, but the specimens from these sites had not been unpacked or studied when this report was prepared for publication.

We know less about the fauna of Bed IV than about that of Beds I and II. We can say with certainty that the fauna is of Middle Pleistocene character and includes such animals as *Elephas recki*, *Hippopotamus gorgops*, *Libytherium olduvaiensis*, *Giraffa jumae*, *Afrochoerus nicoli*, *Tapinochoerus meadowsi*, *Phacochoerus altidens*, *Parmularius rugosus*, *Xenocephalus robustus*, *Bularchus arok*, *Simopithecus jonathani*, etc., as well as many gazelles, antelopes, equidae and rhinoceros. The giant baboon *Simopithecus oswaldi* is now very common. This fauna closely resembles that from Kanjera and Olorgesailie in Kenya, where the cultural material is also of Acheulean facies as it is in Bed IV at Olduvai.

The fossils so far obtained from Bed IV are from the lower and middle parts of the bed. They have been found mainly at human occupation sites. The upper part of Bed IV is mainly a brown aeolian tuff which appears to be unfossiliferous. It is regarded as indicating renewed dry conditions, possibly even sub-desertic. It seems likely that it was this dry period which was responsible for the extinction of so many of the animals which are characteristic of the upper part of the Middle Pleistocene in East Africa. We know from the evidence of sites in Kenya that the fauna of the Upper Pleistocene, in Gamblian pluvial times, is with few exceptions the same as that of the present day.

[1] The suggestion published in 1958 that there was a Chellean stage of culture at the base of Bed II, at sites BK II and SHK II, has proved to be incorrect. This site is now known to belong to the upper part of Bed II.

Table 3. *Vertical distribution of some of the faunal elements in Beds I to IV*

Species	Bed I	Bed II lower	Bed II upper	Bed IV
Hystrix sp.	×	.	.	.
Lepus sp.	×	.	.	.
Lutra sp.	×	.	.	.
Otocyon recki	×	.	.	.
Machairodontinae gen.et sp.indet.	×	.	.	.
Simopithecus sp.	×	.	.	.
Papio sp.	×	.	.	.
Ceratotherium cf. *efficax*	×	.	.	.
Ancylotherium cf. *hennigi*	×	.	.	.
Hippopotamus sp.	×	.	.	.
Ectopotamochoerus dubius	×	.	.	.
Potamochoerus intermedius	×	.	.	.
Promesochoerus mukiri	×	.	.	.
Tapinochoerus sp.	×	.	.	.
Gazella cf. *wellsi*	×	.	.	.
Parmularius altidens	×	.	.	.
Beatragus antiquus	×	.	.	.
Okapia cf. *stillei*	×	.	.	.
Elephas cf. *africanavus*	×	.	.	.
Crocuta aff. *ultra*	×	×	.	.
Panthera aff. *crassidens*	×	×	.	.
Canis africanus	×	×	.	.
Phacochoerus altidens robustus	×	×	.	.
Strepsiceros maryanus	×	×	.	.
Hippotragus gigas	×	×	.	.
Elephas recki (a primitive stage)	×	×	.	.
Deinotherium bozasi	×	×	.	.
Damaliscus antiquus	×	×	×	.
Thos mesomelas	×	×	×	×
Equus oldowayensis	×	×	×	×
Stylohipparion albertense	×	×	×	×
Gorgon olduvaiensis	×	×	×	×
Libytherium olduvaiensis	×	×	×	×
Giraffa gracilis	.	×	.	.
Panthera cf. *tigris*	.	.	×	.
Machairodontinae gen. et sp.indet.	.	.	×	.
Okapia sp.indet.	.	.	×	.
Notochoerus compactus	.	.	×	.
Orthostonyx brachyops	.	.	×	.
Mesochoerus olduvaiensis	.	.	×	.
Pelorovis oldowayensis	.	.	×	.
Parmularius sp.	.	.	×	.
Strepsiceros grandis	.	.	×	.
Alcelaphus kattwinkeli	.	.	×	.
Pultiphagonides africanus	.	.	×	.
Phenacotragus recki	.	.	×	.
Gazella aff. *granti*	.	.	×	.
Aonyx sp.	.	.	×	×
Diceros bicornis	.	.	×	×
Ceratotherium simum	.	.	×	×
Simopithecus jonathani	.	.	×	×
Afrochoerus nicoli	.	.	×	×
Tapinochoerus meadowsi	.	.	×	×
Potamochoerus majus	.	.	×	×
Hippopotamus gorgops	.	.	×	×
Hippotragus niro	.	.	×	×
Damaliscus angusticornis	.	.	×	×
Giraffa jumae	.	.	×	×
Elephas recki	.	.	×	×
Crocuta sp.	.	.	×	×
Bularchus arok	.	.	×	×
Notochoerus hopwoodi	.	.	.	×
Phacochoerus altidens altidens	.	.	.	×
Tapinochoerus minutus	.	.	.	×
Taurotragus arkelli	.	.	.	×
Parmularius rugosus	.	.	.	×
Thaleroceros radiciformis	.	.	.	×
Cercopithecoidea gen. et sp.indet.	.	.	.	×
Papio indet.	.	× ?	.	×

Table 4. *Fossil mammals in the Olduvai beds as at 1951*

1. *Hystrix* sp.
2. *Canis africanus*
3. *Thos mesomelas*
4. *Otocyon recki*
5. *Aonyx* sp.
6. *Panthera* sp.
7. *Felis* sp.
8. *Crocuta* sp.
9. *Simopithecus oswaldi leakeyi*
10. *Papio* sp.
11. *Elephas recki*
12. *E. exoptatus*
13. *Deinotherium bozasi*
14. *Ceratotherium simum*
15. *Diceros bicornis*
16. *Stylohipparion* sp.
17. *Hipparion* sp. (? *Hypsihipparion*)
18. *Equus oldowayensis*
19. *Hippotigris* sp.
20. *Metaschizotherium*
21. *Potamochoerus majus*
22. *Mesochoerus olduvaiensis*
23. *Notochoerus* sp.
24. *Phacochoerus altidens*
25. *Hippopotamus gorgops*
26. *Giraffa* sp.
27. *Libytherium olduvaiensis*
28. *Taurotragus* sp.
29. *Strepsiceros* sp.
30. *Strepsiceros* sp.
31. *Strepsiceros* sp.
32. *Tragelaphus scriptus*
33. *Bularchus arok*
34. *Philantomba* cf. *manticola*
35. *Hippotragus niro*
36. *H.* cf. *equinus*
37. *H.* cf. *niger*
38. *Damaliscus angusticornis*
39. *Parmularius altidens*
40. *Alcelaphus kattwinkeli*
41. *Beatragus* sp.
42. *Gorgon* sp.
43. *Nesotragus* sp.
44. *Gazella praecursor*
45. *G. granti*
46. *Phenacotragus recki*
47. *Pultiphagonides africanus*
48. *Thaleroceros radiciformis*
49. *Pelorovis oldowayensis*
50. *Redunca* sp.
51. *Afrochoerus nicoli*

SUMMARY

The fauna of Olduvai does not belong to one faunal stage but to several. Bed I is Villafranchian and the lower part of Bed II belongs more with Bed I than with the upper part of Bed II. It has fauna almost identical with that of Omo. The upper part of Bed II has much in common with Bed IV, and these two together represent the Middle Pleistocene of East Africa. At the end of Bed IV times a second major faunal break occurs and, thereafter, the fauna of East Africa is, in the main, that of today.

It is thus clear that the earlier concept of an 'Oldowan faunal stage' intermediate between a 'Kaiso stage' and a 'Gamblian stage' is no longer tenable.

FAUNAL LISTS

Table 3 lists certain elements of the fauna as it is now known, particularly of the larger mammals. A fuller list cannot be given at this stage since many of the identifications of the carnivora and rodentia are only generic, while in other groups, such as the Equidae, no reports have been received. The table indicates the beds in which the faunal groups are known to occur, but it must be emphasised that the apparent absence of any particular species or genus from any horizon does not necessarily mean that it could not occur at that level.

For convenience and easy reference the fauna as recorded from Olduvai in 1951 is listed in Table 4. Where necessary, alterations have been made in the names. For example, *Simopithecus leakeyi* is shown as *Simopithecus oswaldi leakeyi* and what was formerly called *Palaeoloxodon recki* is listed as *Elephas recki*. In some cases the specific identification that was given in 1951 has been omitted, because the validity is now doubtful.

CHAPTER VII

THE PROBLEMS OF THE CLIMATIC SEQUENCE

The view has been expressed by a number of geologists that the magnificent fossil-bearing deposits at Olduvai can be interpreted without reference to the possibility of Pleistocene climatic changes. This does not seem to be compatible with the observed facts. The changing sequence of deposits exposed in the gorge, when taken in conjunction with their geographic setting, seems to be incapable of satisfactory explanation without reference to major fluctuations of climate during the time that they were being formed. Even if there was no other supporting evidence elsewhere in East Africa, the Olduvai sequence suggests most strongly that there were prolonged periods when the climate was wetter than it is today, and others when it was fully as dry, or even drier.

Before summarising the evidence for climatic change in Pleistocene times in East Africa, it is necessary to define clearly what we mean by the terms 'pluvial' and 'interpluvial'. These two words are used to refer to major changes of climate which were comparable to the glacials and interglacials of the Pleistocene in Europe and North America. Just as the term 'glacial' does not imply that it snowed all day and every day, so the word 'pluvial' does not mean that it rained every day. In Europe the words 'glacial period' imply a time during which generally colder conditions than those of the same region at the present day, persisted over a prolonged period. This general decrease in over-all temperature resulted in the formation of ice sheets under certain conditions and also of other phenomena indicative of a colder climate. Similarly, the word 'interglacial' implies generally warmer conditions than those of the present day; sometimes warm and wet and sometimes hot and dry.

In East Africa the word 'pluvial' is used to convey the idea that over a prolonged period of time there was a general increase of precipitation relative to that of the same region today. This increase of precipitation and consequent cloud cover was probably accompanied by some reduction in the average temperature. Within such a period there must inevitably have been numerous oscillations, some of them minor and some, perhaps, fairly marked. When the term 'interpluvial' is used it does not necessarily mean that the climate had become desertic, or even very dry. It means rather that the average rainfall over a prolonged period was markedly less than that in the same region today, and certainly markedly less than during the pluvials which preceded and followed it. Areas that are somewhat dry today became drier still, possibly desertic or sub-desertic. Areas where the present-day rainfall may be 60 or 70 inches a year, may have had much less than this, but still remained wet enough for abundant vegetation and adequate water supply to support animal life.

During what is called an 'interpluvial' there may have been a number of occasions when torrential rainfall occurred, and this sort of condition could have been maintained for several decades without being regarded as more than a minor oscillation in a generally interpluvial climate.

From what has been stated above it follows that when we speak of a pluvial facies, when referring to any geological deposit, there may still be evidence of temporary lake recessions and decrease in rainfall within that time. Equally, during what is described as an 'interpluvial' period, torrential river gravels may have formed as a result of a minor increase in precipitation within an over-all drier phase.

It is even more important to remember that the terms must be used in direct relation to the climate of the region which is being studied. A climate which would rank as wet and pluvial for Olduvai today, would rank as dry in some of the high-rainfall zones of the Kenya Highlands.

Before attempting to summarise the evidence which is available today to support the theory of pluvials and interpluvials in East Africa, the history of the study of climatic changes in this region must be briefly outlined. It must also be made quite clear that we do not suggest that the East African evidence in any way proves the existence of comparable climatic fluctuations elsewhere. Only the facts as we believe them to exist in this area are presented for consideration.

Professor J. W. Gregory, the pioneer geologist of East Africa, was the first person to propound the view that during the Pleistocene this region had witnessed major climatic changes. He did not elaborate his theory, but he made it clear that he considered that the evidence available to him in 1918 showed that there had been long periods of much greater rainfall than today.

Mr E. J. Wayland, Director of the Geological Survey Department of Uganda after the First World War, was the first person to consider the problem of East African pluvials and interpluvials in any great detail, and he wrote a number of papers on the subject. While it is possible that a part of his evidence, obtained in Uganda, was not quite as conclusive as he believed it to be, he did adduce many facts that are worthy of more study than has been given to them by his critics.

Our studies in Kenya began in 1926, and it was at once apparent that important fluctuations of climate had taken place during Neolithic and Mesolithic times. As the studies were extended backwards into the Pleistocene, it also became clear that there had been major changes in precipitation during the Upper and Middle Pleistocene periods.

Dr Erik Nilsson of Stockholm also carried out important studies in this field in Kenya, where he worked with us, and in Ethiopia. The result of his observations supported and complemented ours. The first summary of our earlier studies was given in *Stone Age Cultures of Kenya*, published in 1931. It was fully realised, at that time, that a good deal of revision and modification would inevitably be needed as work progressed.

In 1947, at the first Pan-African Congress of Prehistory held in Nairobi, the evidence concerning our local climatic changes was presented and was given careful consideration. As a result of the provisional acceptance of the sequence of events which was indicated by our evidence, the following resolution was passed by the Congress at its final plenary session. 'Certain stratigraphical units should be recognised in East Africa from now onwards. They are: Nakuran, Makalian, Gamblian, Kamasian, Kageran.' This was not a wise resolution, since it took names which had been applied by us in East Africa for pluvials and post-pluvial wet episodes and suggested that they should be recognised as 'stratigraphic units'. The intention, however, was clear. It was a recognition of the fact that a series of geological deposits existed in the area and that these were associated with good evidence of climatic changes. It recognised, too, that these deposits contained fossil faunas in places, which provided an excellent provisional framework for future study.

Unfortunately, many workers in other parts of Africa seized upon this resolution too readily and used the East African terminology somewhat indiscriminately, without having any real, or adequate, correlation with our East African sequence. As a result the whole idea of Pleistocene climatic changes in Africa was brought into disrepute, since our ideas were being applied loosely in regions where the evidence was not yet adequate. This does not mean that the East African evidence for climatic changes during the Pleistocene is unsound.

On the other hand, it does not follow that the climatic changes, of which there is clear evidence in East Africa, will necessarily be encountered in other parts of the continent. The study of each and every area must be made independently and only correlated with results elsewhere if it is found to be scientifically possible. Moreover, it does not follow that any given East African pluvial is necessarily synchronous with a similar, or corresponding, pluvial which can be shown to have existed elsewhere. On the other hand, it is now quite clear that there is some evidence of major climatic fluctuations during the Pleistocene in very many different parts of Africa. The time is yet to come when it will be possible to make really widespread comparisons. When it does, the faunal evidence of geological age will also have to be taken into account.

Those who question the idea of pluvials and interpluvials overlook the fact that climate is a world factor and that, if there were major climatic changes in Europe, America and Asia during the Pleistocene, similar events must have occurred in Africa. It is wholly impossible to accept the view that the climate of Africa remained static, and the same as it is today, over the long period of time when glacials and interglacials were leaving their mark upon other continents.

Even if no evidence could be found at all in East Africa, to indicate climatic changes, it would still be necessary to postulate that they had occurred, simply upon theoretical grounds. However, in Kenya, Tanganyika, Uganda and Ethiopia there is abundant evidence of pluvials and interpluvials. Similar facts are also emerging from many other parts of Africa. In East Africa we do not yet know as much about this subject as we would wish, but a great deal of work has already been accomplished and our theories and ideas are an attempt to interpret the facts.

When discussing the evidence of climatic changes, it is of course necessary to take into account all other possible influencing factors, in fact the total geological picture, including that of tectonic activity. Earth movements are certainly important but they should not be misused. It has, for example, been claimed that climatic changes should not be invoked to interpret the Olduvai deposits (Beds I–IV) because 'these deposits were formed in the Rift Valley'. In fact, the deposits in question antedate the faulting and were cut through by the Rift Valley faults subsequent to their formation. Similarly, suggestions have been made from time to time, concerning hypothetical and repetitive damming and draining of lake basins by volcanic activity, instead of interpreting the observed fluctuation of the lakes as being due to climatic changes. Those who suggest this seem to overlook the fact that there would have to have been simultaneous damming and draining of a very large number of lakes, not once but on a number of successive occasions.

Let us now turn to the available evidence for climatic changes in East Africa. It is not possible to give all the evidence here, but a brief summary of the salient facts is necessary. We recognise in East Africa the existence of a pluvial period at the beginning of the Pleistocene, during the Villafranchian. For this pluvial the name Kageran is used. It is not perhaps the most suitable, since Wayland's original evidence, from the valley of the Kagera River, was not very firmly based. However, this name has precedence and it has been widely used. While, therefore, we still use the name Kageran for the first Pleistocene pluvial which we recognise, we must stress that the best evidence for its earlier phases is to be found at Kanam East and Kanam West in Kenya. The geological formations at these two sites consist of lake deposits containing an excellent Lower Villafranchian fauna by which they are dated. These deposits were formed during a period when the waters of Lake Victoria stood at a height of 300 ± ft. above the present level of the lake. After studying all the available evidence, it is not possible to see how Lake Victoria could have attained and maintained such a high level except under conditions of precipitation very much greater than those of today.

It is often argued that, before the Lake Victoria outlet at Jinja was formed at its present level, rainfall of the same general order as that which is normal in the region today could have filled the enormous closed basin to a depth of 300 ± ft. above the present level. This argument overlooks the fact that several recent attempts to raise the water level of Lake Victoria by a few feet, by building a weir across the only outlet at the Ripon Falls, have failed under conditions of normal rainfall. This has been due to the fact that the evaporation rate is very high indeed relative to the normal yearly inflow from the watersheds of Kenya, Uganda and Tanganyika. But, in 1961–2, as a result of periods of excessive rainfall, combined with a much greater cloud cover and consequent reduction of evaporation, the lake level immediately rose some 4 ft. within a short space of time.

It is clear that unless the rainfall was considerably greater than today and remained so over a prolonged period, with evaporation also lower because of increased cloud cover, the 300 ± ft. lake in the Victoria basin could neither have formed nor have lasted. We know, however, that it existed for a considerable time, as can be seen by the depth of the sedimentary deposits which formed along its margins at this level.

It is not only in the Lake Victoria basin that we have evidence of prolonged increased precipitation during the Villafranchian. Similar evidence can be found in other areas, including the Lake Rudolf basin, where there is a lake beach with a Villafranchian fauna some 700 ± ft. above the present level of the lake and some fifty miles distant from its modern shore line. Similarly, although the evidence in the Lake Albert basin has been less fully studied, it is difficult to see how the earlier part of the Kaiso sedimentary series could have been formed unless precipitation was much greater than it is today.

The fauna of Olduvai Bed I indicates, as we have seen, that the deposit was formed during the Upper Villafranchian, while the Omo fauna and that of the base of Bed II at Olduvai belong to the closing episodes of the Villafranchian. Both these sites have extensive lake beds, and indicate the presence of lakes which could not have formed, or existed for any length of time, under present-day climatic conditions in those areas.

The evidence for a major change of climate between what we call the Kageran pluvial in Villafranchian times and the next pluvial, which we call Kamasian, was based, in the first instance, upon a study of the geology and fauna in the Kanam and Rawe area, on the north-east side of Lake Victoria. There we found deposits which contained an early Middle Pleistocene fauna resting, with strong unconformity, on the earlier Kanam East and Kanam West series in which there were Villafranchian fossils. This unconformity can only be explained by a lowering of the lake level between the deposition of the two series of rocks.

It has been suggested by opponents of the theory of climatic fluctuation that the lowering of the water level after the deposition of the Kanam series was the result of a deepening of the overflow channel, either through erosion or by tectonic activity. While such an explanation for the fall in water level cannot be ruled out, it would be necessary to postulate that the lowered outlet was then again blocked, to a considerable height, in order that the lake could rise to a high level once more. There is no evidence whatever to support such a theory and it seems much more likely that the lowering of the water level was due to a decline in precipitation together with increased evaporation, and that the subsequent rise which led to the deposition of the lower part of the Middle Pleistocene beds at Rawe was due to renewed increase in rainfall.

The upper part of the Rawe series—the so-called fish beds—rest conformably upon the lower part of the series. The study of that deposit shows that the lake was then gradually drying up and that many species of fish were becoming extinct. The exceptions include catfish and lung fish, which are capable of surviving in swamps and in shallow water.

Supporting evidence comes from Olduvai, where the lower part of the Middle Pleistocene is represented by the upper part of Bed II. This consists of lake beds and river sands containing fish, crocodile and hippopotamus, associated with many ungulates of very large size which must have required more abundant vegetation than could grow at Olduvai under present-day conditions. Bed II is followed by the deposition of Bed III, which is a terrestrial deposit formed under different and much drier conditions.

The evidence from both areas thus strongly suggests that the Kamasian pluvial was followed by a drier interpluvial, in the same way that the Kageran was separated from the Kamasian by a long and relatively dry interval.

Next we have the Kanjeran pluvial. This name was first proposed and accepted at the second Pan-African Congress on Prehistory held in Algeria in 1952. By that time it had become clear that what we had previously regarded as merely two distinct peaks of the Kamasian pluvial were in reality two separate pluvials. The name Kamasian was therefore retained for the first of these (as we have seen) and the term Kanjeran introduced for the renewed period of pluvial conditions which followed the deposition of the Rawe fish beds in the Victoria basin and Bed III at Olduvai.

The name Kanjeran derives from the site at Kanjera on the south shore of the Kavirondo Gulf of Lake Victoria, where fossil-bearing lake deposits were discovered by Dr Felix Oswald in 1910. The beds were then thought to be Pliocene but are now known to belong to the Upper Middle Pleistocene. These beds—which were later warped and faulted—rest unconformably upon the Upper Rawe fish beds which, as we have seen, represent the drying up of the early Middle Pleistocene lake with consequent extinction of the fish fauna.

The Kanjeran beds could only have been deposited on the earlier series if the lake had risen to a height of about 100 ft. above its present level. There are those who would like to postulate that the lowering of the lake level and subsequent rise were due to a second lowering and blocking of the outlet. This view ignores the fact that there is evidence of similar climatic fluctuations at the same time, elsewhere in East Africa. Indeed, it is not possible to invoke tectonic activity to account for the events that occurred at that time, since major tectonic disturbance did not occur until after the end of the Kanjeran pluvial. As examples of events similar to those of Kanjera we may cite Olduvai and Olorgesailie.

At Olduvai the lower part of Bed IV rests with marked unconformity upon Bed III. In some places channels were cut so deep at this period that the whole of Bed III was removed and Bed IV rests directly on Bed II. The deposits of the lower and middle part of Bed IV are mainly fluviatile, but there are also some swamp and shallow lake beds. The fauna of these deposits includes many aquatic forms such as hippopotamus, crocodile, fish, *unio* shells, etc., as well as many large extinct herbivorous animals such as *Libytherium*, *Bularchus*, *Afrochoerus*, *Tapinochoerus*, *Hippotragus*, etc. All of these creatures probably required abundant vegetation in order to survive. We therefore conclude that the climate was considerably wetter in the Olduvai region than it is today. At present, this area is waterless for many months at a time. Even during years of excessive rainfall, such as occurred in 1961–2, only a small shallow lake formed in the Balbal depression; and this lasted for only 18 months, which was not nearly long enough for fish, crocodile and hippopotamus to establish themselves. It would require a prolonged period of increased precipitation and increase in cloud cover to give rise to the conditions and the type of fauna which was present in the lower and middle parts of Bed IV.

The evidence at Olorgesailie in Kenya shows that similar conditions existed there at the same time. The annual rainfall around Olorgesailie today and on its watershed results in torrential flows of water at certain seasons of the year, but does not suffice to maintain any permanent sheet of water—even shallow water—for more than a few weeks of the year, because of the very high rate of evaporation. But in the upper part of the Middle Pleistocene, when Acheulean man inhabited the area, there was a lake with hippopotamus and fish which, over a period of time, was sufficiently deep and stable for the formation of considerable beds of diatomite. The fauna included giant forms such as *Libytherium* and *Tapinochoerus*.

The problem of what happened at the close of the Middle Pleistocene is not easy to resolve, but there is considerable evidence which points to a drastic change of climate to dry and possibly even desertic conditions, before the onset of the Gamblian pluvial. In the first place, a large number of animals that are common in the Kanjeran pluvial deposits have become extinct at the time of the Gamblian pluvial, in the Upper Pleistocene, and are replaced by the living species. In East Africa there is no evidence after the Kanjeran of the existence of *Elephas recki*, *Libytherium olduvaiensis*, *Tapinochoerus meadowsi*, *Afrochoerus nicoli*, *Bularchus arok*, and many other giant forms. The disappearance of the Kanjeran fauna at this time suggests, by itself, a considerable climatic break.

There is considerable evidence that towards the close of the Middle Pleistocene, at the end of the Kanjeran pluvial, there were major tectonic movements resulting in faulting and warping on a big scale. This resulted in the formation of a whole series of new lake basins in the floor of the Great Rift Valley, in Ethiopia, Kenya and Tanganyika. The deposits in these basins have been studied in detail, and as a result there can be no doubt that there was a pluvial period during the Upper Pleistocene. This has been termed the Gamblian.

There has been some doubt in the past as to whether there was a major climatic break between the Kanjeran and Gamblian pluvials. The evidence was based mainly on fauna, and it was therefore possible to argue that the changes in water levels were caused by the earth movements known to have taken place at the time. It has even been suggested that the extinction of many species of mammals in East Africa after the Kanjeran pluvial might have been the result of gases released by tectonic activity.

Recently, however, fresh evidence has come to light at Olduvai. This evidence has been summarised by Dr Hay and will be dealt with in greater detail when he has carried out another season in the field. At Olduvai, the upper part of Bed IV consists of a deposit of aeolian tuff lacking fossil remains, which seems to have been formed under desertic conditions. This, taken in conjunction with the extinction of the fauna, points strongly to a dry interpluvial period between the Kanjeran and the Gamblian.

The fluctuations within the Gamblian pluvial cannot be seen clearly at Olduvai, where this pluvial resulted in the erosion of a mature valley. But elsewhere in Kenya and Ethiopia, the effects of the Gamblian pluvial have never been seriously questioned.

In the Lake Victoria basin in Upper Pleistocene times a well-marked beach was formed at about 30 ft. above the present level. Since this was formed after the outlet at Jinja was in existence the climate must have been considerably wetter than it is today in the area, to achieve such a result. The effect of increased rainfall would of course also be accelerated by the decreasing evaporation rate resultant upon increased cloud cover during a pluvial period.

At the close of the Gamblian pluvial there is evidence of a short but very dry period during which windborn deposits were formed in many areas which carried a heavy vegetation both before and after. At this time, too, most of the smaller lakes dried up completely and their fish fauna was exterminated. In the Olduvai region this dry period resulted in the partial filling up of the valley cut in Gamblian times by windblown sands which contain little in the way of fossil fauna save for some land snails of a species which are today found in semi-desertic regions.

The post-Pleistocene climatic changes, which we have termed the Makalian and Nakuran, will not be discussed here, since they have no place in this book.

The following is a brief summary of the evidence for pluvial and interpluvial periods to be found in the Olduvai deposits.

Bed I, while mainly composed of coarse volcanic material in the east and of finer but similar deposits in the centre and west, was formed during a time when the climate was wetter than it is today in the Olduvai region. The base of Bed I is mainly composed of clays with many remains of crocodile and some fish and water birds. These, together with occasional hippopotami, continue throughout Bed I in suitable deposits. Temporary land surfaces occur throughout the bed, often with living-floors upon them. On these the fauna also includes many land mammals. The base of Bed II belongs with Bed I as part of a single pluvial.

After the deposits of the lower part of Bed II were formed, there was a change to drier conditions —an interpluvial—which is shown by a period in which there was deposition of aeolian sands and other sub-aerial deposits. This was followed by the cutting of channels and deposition of torrent gravels, as wet conditions set in once more. The upper part of Bed II consists of fluviatile and lacustrine deposits with a fauna indicative of a renewed wet climate, a second pluvial. This is followed by Bed III, a period of dry climate but not necessarily desertic. Lake beds and river sands and silts are replaced by sub-aerial soils. In places, these are cut into by channels with torrent gravels. There are also some river sands and gravels containing remains of crocodile and fish.

The lower and middle part of Bed IV represents renewed pluvial conditions with a climate which was certainly wetter than that of today. Fish, crocodile and *unio* are common and the fauna of large mammals becomes once more abundant, with many human occupation sites. The closing stages of Bed IV represent another dry period, this time sub-desertic, during which beds of aeolian sand were laid down. These do not contain any fauna.

After Bed IV and the faulting which followed it, the climate once again became wetter than today. A wide valley was cut from west to east, through the earlier deposits, in places right down to the level of the lava. Man lived by the side of this valley during Upper Palaeolithic times. The dry period which occurred at the end of the Upper Pleistocene is represented by the aeolian deposit of Bed V.

The evidence of Olduvai Gorge is thus fully in keeping with that from other parts of East Africa and confirms that major climatic fluctuations took place during the Pleistocene.

CHAPTER VIII

DATING BY THE POTASSIUM-ARGON TECHNIQUE[1]

AGE OF BED I, OLDUVAI GORGE, TANGANYIKA

Olduvai Gorge is justly famous because of its unique geological sequence of Pleistocene deposits, which are exceedingly rich in fossil fauna, as well as a long sequence of stages of evolution of the earlier Stone Age cultures.

In the monograph published in 1951 the view was expressed that, although Bed I differed from Bed II in faunal content, both belonged to the lower part of the Middle Pleistocene. This view was revised by one of us in 1959 as a result of reviewing the fauna collected in the series of detailed excavations during the period 1952–9 inclusive.

Leakey claimed that it was now apparent that the time-interval between Bed I and Bed II was greater than had been previously supposed, and he reverted to the view which he had published in 1935 that Bed I was of Lower Pleistocene age. In view of the extraordinary wealth of fossil and cultural material in the Olduvai deposits, it was not surprising when in 1959 a most important fossil hominid skull—*Zinjanthropus boisei*—was found in Bed I, at site FLK I, in association with faunal remains and Stone Age Culture material of the Olduwan Culture. In 1960, thanks to the Research Committee of the National Geographic Society, the Wenner Gren Foundation and the Wilkie Trust, very extensive further work was carried out, resulting in the discovery of pre-*Zinjanthropus* fossil hominid material at a lower geological level in Bed I.

Fortunately, many of the deposits at Olduvai Gorge are pure volcanic deposits, containing no derived material, and are therefore usable for the potassium–argon dating method. A preliminary examination of deposits in the Gorge and collecting of material were carried out by two of us, Leakey and Evernden, in 1958 and additional material was collected at intervals and sent to Berkeley by Leakey in 1959 and 1960. In 1961 Curtis visited Olduvai Gorge with Leakey and Richard Leakey for further collecting and study. Although many of the samples collected have not yet been studied, we all three feel that a report on those specimens which have a direct bearing on the age of the fossil hominid remains in Bed I should be published, in view of the scientific importance of these early hominids.

The three sites which have yielded hominid remains in Bed I are: (*a*) FLK I, the site which yielded the skull of *Zinjanthropus boisei*; (*b*) FLK NN I, the site which yielded the remains of a 'child', representing a pre-*Zinjanthropus* hominid; and (*c*) MK I, which has not been referred to in the preliminary reports but which has yielded some teeth and parts of a skull of a hominid.[1] All three sites have yielded archaic fauna and tools of the Olduwan culture, and are regarded as of Lower Pleistocene (Villafranchian) age.

The Bed I samples which have been dated are:

KA 412, 437 — Basal bed in the tuff and sediment sequence of Olduvai Gorge. Sample collected immediately west of third fault on south side of gorge at knife-edge descent. This horizon can be correlated with the series of tuffs in which hominid remains have been found by Leakey.

KA 846 — Site MK, 18 in. below hominid remains. Buff coarse-grained tuff 13–14 in. thick.

KA 847 — Site MK, just above hominid layer. Light grey tuff, 20 ft. thick. Appears to be Pelean type ash. On every criterion, this must be considered an uncontaminated volcanic deposit.

KA 849 — Site FLK I. 11 in. crystal tuff overlying *Zinjanthropus* layer

KA 850 — Site FLK NN I. 2 in. ash above very fine-grained 1 in. ash resting on pre-*Zinjanthropus* floor.

KA 851 — Site FLK I. 12 in. tuff immediately under *Zinjanthropus* floor. Sample is rather weathered and soft with numerous Zinjtime root holes.

KA 664, 664R — Sample collected near top of Bed I, some distance from the hominid sites. At time of collection, it was thought that this horizon was older than the hominid sites and it is so described in a paper submitted for publication by the INQUA Congress, Warsaw, 1961. However, more careful field-work has established unequivocally that this horizon is definitely younger than any of the hominid sites.

KA 861 — 19 in. soft lapilli tuff, near top of Bed I and definitely younger than any of the hominid sites.

In the area of the three hominid sites, between 25 and 30 individual tuffs and tuffaceous beds can be distinguished in Bed I. At these sites these aggregate 39–43 ft. in thickness. Most tuffs are approximately a foot in thickness, but at the MK site the tuff immediately overlying the hominid floor is 129 in. In order of abundance, the tuffs fall into the following categories: anorthoclase-bearing vitric-crystal tuffs; anorthoclase-bearing crystal vitric tuffs; oligoclase bearing lapilli tuffs; and tuffaceous sandstones.

Within individual beds faint lines of stratification and

[1] The communications in this chapter are reprinted, by permission of the Editor, in the order in which they appeared in *Nature*, to enable readers who have no easy access to that publication to understand the controversy over the dating of the Olduvai deposits by the potassium–argon technique. Whatever may or may not be the dates of the various deposits in terms of thousands of years, Bed I is by its fauna Upper Villafranchian and with it the lower part of Bed II. Upper Bed II to the end of Bed IV is Middle Pleistocene.

[1] There are, in fact, no skull parts from this site.

evidence of size sorting may sometimes be seen. Some beds, however, show little or no sorting, vary tremendously in grain and lapilli size and exhibit characteristics of nuée ardente deposits. This is true of the tuff overlying the hominid remains at MK site. Some of the crystal vitric tuffs show marked cross-bedding and have probably been reworked. Most beds, however, appear to be primary ash falls.

Vitric shards in most of the tuffs have been more or less altered to clay, but the crystals appear fresh. Dark and strongly altered tops of some tuffs probably marked old soil zones. Numerous root casts below such zones strengthen this interpretation. Angular quartz and quartzite fragments are obvious in only a few of the tuffs. The nearest source of basement complex contaminants of this type is an inselberg of quartzite and gneiss approximately 1½ miles north of the hominid sites. More common visible contaminants are bone fragments. These are usually confined to thin strongly weathered layers: but in some cases, as in the tuff below *Zinjanthropus*, occur throughout the tuff.

Reck believed that the source of these tuffs was to the east, and the rapid thickening and coarsening of the strata in that direction is convincing proof that this is so.

At the FLK and FLK NN sites the hominid remains occur approximately 15–16 ft. above the basalt flow at the base of Bed I, while at the MK site, the hominid remains occur 9 ft. above the basalt. The samples dated, with the exception of KA 851, below the *Zinjanthropus boisei* floor which contained numerous bone fragments, showed no visible contaminants. In four of the six samples, the glass shards have been almost completely altered to clay. In one of these, KA 849, about 20 per cent of the glass shards were fresh, while in KA 847, from just above the hominid remains at MK site, most of the glass shards were fresh. Whether the more intense alteration of the four samples is the cause of their somewhat younger ages is not known, but it appears reasonable.

The run data and derived ages are given in Table 1.

The decay constants used in the computation of ages are $\lambda\beta = 4{\cdot}72 \times 10^{-10}$ yr.$^{-1}$, $\lambda\kappa = 0{\cdot}584 \times 10^{-10}$ yr.$^{-1}$. The standard deviation of all of these runs is 0·1 × 10 yr. or less. This figure is, of course, simply a measure of the reproducibility of our results and does not attempt to deal with geological factors which might cause some scattering of the derived ages.

These samples were critically selected and examined and there seems to be no chance of their being contaminated by older eroded debris. It is to be noted that the hominid-site dates fail to fall in the proper relative order but all are between 1·6 and 1·9 million years. At the present time, the best estimate that we can make of the age of these sites is the average of the several aged, that is, 1·75 million years. This figure is in excellent agreement with KA 412, which is at the same stratigraphic level, but which was sampled a few miles from the hominid sites. The two samples from the top of Bed I give an average age of 1·23 million years, that is, 0·5 million years younger than the hominid sites. It may be of interest to note that we have obtained an age of 360,000 years on a post-Chellean II tuff in Bed II, Olduvai. The conclusion is inescapable that Olduwan Culture and Villafranchian fauna are synchronous in time and that both are approximately 1·75 million years old.

L. S. B. LEAKEY, J. H. EVERNDEN, G. H. CURTIS

Table 1

Top of Bed I						
KA 664	Biotite	1·39 g.	6·96 % K	$1{\cdot}77 \times 10^{-11}$ mole $A_{40}{}^{r}$	93 % $A_{40}{}^{at}$	$1{\cdot}02 \times 10^{6}$ yr.
664R	Biotite	5·99	6·96	8·36	82	1·13
KA 861	Oligoclase	6·4	1·17	1·85	93	1·38
Bottom of Bed I (hominid-containing series of tuffs)						
KA 412	Anorthoclase	29·6	3·29	28·3	67	1·63
KA 437	Anorthoclase	6·68	3·06	6·62	70	1·74
KA 846	Anorthoclase	7·91	2·52	5·55	52	1·57
KA 847	Anorthoclase	5·90	3·18	6·16	70	1·85
KA 849	Anorthoclase	6·50	2·04	4·44	77	1·89
KA 850	Anorthoclase	8·17	3·94	10·20	41	1·78
KA 851	Anorthoclase	7·79	3·69	8·37	61	1·64

AGE OF THE BASALT FLOW AT OLDUVAI, EAST AFRICA

The site of Olduvai in Tanganyika, East Africa, is known all over the world because of its remains of early man, the sequence of human industries and rich associated fauna, and we are all greatly impressed by the research performed under difficult conditions by Dr and Mrs Leakey. Four different beds have been distinguished. Bed I is underlain by a massive flow of basalt, and it should be possible to obtain here an age determination by means of the potassium–argon method. We are greatly indebted to Mr S. E. Ellis of the Mineralogical Department of the British Museum (Natural History), who kindly gave us a rock specimen for that purpose (B.M. 1947/211: collected by K. P. Oakley). This sample is called here O I. As many implements at Olduvai are executed in basalt, we used two pebble tools from Bed I (collected by von Koenigswald, 1951), in order to try to get information for Bed I; these are the samples O II and O III.

If we try to establish the age of Bed I with the help of the fauna, we find that *Elephas planifrons* and *Stegodon*, typical for the Villafranchian of Africa, are missing. There are many archaic types still present such as *Dinotherium* and *Mastodon*, but the dominant species among the elephants is *Elephas recki*, representing the same evolutionary stage as *E. antiquus* in Europe. Beds I and II have many species in common which do not occur higher up. According to Hopwood: 'There is no faunistic evidence to suggest that the lower part of the Olduvai series is of Lower Pleistocene age.' Leakey expressed the opinion that Olduvai 'extends from the beginning of the Mindel to the very end of the Riss glaciation'. After the discovery of *Zinjanthropus*, however, Leakey included Bed I in the Lower Pleistocene, give us the right solution. Adopting the last suggestion we would be forced to revise considerably our ideas about the duration of the Pleistocene, which might be necessary, but should only be done if more age determinations of Pliocene and Villafranchian rocks are available. Therefore, it seems likely to us for the time being that a time-interval does exist between the basalt and the formation of Bed I.

Just after Gentner and Lippolt had finished their report on these measurements, an article was published by Leakey *et al.* on the age of Bed I at Olduvai. Their figures, based on biotite, oligoclase and anorthoclase crystals from various tuffaceous intercalation are higher (1·6–1·9 million years) than ours, though their samples are geologically younger. We thus wonder where this discrepancy comes from.

Potassium–argon ages of basalt samples from Olduvai

	= 4·74 × $10^{-10}a^{-1}$		= 0·58 × $10^{-10}a^{-1}$		
Sample	Weight (g.)	$A_{40}{}^{r}$ (10^{-7} c.c./g.)	$A_{40}{}^{at}$ (%)	K (%)	Age with potassium–argon (10^{6}a)
O I	0·52	0·58	88·6	—	—
	0·51	0·60	87·9	1·13	1·3 ± 0·1
	0·51	0·57	90·6	—	—
O II	0·51	1·91	70·2	—	—
	0·71	1·93	69·6	2·17	2·25 ± 0·16
	0·51	1·94	72·4	—	—
O III	0·52	0·93	89·8	—	—
	0·60	0·86	89·3	1·67	1·4 ± 0·1
	0·61	0·96	88·0	—	—
	0·69	0·92	87·9	—	—

the Villafranchian. Considering the various views expressed without further discussion we expect that Bed I has an age of 0·4–0·8 million years (Zeuner, 1959) and 0·1–0·8 million years (Emiliani, Mayeda and Selli, 1961), respectively.

The basalt bed, however, has a considerably higher age determined by using potassium–argon. The analysis has been carried out on stony material reduced to small pieces. The average diameter of the fragments of the rocks was about 0·8 mm., which guarantees sufficient homogeneity.

The basalt below Bed I has an age (determined using potassium–argon) of about 1·3 million years. The pebble tools from Bed I are not younger but older, and therefore not suitable for our purpose. The age of O III, however, is only a little higher than that for O I and might be regarded as confirming the figures for O I.

We see that the age determined with potassium–argon is about 0·5 million years higher than we had expected it to be. There are three possibilities for this: (*a*) The age determined with potassium–argon is too high, because the amount of argon-40 (rad) is accumulated by inherited argon. This cause can only be excluded by further work on distant samples. (*b*) The sedimentation at Olduvai did not start directly after the formation of the basalt flow but considerably later. (*c*) Bed I has an age of about 1·0 million years and the former estimates were not correct.

It is not easy to say which of the three alternatives may

We can exclude that this is conditioned by the technique of measuring, since we are in good agreement with the age determination of a late Trinil basalt from Java measured by Evernden *et al.* Both laboratories obtained an age of 0·5 million years for this sample, which may be used to date the *Pithecanthropus erectus*.

Thus the difficulty comes from the sample itself. In connection with another problem, we investigated different Pleistocene samples (biotite and sanidine) originating from tuffs and found that their ages, determined with potassium–argon, do not necessarily have to agree with the geological sequence, but may be too high because of inherited argon. There is every appearance that this error may be present just in tuff minerals of low ages.

Furthermore, we have some sedimentological and palaeontological questions concerning an extended time-scale.

For the top of Bed I the average age is given as 1·23 million years, for the lower part as 1·75 million years. Consequently, the total of Bed I should have been formed during a period of (at least) 0·50 million years. One might expect that such a long period should be subdivided by fauna, sedimentation, dry and wet periods, industry; nothing of that kind has been observed. Bed I is about 30 m. thick, which would mean an average rate of sedimentation as low as 0·06 mm. per year. Where are the sediments to support this? By such a slow rate of sedimentation every bone would completely disintegrate before it would have a chance to survive as a fossil, but Bed I contains a rich fauna. Certain places might have been temporarily dry, to allow occupation sites, but no soil horizons have been formed or marked disconformities. Does not the preservation of the bones and the general unity of fauna and industry through the whole of Bed I give the impression of a 'normal' sedimentation?

Perhaps there is a marked time-interval between Bed I and Bed II. If we look at this new time-scale, the top of Bed I is estimated to be 1·23 million years old. According to the same authors a post-Chellean II tuff in Bed II has yielded a date of 0·36 million years. If we estimate the bottom of Bed II to be 0·50 million years old, the supported interval would last nearly 0·75 million years. Here again we might ask why the faunae of Beds I and II are roughly the same, having in common even archaic types of elephants, and why both beds, possessing the same grey colour, seem to rest conformably on each other. Where are the soils or zones of weathering which would certainly have been formed during this considerable interval of time and should separate the beds?

As the lower part of Bed I is said to have an age of 1·75 million years, the whole Pleistocene period would have lasted well over 2·00 million years. This is more than double of what we, cautiously and with good reasons, have estimated.

Having considered the many unanswered sedimentalogical and palaeontological questions which make Olduvai a difficult test case, we prefer, as stated here, to be cautious until more dates for the Lower Pleistocene are available.

G. H. R. VON KOENIGSWALD,
W. GENTNER, H. J. LIPPOLT

AGE OF BASALT UNDERLYING BED I, OLDUVAI

I have received a copy of the communication by Drs von Koenigswald, Gentner and Lippolt, on the age of the basalt flow at Olduvai. This article contains certain mis-statements, which seem to be the result of overlooking reports between 1951 and 1960 and quoting from out-of-date chapters of my book on Olduvai.

Drs Gentner and Lippolt have apparently used three specimens.

(*a*) A specimen collected by Oakley in 1947, which they use without any indication of whereabouts in the Gorge it was collected. It must, I think, represent a specimen from the weathered and exposed surfaces of the basalt on the floor of the Gorge, since in 1947, when Oakley visited Olduvai for a few hours, he had no means of access to the underlying lava at places where it is sealed in by the overlying beds.

(*b*) The other two specimens used were tools of Oldowan type collected when von Koenigswald visited Olduvai with me in 1951. I made it quite clear to him at the time that these were surface specimens and might have come from Bed I, but equally from anywhere up to the top of Bed IV, since, typologically, Oldowan tools continued to be used right to the top of the sequence, although in Bed I they are characteristic, and in Bed II, at the base, predominate over all other types.

These two specimens were only given to von Koenigswald as representative examples of the shape and form of the Oldowan type of tool and certainly not as representing *in situ* Bed I material for dating purposes; so far as Bed I is concerned, they are irrelevant, since they may have been tools used in mid-Pleistocene times and made from mid-Pleistocene lavas, or anything earlier.

The three authors also discuss the age of Bed I on the basis of the fossil fauna, but, in this context, quote nothing more recent than Hopwood's out-dated lists in my 1951 volume on Olduvai. I have made it clear on a number of occasions, since then, that this is out of date and inaccurate.

The three authors correctly state 'after the discovery of *Zinjanthropus*, however, Leakey included Bed I in the Lower Pleistocene', but they ignore the fact that I have given reasons in various places for doing so, and that, in fact, I suggested a Lower Pleistocene date as far back as 1935, although I temporarily withdrew this in 1951 because Hopwood had wrongly identified some of the fauna and so misled me.

A lengthy three-volume report on the fauna and other aspects of the Olduvai work will shortly be going to press. I must state, however, that there is no trace whatsoever of *Elephas* (*antiquus*) *recki* in Bed I where the characteristic elephant is *Elephas* cf. *africanavus* of the Villafranchian of North Africa. Even at the base of Bed II I have not, so far, found *Elephas* (*antiquus*) *recki in situ* but only *Elephas exoptatus* (or cf. *meridionalis* of Arambourg).

Indeed, far from what Hopwood claimed, the fauna of Bed I, in so far as its very numerous large mammals are concerned, is more than 95 per cent extinct.

There is abundant faunal as well as geological evidence to indicate a time-break between Bed I and Bed II and, moreover, I have indications that the base of Bed II is different from the top of Bed II; with the base of Bed II probably of the same age as the Omo fossil beds. There is also a distinct difference between the fauna of the lower and middle part of Bed I and the top of Bed I.

Bed I is not '30 metres thick'. It has such a thickness in places and varies from as little as 18 ft. to more than 100 ft. In the area of the *Zinjanthropus* sites it is about 40 ft. thick. The authors repeatedly restate the wholly out-dated suggestion that the fauna of Bed I and Bed II is roughly the same, which is quite incorrect.

Samples of the lava, which underlies the Olduvai deposits at site FLK I and other sites, are being studied by Curtis and Evernden and I am asking them to reply independently to this aspect of the communication by von Koenigswald, Gentner and Lippolt.

L. S. B. LEAKEY

In Dr Leakey's reply to our communication I miss his answer to our main question: What is his date for the basalt of Olduvai?

After Dr Curtis's cautious remarks early in 1961 about the age of determinations from Olduvai—'However much of that section is contaminated . . . I think the thing to do is to get a sequence of dates and throw out those that are vastly anomalous'—I did not expect such categorical statements shortly afterwards. I am fully aware of the difficulties, but just as the results seem to indicate a so much greater age than expected I feel that all sources of possible errors and the sedimentological implications should be fully discussed before they are accepted.

So far as the fauna is concerned, I am surprised that Dr Leakey now rejects Dr Hopwood's determinations which he included in 1951 in his book *Olduvai Gorge*. Dr Leakey stated that certain species are restricted to Bed I; Beds I and II still have in common such archaic genera as *Dinotherium*, *Mastodon* and even *Australopithecus*; this could scarcely be expected if such a major break existed between Bed I and Bed II as suggested by Dr Leakey.

It is, in connexion with the time-factor, interesting to learn the Oldowan tools, typologically, continue to the top of the sequence. But who is responsible for the Oldowan culture *s. str.*? Quoting Dr Leakey: 'In my report to *Nature* on the discovery of the skull of *Zinjanthropus boisei* at Olduvai Gorge in 1959, I made it clear that I believed that the very close association of quantities of primitive Oldowan stone tools with the skull of *Zinjanthropus* . . . suggested that it represented the makers of this culture. I admit, that I am no longer quite so certain, but I still think that in all probability . . . .' G. H. R. VON KOENIGSWALD

The age of 1·75 million years obtained by us for early man at Olduvai Gorge has been questioned by von Koenigswald, Gentner and Lippolt on the basis of their determination of an age of 1·3 million years for the basalt underlying the strata containing the hominid remains.

While at the Gorge in April 1961, one of us (G. H. C.) examined the basalt carefully for dating purposes. In most exposures, however, the basalt shows effects of chemical alteration and is clearly not suitable for obtaining reliable ages. Nevertheless, three specimens were sent here for examination. Thin-section examination revealed that none of these was suitable for dating; but with the appearance of the above-mentioned communication we chose the best two specimens and made age determinations in order to compare the numbers.

Specimen KA 927 was collected by Leakey at Site FLK I and specimen KA 933 was collected by Curtis at the third vault. KA 927 is an olivine basalt exhibiting subophitic or intergranular texture with numerous phenocrysts of olivine up to 1 mm. in length in a groundmass of titanaugite and plagioclase. The plagioclase laths average about 0·4 × 0·1 mm. but vary considerably, while the titanaugite is more nearly equidimensional and averages about 0·2 mm. in diameter. Grains of ilmenite compose about 5 per cent of the rock. Some of the sparse vesicles are filled with calcite, and a carbonate is intergrown with plagioclase and titanaugite in the groundmass in a manner suggesting that it is a primary mineral. Most of the margins of the titanaugite are chloritized, and chlorite or a serpentine mineral replaces the margins and cleavage cracks in most of the olivine. Even some plagioclase grains are extensively chloritized.

Specimen KA 933 was probably an olivine basalt or olivine trachyte, but the few phenocrysts of olivine have since been completely converted to iddingsite. In other respects, however, the specimen appears less altered than KA 927 and no chlorite is present. The rock is exceedingly fine-grained though holocrystalline. The average plagioclase is approximately 0·05 mm. in length and less than 0·01 mm. in width. A few plagioclase phenocrysts attain 1·5 mm. Intergranular clinopyroxene averages approximately 0·01 mm. in diameter, but a few grains reach 0·2 mm. Dark brown and black equidimensional grains of magnetite and haematite averaging 0·01 mm. in diameter are so abundant that they render the thin-section opaque in 30μ thickness. Only thin edges of 10μ or less in thickness can be examined in transmitted light. The grains of magnetite and haematite occur along margins, in centres, and in zones within the plagioclase. A striking feature of the rock is the presence of biotite both in the groundmass and extending out from the sides of very numerous micro-vesicles. Some biotite flakes are up to 0·5 mm. in length. Many vesicles are partly coated with calcite.

The proportions and types of minerals are different in the two specimens, and it appears unlikely that they are specimens of the same flow. The difference in potassium contents supports this view. As two superimposed flows can be seen at one of the exposed faults in the Gorge, the occurrence of two different types at widely spaced localities is not surprising.

The data that we have obtained on these two specimens are given in Table 1.

Table 1

KA	Weight (gm.)	Percentage K	Percentage A_{40}^{at}	Ar_{40}/K_{40}	Age (m.y.)
927	32·19	0·876	92	$2{\cdot}37 \times 10^{-4}$	4·1 ± 0·5
927 (repeat)	16·46	0·876	97	2·32	4·0 ± 1·0
927 (repeat)	18·34	0·876	80	2·56	4·4 ± 0·2
933	21·16	1·472	90	1·00	1·7 ± 0·2
933 (repeat)	18·77	1·472	89	1·00	1·7 ± 0·2

We conclude that the basalts at Olduvai are at least four million years old but that they are unreliable for dating purposes. There is thus no conflict between the dates obtained on the overlying tuffs and the best estimate of the age of the underlying basalts. There are good reasons why one should question the reliability of dates obtained on these basalts. The location of potassium in basalts is not known with certainty. Some is contained within the plagioclase and trace amounts go into the pyroxene. Probably in most cases potassium is concentrated in the final intergranular hydrous siliceous fluid where it either enters the albitic rims of the plagioclase or, in the case of specimen KA 933 above, forms late potassium-bearing minerals in the interstices of other minerals or in the vesicles. Any later chemical alteration of the rock attacks these grain-boundary areas first and the resultant loss of radiogenic argon may considerably reduce the age obtained. Several basalts which we have attempted to date from elsewhere, such as the Columbia River area and the Berkeley Hills, have yielded anomalously low ages, and we can only explain them along the foregoing line of reasoning. In view of the difficulty that we had in obtaining moderately fresh basalt from Olduvai Gorge, it appears unlikely to us that

Gentner's specimen, obtained from the British Museum by von Koenigswald and collected by Oakley without this objective in view, is as reliable for dating as ours.

Equally significant may be the extremely fine-grain size of KA 933. Though average dimensions are given above for grain sizes, much of the rock is so fine-grained that individual crystals appear only as specks of light, that is, sufficient only to attest to the crystalline rather than glassy nature of the groundmass. The effective grain-size for argon diffusion is no doubt much greater in volcanic glass than in this rock. The rock carries about 5 per cent biotite so that only about 20 per cent of the potassium of the rock is in this mineral with the remainder probably being in the albitic rims of the plagioclase crystals. These crystals have an average width of less than 0·01 mm. with the result that most of the potassium is within less than 1μ of a grain boundary. Even this unattractive situation is rendered worse by the fact that these crystals are charged with magnetite and haematite. It is not certain that such inclusions will adversely affect argon retention, but it is certain that they will reduce the average distance to grain boundaries to less than one micron and will introduce strains into the feldspar lattice. Quantitative treatment of the indicated diffusion problem is impossible due to the uncertainties of some of the parameters, but approximate solutions indicate that the diffusion coefficient necessary to explain the observed losses is probably reached at the surface temperatures of the Earth.

Some of the arguments used by von Koenigswald *et al.* require further discussion and clarification. Gentner raises the question of inherited argon in the samples of tuff used by us. While one cannot be absolutely certain that inherited argon is not a problem in some rocks, certain definite statements can be made. All available data on argon diffusion in feldspars indicate that argon retention in feldspar crystals suspended in molten rock is essentially zero over time spans of even a few days. We can state that the few volcanic sanidines of historic age dated by us have yielded ages consistent with the concept of zero argon content at the time of eruption. Both the 1912 eruption of Katmai and the 1304 eruption of Ischia yielded zero potassium/argon ages. Also, dates of late or post-Pleistocene events have given reasonable ages. A late Gamblian tuff from Lake Naivasha in Kenya gave 28,000 years and a prehistoric post-glacial pumiceous rhyolite dome near Mono Lake, California, gave 5,600 years.

If the possibility of faulty stratigraphic correlation can be eliminated, the most probable source of error in any tuff lies in the possibility of contamination from older admixed debris, whether erosional debris or country rock torn from the sides of the conduit at the time of eruption. The Olduvai tuffs were examined carefully for such material before being used. The mineral used would virtually eliminate any possibility of significant conduit contamination, and the nature of the deposits (that is, primary tuff falls) argues strongly against any admixture of eroded debris. We see no possibility of explaining the tuff ages in terms of admixed older anorthoclase crystals.

The only sample of post-Pleistocene feldspar that has yielded an age far older than that assigned to it by field investigators is the concentrate of Laacher See sanidine sent to us by Dr Frechen, University of Bonn. We have made several runs on this material, and all ages obtained are several hundred thousand years rather than the 11,000 years suggested by its assignment to Alleröd. As we had nothing to do with the collection or concentration of this material, we are in no position to explain the gross discrepancy. However, we feel that the diffusion data and the dates from other young samples argue so positively for total argon loss at temperatures of 1000° C. that we must conclude that the Laacher See sample is either contaminated or misplaced stratigraphically.

von Koenigswald appears concerned about the slow rate of sedimentation in Bed I and believes that every trace of bone would have disintegrated before becoming sufficiently buried. This opinion is based on a complete misconception as to the nature of sedimentation at Olduvai. Virtually all the layers in Bed I are primary tuffs, that is, ash falls directly from the air. They instantly buried everything lying on the land surface or in the shallow margins of the lake. One of these ash-fall units is 12 ft. thick.

Again in response to an objection of von Koenigswald, it is to be noted that the tops of many of these ash-fall layers show well-developed soil profiles, some of them quite deep (up to 4 ft.) and indicating protracted periods of time of development. Each of these soil profiles indicates a disconformity. The soil profiles are numerous, well-developed, and plainly visible in quarry faces though not in weathered canyon-wall views. It may come as a surprise to many to learn that the stratigraphy of such an important locality as Olduvai Gorge is very imperfectly known. The only reference describing Bed I (*Olduvai Gorge Monograph*, Leakey *et al.*) does not even mention the crystal content of the tuffs though this content may make up more than 30 per cent of the rock. This reference does not mention the existence of soil profiles and virtually implies absence of them. The large cuts made in conjunction with the *Zinjanthropus* excavations now clearly display an extensive development of soil profiles. This lack of adequate description of the Olduvai section will be remedied shortly as the tuffs are being studied by a competent volcanologist.

In conclusion, it can be stated that the basalts at Olduvai are at least four million years old, and that the result obtained by Gentner is to be interpreted as the result of using unreliable material. In addition, the soil profiles sought by von Koenigswald are present in profusion, and his bone destruction hypothesis is invalid on the grounds of inadequate understanding of the mechanism of tuff accumulation at Olduvai. All data so far available agree that there is no alternative to accepting an age for *Zinjanthropus* and for Lower Pleistocene of approximately two million years.

G. H. CURTIS, J. F. EVERNDEN

CHAPTER IX

NOTE ON FOSSIL HUMAN DISCOVERIES AND CULTURAL EVIDENCE

In this first volume I have tried to set out the present state of our knowledge of the background evidence, against which the very important fossil human remains from Olduvai and the Stone Age cultural sequence there can be assessed. Preliminary reports upon the fossil hominid discoveries have appeared from time to time since 1958, both in *Nature* and in the popular press, and scientists are therefore already aware of the general picture. The study of the fossil human remains and of the stone tools, together with a detailed report upon the excavations of sites will appear in subsequent volumes.

Volume 2 by Professor Phillip V. Tobias, of the Medical School, Johannesburg, is now in the final stages of preparation, and will be published during 1967. It deals with the skull of *Zinjanthropus boisei* in very considerable detail.

Three further volumes are in preparation. Mrs M. D. Leakey's deals with the living sites and cultural sequence of Bed I and Bed II, and will include chapters by Dr Leakey on the fauna associated with each living floor.

Dr Richard Hay's volume, dealing with the details of the geology of Olduvai Gorge, is being prepared.

Another volume by Professor Phillip Tobias, also now in active preparation, will be a careful study of *Homo habilis*, the LLK skull, and a variety of other Olduvai hominids.

At the time of writing this report, i.e. up to the end of December 1961, fossil hominid remains had been found at three different levels in Bed I. These are as follows:

(*a*) *At site MK I* a few teeth have been found at a level which is older than the FLK I sites.

(*b*) *At site FLK NN I* a living-floor was found which yielded remains of a juvenile and also of an adult.

The juvenile is represented by parts of the skull, part of a lower jaw, one upper molar, a collar bone and parts of a hand.

The adult is represented by an almost complete foot and a few hand bones.

(*c*) *Site FLK I*, where the skull of *Zinjanthropus boisei* was found on 17 July 1959. This site has also yielded a few fragments that represent a quite different and distinct individual, probably of the same type as the juvenile above.

There is general agreement that *Zinjanthropus boisei* represents an australopithecine hominid, and it is certain that this australopithecine was found on a living-floor with tools of the Oldowan culture. It is by no means so certain, however, that *Zinjanthropus* made the culture associated with him. The fragments of a second individual on this floor appear to represent a wholly different type of hominid—one that is comparable with the juvenile from site FLK NN I. It seems to represent a very early and primitive hominid leading towards *Homo*.

Hominid remains had also been found at two sites in Bed II. These consisted of two deciduous teeth found at site BK II in 1954 and an incomplete calvaria found at site LLK II in 1961. The BK II site was originally thought to be near the base of Bed II and, in a preliminary note published in it was described as such. New evidence shows that the deposit at BK II consists of fluviatile sands and gravels filling an old channel in the upper part of Bed II.

The site at LLK II is also near the top of Bed II, at a level which yielded cultural material in 1931, then regarded as belonging to a late stage of the Chellean. When the skull was found, it was therefore considered to represent one of the makers of the Chellean culture. During the last two years' work it has become apparent that the cultural sequence in Bed II is more complex than first considered, it is therefore necessary to be cautious in assigning this skull to any particular cultural stage.

The available evidence, which will be set out in volume 3 indicates that the Oldowan culture does not end at the top of Bed I, but continues in a more evolved form into the lower part of Bed II. The whole cultural sequence is now being revised on the basis of detailed excavations at successive levels in Beds I and II and the results will be published in volume 3.

Since no detailed excavations in Beds III and IV have been carried out on the same basis as those in Beds I and II, with the exception of three uncorrelated sites, the Beds III and IV cultural material will not be described in volume 3.

APPENDIX 1

PRELIMINARY NOTES ON THE STRATIGRAPHY OF BEDS I–IV, OLDUVAI GORGE, TANGANYIKA

by Dr Richard L. Hay

ABSTRACT

Bed I at Olduvai Gorge is a conformable sequence of lava flows and varied sedimentary deposits that extend upward from a welded tuff overlying the Precambrian basement to the top of a widespread marker bed. Bed II is a sequence of lacustrine clays and laterally equivalent fluvial, eolian and pyroclastic deposits. Bed III comprises alluvial deposits and a laterally equivalent assemblage of fluvial, lacustrine and eolian beds. Bed IV can be widely subdivided into a lower unit of fluvial clays, sandstones and conglomerates and an upper unit of eolian tuffs. Although moister than that of the present day, the climate was relatively dry throughout much of the time that these beds were deposited. Semi-desert or desert conditions rather like those of the present day may have prevailed at least twice. Tectonic movement seems to have taken place between the deposition of Beds III and IV.

A stratigraphic and environmental framework more detailed than that of Reck (1951) and Pickering (1960) is here presented for the Pleistocene succession of Olduvai Gorge (Fig. 1), which contains hominid remains of great antiquity (Leakey, Curtis and Evernden, 1961) and an unsurpassed sequence of Palaeolithic culture levels (Leakey, 1951; 1963). The need for geological information about the Olduvai succession became clear in recent controversies over the potassium–argon dates and geologic histories of Beds I and II (Straus and Hunt, 1962; Von Koenigswald, Gentner and Lippold, 1961; Leakey, Curtis and Evernden, 1962). Some, at least, of this argument would not have arisen if the geology had been properly understood. This paper summarises briefly the principal results of eight weeks' geologic field-work at Olduvai Gorge during the summer of 1962, which was supported by the Miller Institute for Basic Research in Science and the National Geographic Society. L. S. B. Leakey kindly provided his camp facilities and guided me in the field during the early stages of this study. I also had the benefit of reading R. Pickering's unpublished report on Olduvai Gorge. Field-work has been supplemented by extensive microscopic and X-ray study of the rocks. The stratigraphic synthesis of Fig. 2 is based on approximately fifty measured sections and the lateral tracing of key horizons. This report is primarily intended to clarify the major stratigraphic relationships; the geology will be described more fully in a subsequent paper.

The succession in Olduvai Gorge was divided by Reck (1951) into a basal series of basalt flows and mappable units termed Bed I, Bed II, Bed III, Bed IV and Bed V. Bed V overlies the older beds with pronounced angular unconformity and will not be considered further. The subdivision and nomenclature of Reck will be continued here with few modifications, unlike R. Pickering's forthcoming report (unpublished), in which Reck's terms are abandoned.

The basalts comprise a lower flow, 35–40 feet thick, which has a typical aa surface structure, and an overlying series of thinner, pahoehoe flows. The aa flow is a biotite-bearing olivine basalt, and the pahoehoe flows are biotite-free olivine-rich basalt. The original ropy surface is widely preserved over the basalts, as Reck has noted (1951), and the primary lava tumuli form a hummocky surface having a local relief of 5–20 feet. Five feet of trachyte tuff and claystone were exposed beneath the basalt flows near the third fault in the summer of 1962, and four more feet of tuff and claystone were exposed by excavating. These beds below the basalts resemble tuffs and claystones of Bed I above. One tuff, 8 feet below the basalts, is rich in coarse anorthoclase crystals of two habits and compositions.

Bed I as defined by Reck is a series of tuffs whose lower and upper limits are the basalt flows and the 'lacustrine marls' of Bed II. This description seems to fit best the relationships at sites HWK and FLK, where Marker Bed B of Fig. 2 would be the uppermost stratum of Bed I, as Leakey has considered it (1962, personal communication). It varies in thickness and lithology, and pinches out locally in the area to the east, where it can easily be confused with a widespread reworked tuff 30–50 feet lower in the sequence (Marker Bed A of Fig. 2). Leakey, Curtis and Evernden (1961) gave a thickness of about 40 feet for Bed I at site MK, indicating that here they took Marker Bed A as the top of Bed I. The maximum thickness of 40 metres for Bed I given by Reck suggests that he measured up to the top of Marker Bed B in the vicinity of the second fault (see Fig. 2). In the present report, Marker Bed B is taken as the top of Bed I.

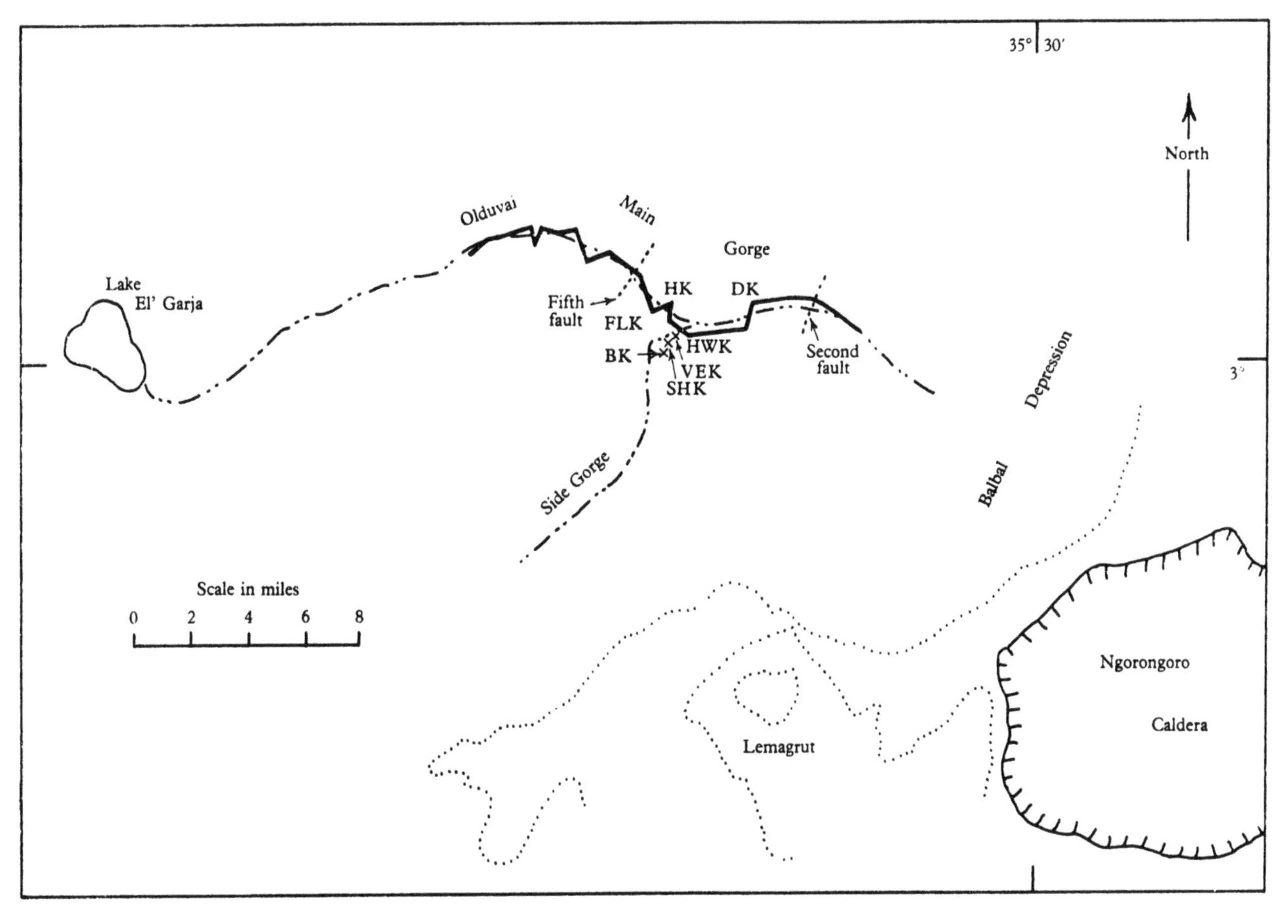

Fig. 1. Sketch-map of Olduvai Gorge and its environs, after Leakey (1951), showing the locations of the second and fifth faults and archaeological sites DK, HWK, FLK and HK. Heavy line along the Olduvai Main Gorge is the line of section for Fig. 2.

A conformable series of rather uniform tuffs and clays 40–75 feet thick underlie Marker Bed B and

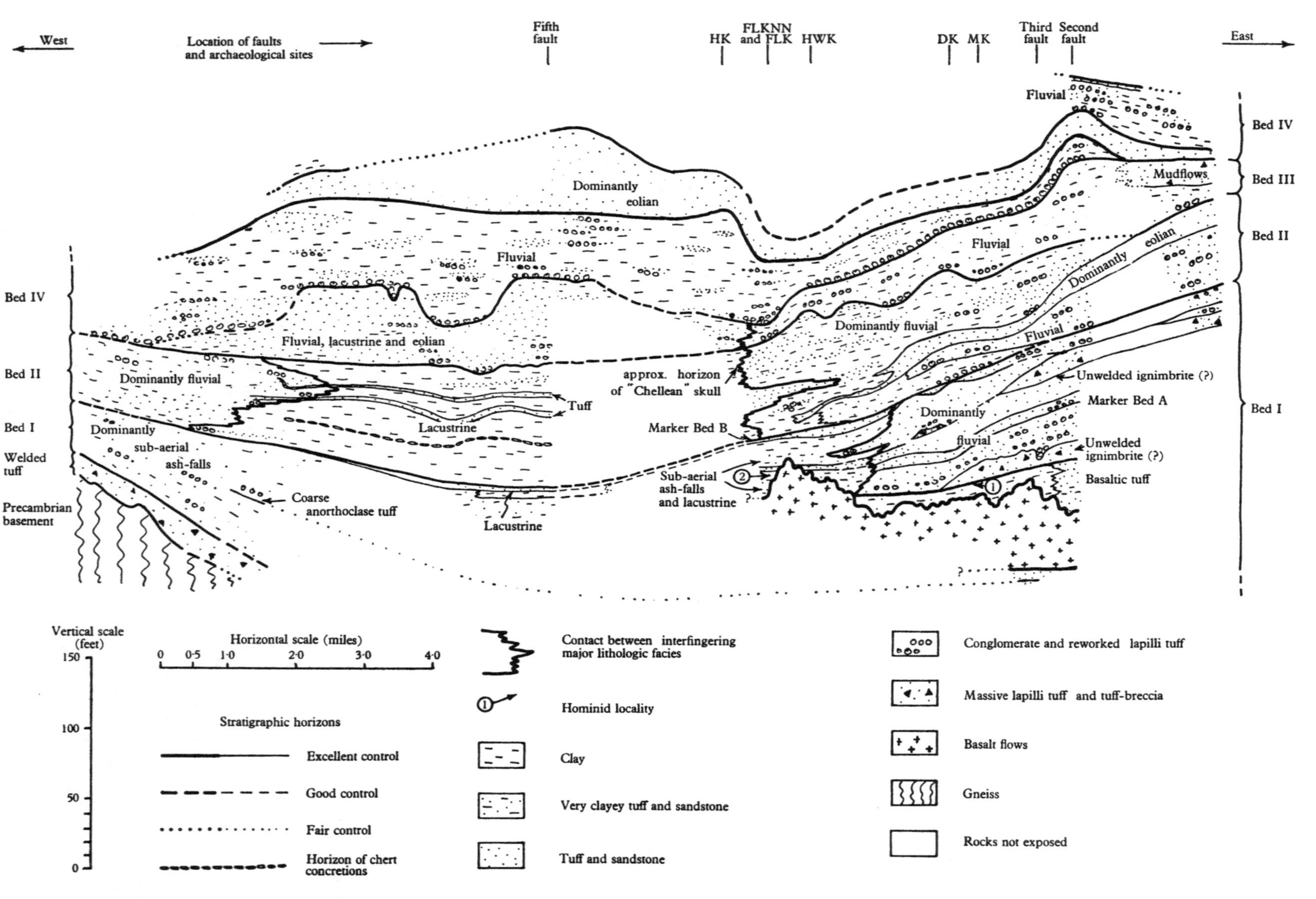

Fig. 2. Stratigraphy of Beds I–IV along Olduvai Main Gorge, based upon measurements along the line of section shown in Fig. 1. The sequence is reconstructed approximately as it would have appeared prior to faulting. Heavy lines separate major stratigraphic units, and light lines are used to delineate subdivisions of major units. Hominid locality 1 represents the hominid site at MK (3), which lies at approximately the same horizon as small stone tools and a structure built of basalt blocks at site DK (5). Locality 2 represents the *Zinjanthropus* site; the pre-*Zinjanthropus* child was found stratigraphically about 2 feet lower.

overlie a trachyte welded tuff to the west of the fifth fault. These tuffs resemble those of Bed I to the east, except for a greater degree of weathering and the presence of conglomerates of welded tuff, gneiss and quartzite debris supplied from the west or north. A coarse anorthoclase tuff near the middle of this sequence appears to be the same as that 8 feet below the basalt flows 12 miles to the east, suggesting that Bed I as defined by Reck is equivalent to only the upper part of this conformable sequence to the west of the fifth fault. In order to make Bed I a recognisable stratigraphic unit to the west of the fifth fault, it is redefined to include the entire sequence of tuffs between the welded tuff and the top of Marker Bed B. By this definition, the basalt flows constitute a member of Bed I in the area to the east.

In its easternmost exposures, Bed I consists largely of stream-worked trachyte tuffs, lapilli tuffs and conglomerates. It also contains two unsorted trachyte lapilli tuffs, probably ignimbrites (i.e. the deposits of Pelean eruptions), which can easily be traced within the eastern, fluvial facies of Bed I. Stream-channel alignments generally depart little from S. 60° E., suggesting that Ngorongoro could have been the volcanic source of most or all of the trachytic materials. The stream-laid deposits interfinger westward with dominantly land-laid trachyte tuffs and tuffaceous clays that are extensively penetrated by root channels. A few diatoms, oolites and thin-walled pelecypods in some of the tuff and clay suggest temporary floodings by a lake. Hominid remains (see Fig. 2) and artefacts within Bed I are largely or entirely confined to these sub-aerial and shallow-water deposits. Falls of volcanic ash and deposits of lacustrine clay have contributed to the preservation of artefacts and fossils in undisturbed context on occupation areas.

A relatively small thickness of purely lacustrine clay underlies Marker Bed B near the fifth fault. Farther to the west, Bed I is mostly a weathered accumulation of ash deposited on the land surface. Caliche layers in the more weathered of these ash deposits suggest a climate that was not sufficiently humid for calcium carbonate to be leached from the soil. Rosettes of gypsum crystals, now replaced by calcite, occur at a few horizons above the basalts in the area to the east. The gypsum, like the caliche, probably indicates a relatively dry climate.

Bed II conformably overlies Bed I, and the two are difficult to separate where Marker Bed B pinches out to the east of site DK and near the west end of the gorge. Purely lacustrine deposits as much as 90 feet thick occupy the axis of a small sedimentary basin transected by the Olduvai Main Gorge (Fig. 1). Unfossiliferous green clays form most of the lacustrine sequence which also includes trachyte tuff, dolomite, oolitic limestone and chert nodules. The lack of fossils and presence of dolomite beds and of authigenic potash feldspar in clays and tuffs suggest to me that the lake was moderately or strongly saline and alkaline for much of its history, and it probably lacked a permanent outlet. In most years, evaporation must have exceeded inflow to the lake. Stream-worked tuffs form most of Bed II to the west of the lake, but there are substantial proportions of sub-aerial tuff and non-volcanic sandstone and conglomerate.

Bed II to the east of site FLK is a diverse sequence of fluvial, eolian and pyroclastic deposits 50–90 feet thick. Several horizons are moderately or deeply weathered and channelling is visible in many places, particularly along the Side Gorge. Clays, sandstones and conglomerates of fluvial origin form most of the eastern sequence, but some of the clays and sandstones appear to have been deposited by the lake when it spread eastward beyond its usual limits. Detritus of the sandstones and conglomerates is largely trachytic, and stream-channel trends suggest that most of it may have been supplied by Ngorongoro. Trachyte, nephelinite, and olivine basalt cobbles in conglomerates near sites HWK and VEK were derived from the direction of Lemagrut.

Trachyte tuffs of ash-fall origin and pyroxene-rich tuffs redeposited by wind together constitute roughly a fifth of the eastern sequence. The eolian tuffs consist of mineral grains and rock fragments that were rounded and sorted by wind action, and the thickness and extent of these eolian tuffs suggest a climate possibly not unlike that of the present time at Olduvai Gorge. At about the time these eolian tuffs were deposited, the lake to the west temporarily shrank. Lacustrine chert nodules were then eroded at its margin, and mud cracks formed over much of its floor.

Stone artefacts and well-preserved fossils are relatively abundant in beds deposited along the eastern and southern margin of the lake (e.g. sites HWK, FLK, SHK and BK). In the lower part of Bed II there are tools of chert that was obtained from the margin of the lake during its brief period of desiccation. The skull from LLK II reported by Leakey (1961) was found 15–20 feet below the top of Bed II in land-laid tuffs deposited near the margin of the lake.

Bed III is separated by disconformities from both Bed II and Bed IV. An eastern facies of Bed III consists largely of stream-laid conglomerates, clayey sandstones, and sandy claystones; it contains minor proportions of tuff and mudflow deposits. These beds were weathered as they accumulated, and most of them are reddish brown, well consolidated, and penetrated by root channels. Irregular nodules and anastamosing twig-like structures of calcium carbonate are concentrated in horizontal layers at many places in the sandstones and claystones. These calcite bodies seem to have been formed at shallow depth as the beds were being deposited, for they are locally channelled by fluvial conglomerates. Volcanic debris forms most of the sandstones and conglomerates, and channel orientations suggest a southerly source—probably Lemagrut—for most of the detritus.

To the west, Bed III consists of yellowish grey sandstones and smaller amounts of conglomerate, clay and dolomite. South-east of site HK these beds intergrade with the reddish brown alluvial deposits, a fact evidently not recognised by Reck. Sandstones are dominantly of volcanic detritus, but many of them also contain a small to moderate proportion of gneissic debris and calcite oolites. The oolitic sandstones probably accumulated in and along the margin of a lake. Large-scale cross-bedding suggests that some of the oolitic sandstones are eolian dune deposits. Conglomerates and many of the sandstones are lenticular, and some of them fill steep-walled stream channels. The conglomerates are formed largely of quartzite, gneiss and welded-tuff pebbles derived from the west or north. Dolomite is present both as lacustrine beds and as dolomitised caliche layers.

Dolomite and calcareous structures in the weathered alluvium suggest a climate in which evaporation exceeded precipitation, either seasonally or throughout the year. Climatic significance of the reddish brown colour in the weathered alluvium is controversial, as Pickering (1960) has pointed out. More important is the dominance of montmorillonite and illite (clay mica) as the clay minerals in these beds, which indicates that leaching was mild by comparison with that on the rainy, upper slopes of Ngorongoro, where gibbsitic reddish brown latosols have been formed. Stone artifacts occur in the reddish brown alluvial deposits and in a few fluvial layers of the mixed lacustrine-fluvial-eolian assemblage.

Bed IV comprises a widespread lower unit of clays, sandstones and conglomerates and an equally widespread upper unit of eolian tuffs. Near the mouth of the gorge the eolian tuffs are overlain by conglomerates, sandstones and clays. The lower member of Bed IV, as much as 80 feet thick, consists almost exclusively of Precambrian debris which coarsens westward along the gorge. Lenticular shape of the sandstones and conglomerates, and root channels and evidence of sub-aerial weathering in the clays suggest to me that these beds were deposited in a stream-channel and flood-plain environment, rather than in a lake as Reck believed. Catfish bones and pelecypods (Unionidae) are common in the sandstones and conglomerates thought to have been deposited in stream channels, and stone artefacts are widespread in both stream-channel and flood-plain deposits.

The eolian tuff member, as much as 60 feet thick, is formed of aegerine nephelinite ash particles that have been rounded and redeposited by wind. Cross-bedding characteristic of sand dunes can be seen in a few places. Dense, travertine-like layers of caliche, known in East Africa as *steppe limestone*, are interbedded with the tuffs. Neither stone artifacts nor fossils occur in these beds, as far as I am aware, and these eolian tuffs probably accumulated in a desert or semi-desert environment, possibly even drier than that which prevails here today. The overlying beds are generally similar to the fluvial deposits underneath the eolian tuffs and catfish bones were noted in a conglomerate. Reck does not mention the eolian tuffs of Bed IV, possibly because he confused them with the similar deposits of Bed V, which they often underlie.

The earth's crust here may have been slightly warped before Bed IV was deposited, for the direction of stream flow in the vicinity of Olduvai Gorge seems to have been reversed between Beds III and IV. Bed III received most of its debris from a volcanic source to the south and possibly south-east, whereas most of the debris at the base of Bed IV nearly as far east as the third fault was derived almost exclusively from exposures of welded tuff and Precambrian basement to the west and north. Crustal warping would also account for the channelling of Bed III by the base of Bed IV and the abrupt disappearance or displacement of the lake which had existed in approximately the same place for much of the time that Beds I, II and III were deposited.

Only a few points concerning the geochronology of Olduvai Gorge will be mentioned here, as Professors Curtis and Evernden will shortly publish many new dates. The 4·4 million-year date given by a basalt sample (Leakey, Curtis and Evernden, 1962) implies a hiatus of 2·7 million years between the basalt and the overlying tuffs of Bed I, which have an average age of about 1·7 million years (Leakey, Curtis and Evernden, 1961). Geologic evidence to the contrary suggests that the basalts were extruded and buried over a relatively short period of time. Evernden and Curtis have given me permission to state that their published figure of 4·4 million years on sample KA 927 is grossly erroneous because of procedural difficulties of which they were unaware at the time. Their statement is as follows:

The argon extraction procedure used for the published date (Leakey *et al.* 1962) involved the freezing out of CO_2 if it were present in the gas obtained from fusion of the sample. KA 933 had no such contaminant, but much disseminated carbonate in KA 927 resulted in large quantities of CO_2. Isotopic fractionation of the argon incorporated in the CO_2 (preferential selection of A_{36}) resulted in a residual argon sample enriched in A_{40}. High percentages of atmospheric argon in these samples magnified the effects as regards determination of the radiogenic argon content. Modification of procedure has resulted in ages on KA 927 of approximately 1·8 million years. As noted in our paper (Leakey *et al.* 1962), the character of KA 933 was such as to suggest probable fractional loss of argon so that we feel that there is no disagreement between the ages of basalt samples KA 933 and KA 927 (1·7 and $1{\cdot}8 \times 10^6$ years, respectively).

The biotite sample giving an age of 1·0 to 1·1 million years (Leakey *et al.* 1961) was probably obtained from Bed II rather than Bed I as reported, for only a few scattered flakes of biotite were noted in samples from Bed I, but biotite is locally common in and below the eolian tuff unit of Bed II. This change in stratigraphic assignment accords with the fact that the land-laid tuffs of Bed I which overlie the basalt flows are not weathered as severely as they should be if they had accumulated over a period of 700,000 years. Moreover, there appear to be no significant erosional gaps within this sequence of beds.

The sample from Bed II earlier dated as 360,000 years old (Leakey *et al.* 1961) has been redone by an improved procedure, and Evernden and Curtis have allowed me to state that the present estimate of its age is 490,000 years. Thus, Bed II seems to span at least half a million years. This figure seems a reasonable minimum, in view of the fluvial channelling, horizons of moderate to deep weathering, and complex geologic history of Bed II.

IMPLICATIONS FOR ANTHROPOLOGY

(1) The climate was relatively dry, at least seasonally, throughout most or all of the lengthy period of hominid occupation recorded by Beds I–IV. Hominid occupation may have been first interrupted by desert conditions which prevailed during the deposition of Bed IV.

(2) For the period of Beds I and II, hominid occupation is recorded principally along the southern and eastern shore of a lake which was alkaline and rather strongly saline for most of its duration. Streams draining large active volcanoes to the east and south may have been the principal source of fresh water for this area.

(3) Stone artefacts in the upper part of Bed I are only slightly older than those in the lower part of Bed II.

(4) The Oldowan cultural sequence from Bed I probably represents a shorter period of time than stages 1–5 of the Chelles-Acheul culture, which span Bed II (Leakey, 1951).

(5) None of the stone artefacts or hominid fossils collected from Bed I are appreciably older than 1·7 million years.

REFERENCES

Leakey, L. S. B. (1951). *Olduvai Gorge.* Cambridge University Press.

Leakey, L. S. B. (1961). *Nature, Lond.* **189**, 649.

Leakey, L. S. B. (1963). *National Geographical Magazine,* **123**, 132.

Leakey, L. S. B., Curtis, G. H., and Evernden, J. F. (1961). *Nature, Lond.* **191**, 478.

Leakey, L. S. B., Curtis, G. H. and Evernden, J. F. (1962). *Nature, Lond.* **194**, 610.

Pickering, R. (1960). *Comm. Tech. Co-op. Africa South of Sahara, Pub.* 44, 77.

Reck, H. (1951) In L. S. B. Leakey, *Olduvai Gorge.* Cambridge University Press.

Straus, W. L. and Hunt, C. B. (1962). *Science,* **136,** 293.

von Koenigswald, G. H. R., Gentner, W. and Lippold, H. J. (1961). *Nature, Lond.* **192**, 720.

APPENDIX 2

DESCRIPTIVE LIST OF THE NAMED LOCALITIES IN OLDUVAI GORGE

by M. D. Leakey

BK Bell's Korongo[1]

This site lies on the right bank of the side gorge some 2½ miles above the confluence with the main gorge and a little upstream of DC. It consists of two korongos linked by a short cliff. There are good exposures of the upper and middle part of Bed II but the base is not visible. There are also some exposures of Beds III and IV in the upper part of both the gullies included in the site. Excavations were carried out here during 1952–8 and yielded a very rich culture and many fossils, including two hominid teeth.

In the eastern part of the site, a step trench has been cut down to the floor of the gorge near to the 1953–8 camp-site. A smaller excavation in the area near this trench, also yielded many stone tools and fossils.

Bos K Bos Korongo

This site lies on the right bank of the main gorge, slightly upstream of 'Hand-axe Cliff' and about 1 mile below the fifth fault. Extensive exposures of Bed IV have yielded some implements and also a crushed, rather incomplete *bos* skull, which was found during 1962.

CK Camp Korongo

This is a long gully on the left bank of the gorge between the second and third faults and adjacent to the 1931 camp. Beds IV, III, II and I are all exposed. It was at this site that the first hand axes ever recognised at Olduvai were discovered by L. S. B. Leakey in Bed IV, shortly after the arrival of the 1931 Expedition. Excavations undertaken that year yielded implements from the basal gravel of Bed IV and also from a level some 10 ft. higher up in the sequence.

CMK Catherine Martin Korongo

This site is situated on the left bank of the side gorge, about 3¼ miles upstream of the confluence. Exposures of Bed IV are rich in fossil bones and artefacts, which occur at several different levels. The site was discovered in 1935.

Croc. K Crocodile Korongo

This lies on the right bank of the main gorge, downstream of Bos K and about ¾ mile from FLK. Somewhat limited and patchy exposures of Bed IV have yielded a few tools and flakes together with an incomplete crocodile skull which was found on the surface during 1962.

Dal. K Dalmatian Korongo

This is the next gully on the left bank of the main gorge upstream of PLK. Bed IV exposures are of considerable depth and have yielded a few fossil bones.

DC Donald's Cliff

This site lies on the right bank of the side gorge between SHK and BK. The cliff exposes the upper part of Bed II, a small part of Bed III and some Bed IV. A fossil elephant tusk and jaw were found here in Bed II when the site was first discovered by D. MacInnes during 1932.

DK I Douglas (Leakey) Korongo

This site lies on the left bank of the gorge, about 1¾ miles upstream of the third fault. It consists of

[1] All sites have been designated cliffs or 'korongos' (the Swahili word for gully).

a long series of exposures in Bed I. These exposures are usually in the form of small cliffs, the tops of which are formed by the thick deposit of a hard tuff which also occurs above the fossiliferous horizon at sites MK I and WK I.

During 1962 it was found that the slopes below the small cliffs were particularly rich in fossil bones and artefacts, eroding from the sandy clay below the tuff. Excavations carried out in four areas yielded a considerable number of artefacts and a large amount of broken-up fossil bones. An occupation level was found in this area, sometimes resting on the surface of the basalt. On part of the floor a loosely piled circle of stones was uncovered, which may have been an artificial structure.

(Exposures of Beds IV, III and the upper part of II in this area are described under site EF–HR.)

EF–HR Evelyn Fuchs–Hans Reck

This site was discovered by Professor Hans Reck during 1931 and was the first site to yield artefacts *in situ* in Bed II. The higher part of the gully, in which Beds IV and III are exposed, was explored on the same day by V. E. Fuchs, who found and reported artefacts in these levels. Since neither discovery had priority, the site was named with the joint initials of both discoverers.

In the map published in the 1951 book on Olduvai Gorge (which was based on a sketch-map supplied by Professor Reck shortly before his death), the site was placed downstream of MK, whereas, in fact, it lies higher up the gorge on the left bank about 1¼ miles above the third fault in the same general area as the exposures of Bed I known as DK I.

Elephant Korongo

This site lies on the right bank of the gorge and consists of a long gully which terminates just east of the third fault. A well-marked cattle track descends the western side of the gully and turns across the line of the fault before continuing to the floor of the gorge. The site was found in 1931 when deposits in the upper part of Bed II yielded some excellent Chellean implements in association with parts of an elephant skeleton.

FC Fuch's Cliff

This site was discovered in 1931 and lies on the left bank of the side gorge opposite, and a little upstream of MNK, about 1 mile from the confluence. In this area, Bed II is very rich in fossils and stone tools and at least four levels of culture have been established. In the western part of the site the 'chert horizon' in the base of Bed II is well exposed. Bed III is present and the upper part has been deeply channelled prior to the deposition of Bed IV. Some preliminary excavations were carried out here in 1961, and during 1962 a nearly complete fossil skeleton of a large snake was found in the basal sands of Bed IV.

FK Fuch's Korongo

This is the next gully, upstream, from site HK, also on the left bank of the gorge. A great depth of Bed IV is exposed, together with the buff-coloured deposits which correspond in this area, and also further upstream, to the Red Bed III, as it is seen to the east and south. The topmost horizon of Bed II is also exposed in a small gully adjacent to FK. The extensive exposures of Bed IV at this site have yielded very little cultural or faunal material, with the exception of the skull of a very large *bos* found during 1962 in the lower part of Bed IV.

FLK Frida Leakey Korongo

The site FLK lies on the right bank of the main gorge, just above the confluence of the two rivers. It was discovered in 1931 and was the first site at which stone tools were found *in situ* in Bed I. A small excavation was carried out at that time.

The whole FLK area was found to be rich in fossil bones and stone tools, not only in Bed I, but also at several levels in Bed II. For convenience, the area was subdivided into FLK and FLK south (FLKS), FLK being the area where the stone tools were found *in situ*, while FLK south lay to the south and east, along the tongue of land where the two gorges meet.

In 1959 the skull of *Zinjanthropus boisei* was found at FLK, within a few yards of the 1931 excavation. Major excavations were carried out at the site during 1960–1. During that season, another site with hominid remains in Bed I was discovered a little to the north, which became known as FLK NN. A third site, near the top of Bed I, excavated during 1960–1 and 1962, is known as FLK N, and proved exceedingly rich in cultural and faunal material. A trial excavation in the upper part of Bed II, carried out in 1960–1, yielded an assemblage of Chellean artefacts.

GC

This site consists of shallow exposures on the right bank of the side gorge opposite CMK. Beds II and IV are exposed, the contact of the two beds being very rich in fossils and stone tools. It was discovered in 1935.

GRC

This is a small cliff lying on the left bank of the side gorge, upstream of FC and about 1¼ miles above the confluence of the two gorges. Beds III and II are exposed, with occasional small patches of Bed IV. Discovered in 1935, the upper part of Bed II here contains fossil bones and some artefacts.

GTC

A low cliff, situated on the left bank of the side gorge, immediately upstream of MRC; Bed IV is exposed. It was discovered in 1935.

Ha. C Hand-axe Cliff

This lies in the main gorge upstream from Rhino korongo and also on the left bank. There are exposures of Bed IV which yielded a number of hand-axes during 1962.

Heb. G Heberer's Gully

This is a small gully on the south face of HWK. Bed IV and the upper part of Bed III are exposed. Hand-axes and fossil bones were found eroding from the base of Bed IV, as well as from a second horizon slightly higher in the sequence, by L. S. B. Leakey when visiting the area with Professor Heberer during 1962. Excavations undertaken at the site under the supervision of Dr J. Waechter during June, July and August 1962 revealed a rich Acheulean floor level with associated fauna in the lowest sand of the Bed IV series.

HG Hoopoe Gully

This site was named by Dr R. Pickering in 1961. It is a deep gully, exposing Beds IV, III and II, and is situated about ½ mile from KK, on the same side of the gorge. Bed IV exposures here show a very complete sequence with the basal gravel in very coarse and massive form. A few fossil remains have been found at this level.

HK Hopwood's Korongo

This site is situated on the left bank of the main gorge, almost due north of FLK, between Dal.K and FK. It was found in 1931; the exposures are mainly of Bed IV, which is here of considerable thickness, as also in the adjoining gullies. Bed III is represented by the grey-buff deposit usual in this area and a small exposure of the upper part of the Bed II series is also visible. The main site at HK is on an 'island' in the centre of the gully. Here a very rich Acheulean living-site was first seen by means of binoculars from the opposite side of the gorge. This was described in the 1951 book.

HWK Henrietta Wilfrida Korongo

This site lies on the right bank of the side gorge near the confluence of the two rivers. It covers a considerable area of extensive exposures, in which Bed II is particularly rich in fossil bones and artefacts.

The HWK erosion gullies centre round a low saddle connecting two pinnacles of deposits, mainly of Bed III, which form prominent landmarks and are known as the 'castle' and the 'tower'. The saddle between the two forms a small watershed, giving rise to a number of minor erosion gullies.

For convenience in working the area, it has been subdivided as follows:

HWK Main. This is a long narrow gully extending from the saddle down to the floor of the gorge, in a generally north–south direction. At its mouth, the lava underlying the lake sediments rises in a small mound to within a few feet of the top of Bed I. To the south, the head of the gully is flanked to the east and west respectively by the 'castle' and the 'tower'.

During 1931 tools discovered in this gully were considered to be *in situ* in Bed I, but are now known to have been found in the base of Bed II. Excavation during 1959 at the same site yielded the greater part of the skeleton of *Giraffa gracilis* associated with Oldowan choppers, etc.

To the west of the HWK Main gully lies a ridge which leads up to a wind gap. This area is partly covered by vegetation, but also has considerable exposures of Bed II which have yielded a number of artefacts *in situ*.

The 'tower', mostly formed by a pinnacle of Bed III, retains residual deposits of Bed IV at its top, where Acheulean hand-axes were found in 1931. To the south of the 'tower' lies Heberer's Gully, where excavations in Bed IV were carried out in 1962 (see Heb. G).

HWK Central consists of a short gully lying to the east of the 'castle' and running in a southerly direction to meet Heberer's Gully. Stone tools and fossils were found here in 1931.

The HWK East gully derives from the point where the HWK central gully and Heberer's Gully meet. In the upper part, it runs mainly east but, approximately ¼ mile from the head, turns sharply to the north, to drain into the gorge. Exposures of the lower part of Bed II in this gully have long been known to be particularly rich in fossil bones and stone tools. Detailed excavations at this level were begun in 1962 and revealed a number of different implementiferous horizons in the lower 20 ft. of Bed II.

JK, JK 1 and JK 2 Juma's Korongos

These sites, situated on the left bank of the gorge, approximately 2 miles upstream of the third fault, consist of a long cliff with associated gullies. At JK, the complete sequence of deposits from Beds IV to I is exposed, whereas at JK 1 and JK 2 only Beds IV, III and the upper part of Bed II are cut through.

The exposures of Bed IV at sites JK 1 and 2 yielded a large number of Acheulean implements with an associated fauna from three separate levels during the 1931–2 season. In 1962, at the invitation of Dr and Mrs Leakey, Dr Maxine Kleindienst carried out a four-month season at these sites, during which an assemblage of artefacts of non-Acheulean facies were discovered in Bed III and many implements and fossils in Bed IV.

Kar. K Karonga Korongo

This lies on the left bank of the main gorge, approximately 2¼ miles above the fifth fault scarp. There are exposures of Beds IV, III and II. A particularly well-preserved *bos* skull was found here by G. Karonga in 1962, eroding from the lower part of Bed IV, which is also rich in hand-axes and other artefacts.

Kestrel Korongo

A deep, sheer-sided gully cutting back into the plains from the left bank of the gorge, below the fifth fault. The nearly vertical exposures cut through the most extensive known deposits of the upper part of Bed IV. The site was named by Dr R. A. Pickering in 1961 but had originally been known as Teale's Korongo during the first season at Olduvai, in 1931, when four artefacts were found there. This name, however, fell into disuse when no further remains were found at the site during the ensuing years.

Kit. K Kitibi Korongo

This site is situated slightly to the west of the first fault, on the left bank of the gorge, and was found in 1962. There are exposures of Bed III and the upper part of Bed II.

K.K. Kudu Korongo

A gully on the right bank of the gorge, immediately downstream of HWK East. Beds IV to I are well exposed. The lower part of Bed II is exceedingly rich in fossil bone and stone tools, which occur in the same levels as at HWK East.

LK Leakey's Korongo

This lies on the left bank of the gorge and is the first gully upstream of the third fault. Exposures of Bed IV yielded Acheulean hand-axes during 1931. Beds III, II and I are also exposed.

Long Korongo

This is a very long gully lying on the right bank of the gorge, immediately east of the fourth fault. At the mouth Beds IV to I are well exposed and contain some artefacts and fossils, particularly in Bed II. The higher part of the gully follows the line of an old drainage channel cut prior to the deposition of Bed V, which became filled-in with Bed V deposits and has later been re-excavated. There are a number of small exposures of Bed III towards the head of the gully.

LLK Louis Leakey Korongo

This is a small exposure cutting through Beds IV, III and the upper part of II. It lies on the left bank of the side gorge, just above the confluence and less than ½ mile from the FLK I sites.

In December 1961 L. S. B. Leakey discovered a human calvaria eroding from the floor of this gully. Although it was not *in situ* the matrix adhering to the skull could be correlated with a bed of fine sandy tuff and indicated that it was derived from the upper part of Bed II.

MK MacInnes Korongo

This site is situated on the left bank of the gorge about 1 mile upstream of the third fault. A series of steep-sided gullies cuts through deposits of Beds IV to I.

In Bed I there is an horizon which is particularly rich in fossil remains and which has also yielded some artefacts. It lies low in the sequence, below a deposit of massive tuff and not far above the underlying basalt. The first known remains of aquatic animals from Bed I, such as crocodile, hippopotamus, fish, etc., were recovered by Dr MacInnes from this level in 1931.

During 1959 the same level yielded a hominid lower molar tooth. Trial excavations were carried out the following season and resulted in the recovery of two further hominid teeth.

MCK Margaret Cropper Korongo

This lies on the right bank of the gorge, to the east of the fourth fault. There are exposures cutting through Beds IV, III, II and I. In this area, deposits of Bed II contain fossil bones and stone tools at several different levels, and the basal sands of Bed IV are also rich in stone implements.

During 1962, on the north face of the saddle which separates this gully from the Long Korongo, Miss M. Cropper found an almost complete skeleton, mandible and part of the skull of a species of a very large baboon allied to *Simopithecus*, eroding from the upper part of Bed II, within a few feet of the base of Bed III.

MJTK M. J. Tippett Korongo

Here deposits of Beds II and I are exposed in a gully on the left bank of the gorge. The site is opposite MCK and lies about ¾ mile from the camp. Mr M. J. Tippett found an incomplete skull and horn cores of a large antelope in Bed II during 1962. Micro-mammalian fauna can also be found some 18 ft below the Bed I 'marker bed'

MLK Mary Leakey Korongo

This site is situated on the right bank of the main gorge, approximately 1 mile west of the fifth fault, it includes the area of earth movement known as the 'landslide' and was found in 1957. There are deep exposures of Beds IV to I. Artefacts and fossils were first located at the main site. During 1962 a well-preserved skull of a very large monkey was found by Philip Leakey eroding from the base of Bed IV on the north-west face of a long ridge which juts out into the gorge, to the west of the original site.

MNK Mary Nicol Korongo

This site is situated on the right bank of the side gorge, approximately 1 mile above the confluence. A number of discontinuous exposures give an aggregate section through Beds IV, III, II and the top of I.

At this site, as at FC on the opposite bank of the side gorge, and also in many parts of the main gorge upstream of the fifth fault, the chert bed can be seen at a low level in the Bed II deposits, about 12 ft. above the marker bed at the top of Bed I. This yielded the raw material for making small flake tools and choppers during Bed II times, such as those found at HWK.

The site was discovered in 1935 by Mrs Leakey and yielded stone tools and faunal remains at the junction of Bed IV and Bed III, where there are also numerous fossil *Unio* shells. A small fragment of a fossil human skull was found at this site in 1935. The fault which cuts through the FLK area also affects this site but with the throw reversed. A very rich cultural level also occurs in Bed II.

MRC

This is a series of exposures on the left bank of the side gorge opposite to and upstream of BK. It was found in 1935, when tools and fossils were discovered in Bed IV in the western part of the area. Bed II is also exposed in places. A car track leads out of the gorge on the eastern side of the gully, and here, in 1957, J. H. E. Leakey found the type mandible of *Simopithecus jonathani* in Bed IV gravels.

NGK

This is a long cliff, associated with a small gully, on the left bank of the side gorge, approximately 4 miles upstream from the confluence. Red tuffaceous beds are exposed. These are mostly barren. A small fault can be seen, with the downthrow to the west.

PDK Peter Davies Korongo

This lies on the right bank of the gorge, opposite site JK 2. A series of deep exposures cut through Beds IV to I, which are largely non-fossiliferous in this area. The greater part of two hippopotamus skeletons were found in Bed I during 1959, one of which was subsequently excavated by Dr A. Sutcliffe.

PEK Peter E. (Kent) Korongo

This is a cliff with very small gullies, situated on the right bank of the side gorge, between the two large gullies VEK and MNK. Beds I and II are exposed.

PLK Philip Leakey Korongo

This is the first long Korongo on the left bank of the gorge above the confluence. Beds IV, III and the upper part of II are exposed. Bed III is here considerably reduced in thickness. Trenches cut during 1962 revealed a lateral change from the usual red facies of Bed III to the brownish-buff deposits which correspond to Bed III in exposures higher up the gorge, at such sites as HK, FK, etc. The type specimen of *Bularchus arok* was found at this site in 1935 and was probably derived from the top of Bed II, where a hand-axe site also occurs.

RK Reck's Korongo

A very small gully on the left bank of the gorge, immediately upstream of the third fault, sometimes referred to as 'Reck's Man site' since it was here that he found the crouched human burial in 1913.

RHC Richard Hay Cliff

An almost vertical cliff situated on the left bank of the gorge approximately ¾ mile from the fifth fault, facing south-east. Beds IV, III and II are exposed. Bed II deposits in this area include a considerable depth of fine-grained green clays, laid down in deep water. The band of chert nodules noted elsewhere in Bed II is well developed and a second much thinner chert band also occurs at a slightly higher level in the bed. Beds II and III have yielded no remains at this site, but a rich Acheulean floor occurs within Bed IV near the top of the main cliff.

Rhino Korongo

This is on the left bank of the main gorge, upstream of FK, but with one unnamed gully intervening. Exposures of Bed IV yielded the greater part of a skull and the complete mandible of a fossil rhinoceros during 1962. A number of well-made Acheulean hand-axes were also found, probably derived from the lower sand of Bed IV.

SC Sam's Cliff

This is a small cliff on the left bank of the side gorge, opposite DC and about 2 miles upstream from the confluence. There are some exposures of Beds II and III and of a small part of Bed IV.

SHK Sam Howard Korongo

This site lies on the right bank of the side gorge about 1½ miles above the confluence. It consists of a long narrow gully cutting back into the plains and a cliff facing the side gorge. Discovered in 1935, it then yielded a hippopotamus skull at a high level in Bed II and a series of associated skeletal and cranial material of a small gazelle.

Excavations were carried out at the corner between the gully and the cliff from 1952 to 1958 and yielded many thousands of stone tools and fossils. In a preliminary report these were regarded as representing Chellean stage 2. It is now clear that the levels containing artefacts here and at BK II belong to a late stage of Bed II.

SWK Sam White Korongo

This is a small, narrow gully on the right bank of the side gorge, upstream of MNK. It was discovered in 1935. Bed II is well exposed and has yielded fossils and culture. There is also a small secondary exposure of Bed IV, separated from the main gully by deposits of Bed V.

TK Thiongo Korongo

This is a large gully on the left bank of the gorge, situated less than 1 mile east of the camp. Beds IV, III and II are exposed over considerable areas. During the 1931–2 season, a Chellean site was explored in a high level of Bed II, whilst hand-axes, and other artefacts, were recovered from the base of Bed IV.

A small side gully, known as 'Fish Gully', on the west side of the main TK gully, which had yielded archaeological material during 1931–2 was partially excavated under the supervision of Dr J. Waechter, of the London Institute of Archaeology, during 1962. Five separate cultural levels were revealed within Bed IV, including a particularly rich living-floor associated with plentiful fish remains.

Th. C Thiongo Cliffs

This site consists of a series of small exposures of Bed I and is situated on the right bank of the gorge, less than ½ mile upstream of the third fault. It was found during 1935 when it yielded foot bones of a large Chalicothere.

VEK Vivian Evelyn (Fuchs) Korongo

This consists of a long gully on the right bank of the side gorge, just upstream of its confluence with the main gorge. There is a car track on the eastern side, which leads out on to the plains. The exposures cut through Beds IV, III, II and I. Bed IV in this area is of variable thickness, being considerably deeper at the head of the gully (where artefacts were collected in 1931) than lower down on the west side, where it is reduced to less than 15 ft. in thickness and is comparable to the exposure at Heberer's Gully. During June 1962 Miss M. Cropper discovered some fragments of a thick human skull lying on the slopes of Beds IV and III in this area.

Lower down the gully, also on the west side, exposures of Bed III (which is usually almost barren) contain a horizon of cemented coarse sand, in which can be seen artefacts and fossil bones, including remains of crocodile. During 1931, at the same site, a number of hand-axes, were found immediately underlying the Beds II–III junction.

Vth FK Fifth Fault Korongo

This site lies on the left bank of the main gorge, immediately west of the fifth fault. Deep exposures, cut through Beds IV, III, II and I. This is the first site west of the FLK group where erosion has cut sufficiently deep to expose Bed I and the lower part of Bed II.

A concentration of antelope remains eroding from an earthy clay, which lies approximately midway in the Bed IV series, was found here by J. H. E. Leakey in 1962. Excavation yielded thirteen skulls and skeletons of a single species resembling the blesbok. An incomplete *bos* skull was also found on the opposite side of this gully eroding from Bed IV.

WK Wayland's Korongo

This is a steep-sided gully exposing Beds IV, III, II and I. It is situated on the right bank of the gorge, immediately east of the Long Korongo. Bed IV deposits have yielded hand-axes. In Bed I, the grey tuff and underlying fossiliferous horizon which occurs at MK I and DK I is exposed near the base of the series. (Note: in the volume on Olduvai Gorge published in 1951 the position of this gully was wrongly shown on the sketch-map as being on the left bank instead of on the right bank of the gorge.)

REFERENCES

Andrews, C. W. (1924). Notes on the occurrence of a species of Chalicothere in Uganda. *J. E. Afr. Ug. Nat. Hist. Soc.* **20**, 22–3.

Arambourg, C. (1943). Observations sur les Suides fossiles du Pleistocène d'Afrique. *Bull. Mus. Hist. Nat., Paris,* (2), **15**, 471–6.

Arambourg, C. (1947). Contribution à l'étude géologique et paléontologique du lac Rudolphe et de basse vallée de l'Omo. *Mission Scientifique de l'Omo.* Paris.

Arambourg, C. (1949). Les Gisements de vertèbres Villafranchien de l'Afrique du Nord. *Bull. Soc. Géol. Fr.* **5**, 19.

Broom, R. (1937). On some new Pleistocene mammals from the limestone caves of the Transvaal. *S. Afr. J. Sci.* **33**, 750–68.

Broom, R. (1948*a*). Some South African Pliocene and Pleistocene Mammals. *Ann. Transv. Mus.* **21**, 8–11.

Broom, R. (1948*b*). Some South African Pliocene and Pleistocene Mammals. *Ann. Transv. Mus.* **21**, 25–32.

Cooke, H. B. S. See Shaw and Cooke (1941).

Dietrich, W. O. (1926). Fortschritte der Säugetierpaläontologie Afrikas. *Forsch. Fortsch. dtsch. Wiss.* **2**, 121–2.

Dietrich, W. O. (1941). Die Säugetierpaläontologische Ergebnisse der Kohl-Larsen'schen Expedition 1937–1939 im nördlichen Deutsch Ostafrika. *Cbl. Min. Geol. Paläont. Stuttgart,* 1941 B, 217–23.

Dietrich, W. O. (1942*a*). Altesquartare Säugetiere aus der südlichen Serengeti, Deutsch Ostafrika. *Paläontogr.* **94** (A), 43–133.

Dietrich, W. O. (1942*b*). Zur Entwicklungsmechanik des Gebisses der afrikanischen Nashörner. *Cbl. Min. Geol. Paläont. Stuttgart,* 297–300.

Dietrich, W. O. (1950). Fossile Antilopen und Rinder Äquatorialafrikas. (Material der Kohl-Larsen'schen Expeditionen.) *Paläontographica. Stuttgart,* 99 A, 1–62.

Emiliani, C., Mayeda, T. and Selli, R. (1961). *Bull. Geol. Soc. Amer.* **72**, 5, 679.

Evernden, J. F. and Curtis, G. H. (1961). *Proc. INQUA Cong. Warsaw.*

Ewer, R. F. (1954). Some adaptive features in the dentitions of Hyaenas. *Ann. Mag. Nat. Hist.* **7**, 188–94.

Ewer, R. F. (1956). The Fossil Carnivores of the Transvaal Caves: Canidae. *Proc. Zool. Soc. Lond.* **126**, 97–119.

George, M. (1950). A Chalicothere from the Limeworks quarry of the Makapan Valley, Potgietersrust District. *S. Afr. J. Sci.* **46**, 241–2.

Greenwood, M. (1955). Fossil Hystricoidea from the Makapan Valley, Transvaal. *Palaeontologia Africana,* vol. III.

Hilzheimer, M. (1925). *Rhinoceros simus germano-africanus* n.sub-sp. aus Oldoway. *Wiss. Ergebn. Oldoway-Exped.* 1913 (N.F.), **2**, 45–79.

van Hoepen, E. C. N. (1930). Fossiel perde van Cornelia O.V.S. *Paleont. Navors. Nas. Mus. Bloemfontein,* **2**, 13–24.

Hopwood, A. T. (1934). New fossil mammals from Olduvai, Tanganyika Territory. *Ann. Mag. Nat. Hist.* (10), **14**, 546–50.

Hopwood, A. T. (1936). New and little known fossil mammals from the Pleistocene of Kenya and Tanganyika Territory. *Ann. Mag. Nat. Hist.* (10), **17**, 636–41.

Kent, P. E. (1941). The recent history of the Pleistocene deposits of the plateau north of Lake Eyassi, Tanganyika. *Geol. Mag. Lond.* **78**, 173–84.

Leakey, L. S. B. (1935). *Stone Age Races of Kenya.* Oxford University Press.

Leakey, L. S. B. (1936). *Stone Age Africa.* Oxford University Press.

Leakey, L. S. B. (1943). New fossil Suidae from Shungura, Omo. *J. E. Afr. Ug. Nat. Hist. Soc.* **17**, 45–61.

Leakey, L. S. B. (1951). *Olduvai Gorge.* Cambridge University Press.

Leakey, L. S. B. (1958*a*). Note on an aberrant lower third molar of *Sus cristata. S. Afr. J. Sci.* **54**, no. 6, 134.

Leakey, L. S. B. (1958*b*). Some East African Pleistocene Suidae. *Fossil Mammals of Africa,* no. 14.

Leakey, L. S. B. (1959*a*). *Nature, Lond.* **181**, 1099–1103.

Leakey, L. S. B. (1959*b*). In: *Proceedings of the 3rd Pan-African Congress of Prehistory at Leopoldville.*

Leakey, L. S. B. (1959*c*). *Nature, Lond.* **184**, 491–3.

Leakey, L. S. B. (1959*d*). *Illustrated London News,* 12 September 1959.

Leakey, L. S. B. (1961). *Nature, Lond.* **189**, 649.

Leakey, L. S. B., Evernden, J. F. and Curtis, G. H. (1961). *Nature, Lond.* **191**, 478–9.

Leakey, L. S. B. and Whitworth, T. (1958). Notes on the genus *Simopithecus* with a description of a new species from Olduvai. *Coryndon Memorial Museum Occasional Papers,* no. 6.

Meester, J. (1955). Fossil Shrews of South Africa. *Ann. Transv. Mus.* **22**, 271–8.

Pohle, H. (1928). Die Raubtiere von Oldoway. *Wiss. Ergebn. Oldoway-Exped.* 1913 (N.F.), **3**, 45–54.

Reck, H. (1925). Aus der Vorzeit des innerafrikanischen Wildes. *Leipzig. illustr. Ztg.* **164**, 451.

Reck, H. (1933). *Oldoway die Schlucht des Urmenschens.* Leipzig.

Reck, H. (1935). Neue Genera aus der Oldoway-Fauna. *Cbl. Min. Geol. Paläont. Stuttgart,* 1935 B, 215–18.

REFERENCES

Remane, A. (1925). Der Fossil Pavian (*Papio* sp.) von Oldoway nebst Bemerkungen über die Gattung *Simopithecus*. C. W. Andrews, *Wiss. Ergebn. Oldoway-Exped.* 1913 (N.F.), **2**, 85–90.

Schwarz, E. (1932). Neue diluviale Antilopen aus Ostafrika. *Cbl. Min. Geol. Paläont. Stuttgart.*

Schwarz, E. (1932). Neue diluviale Antilopen aus Ostafrika. *Cbl. Min. Geol. Paläont. Stuttgart.*

Schwarz, E. (1937). Die Fossilen Antilopen von Oldoway. *Wiss. Ergebn. Oldoway-Exped.* 1913 (N.F.), **4**, 8–104.

Shaw, J. C. M. and Cooke, H. B. S. (1941). New fossil pig remains from the Vaal River gravels. *Trans. Roy. Soc. S. Afr.* **28**, 293–9.

Simpson, G. G. (1945). The Principles of Classification and a Classification of Mammals. *Bull. Amer. Mus. Nat. Hist.* **85**.

Singer, R. and Boné, E. (1960). Modern giraffes and the fossil giraffids of Africa. *Ann. S. Afr. Mus. Capetown*, **45**, part 4.

Stresemann, E. (1928). Ein Raubadler aus den Oldoway-Schlichten. *Wiss. Ergebn. Oldoway-Exped.* 1913 (N.F.), **3**, 69.

Zeuner, F. E. (1959). *The Pleistocene Period.* Hutchinson.

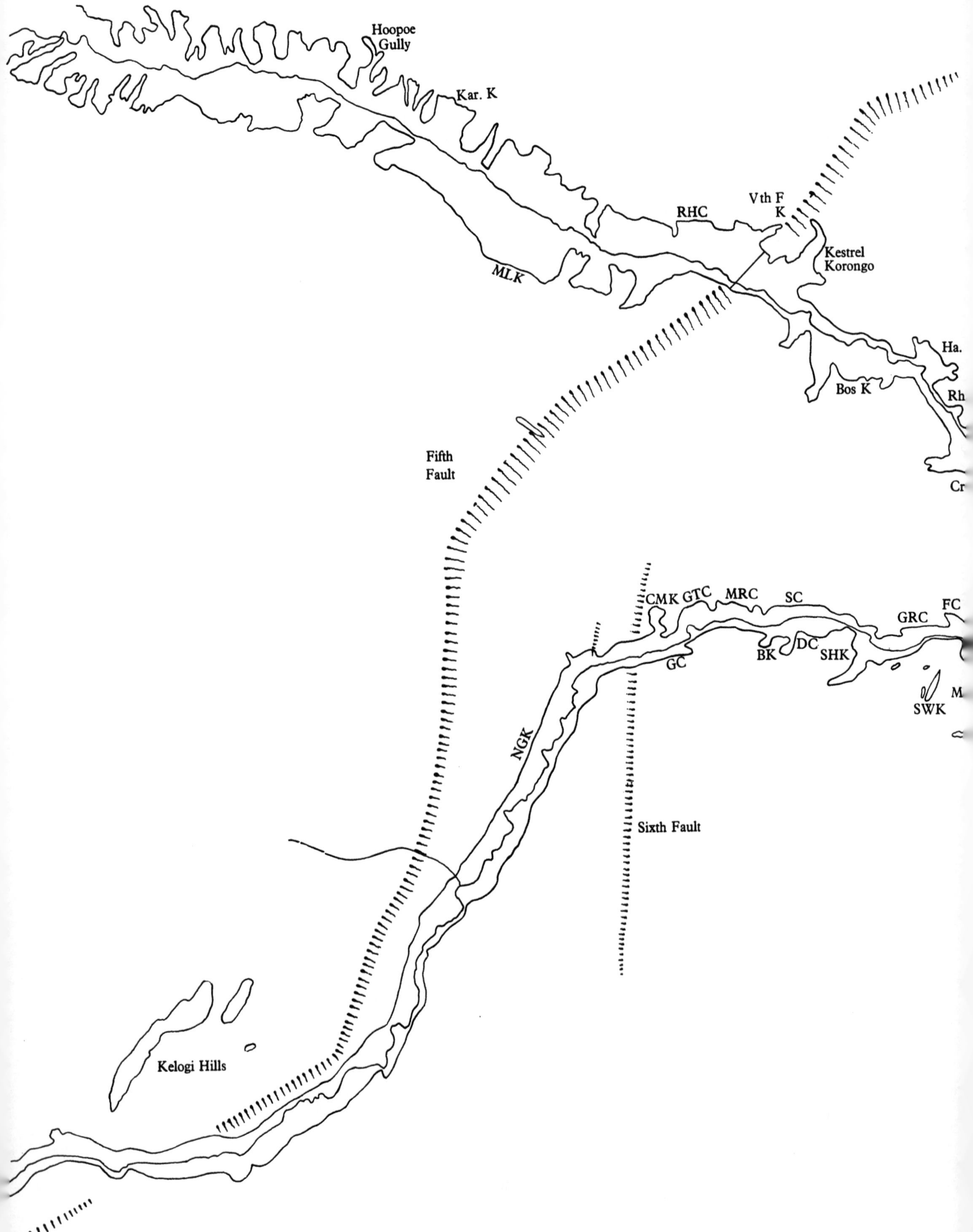

Hoopoe Gully

Kar. K

Vth F K

RHC

Kestrel Korongo

MLK

Ha.

Bos K

Rh

Fifth Fault

Cr

CMK GTC MRC SC

FC

GRC

BK DC SHK

GC

M

SWK

NGK

Sixth Fault

Kelogi Hills

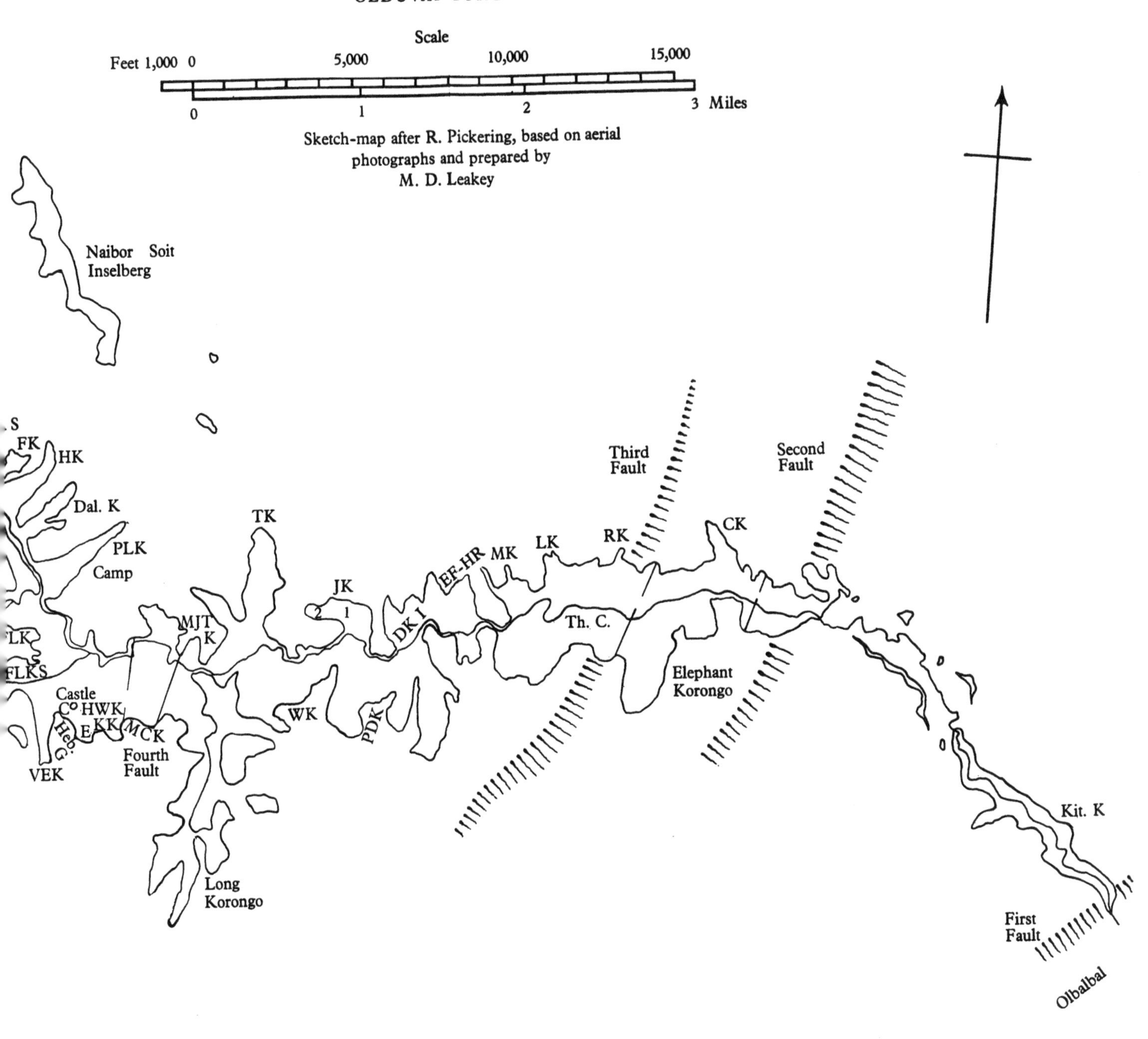
OLDUVAI GORGE
Scale
Feet 1,000 0
5,000
10,000
15,000
0
1
2
3 Miles
Sketch-map after R. Pickering, based on aerial photographs and prepared by M. D. Leakey
Naibor Soit Inselberg
S
FK
HK
Dal. K
PLK
Camp
TK
JK
2
1
MJT K
LK
FLKS
Castle
Co
HWK
E
KK
MCK
Heb.
G.
VEK
Fourth Fault
WK
PDK
DK I
EF-HR
MK
LK
RK
Third Fault
Second Fault
CK
Th. C.
Elephant Korongo
Long Korongo
Kit. K
First Fault
Olbalbal

PLATES

1. The 'marker bed' at the top of Bed I

2. 'Desert roses' *in situ* near the top of Bed I

3. General view of the FLK area from the camp road, with the *Zinjanthropus* site indicated

4. Bed III at site HWK

5. Torrent gravels in Bed II and unconformity in Bed III

6. Bed III torrent gravels

7. Boulder gravel in Bed II

8. Bed IV at site FK

9. Gravel beds of Va near the camp

10. Close-up of gravel beds of Va

11. Bed V unconformable on Bed I at the third fault

12. Excavating a small mammal mandible

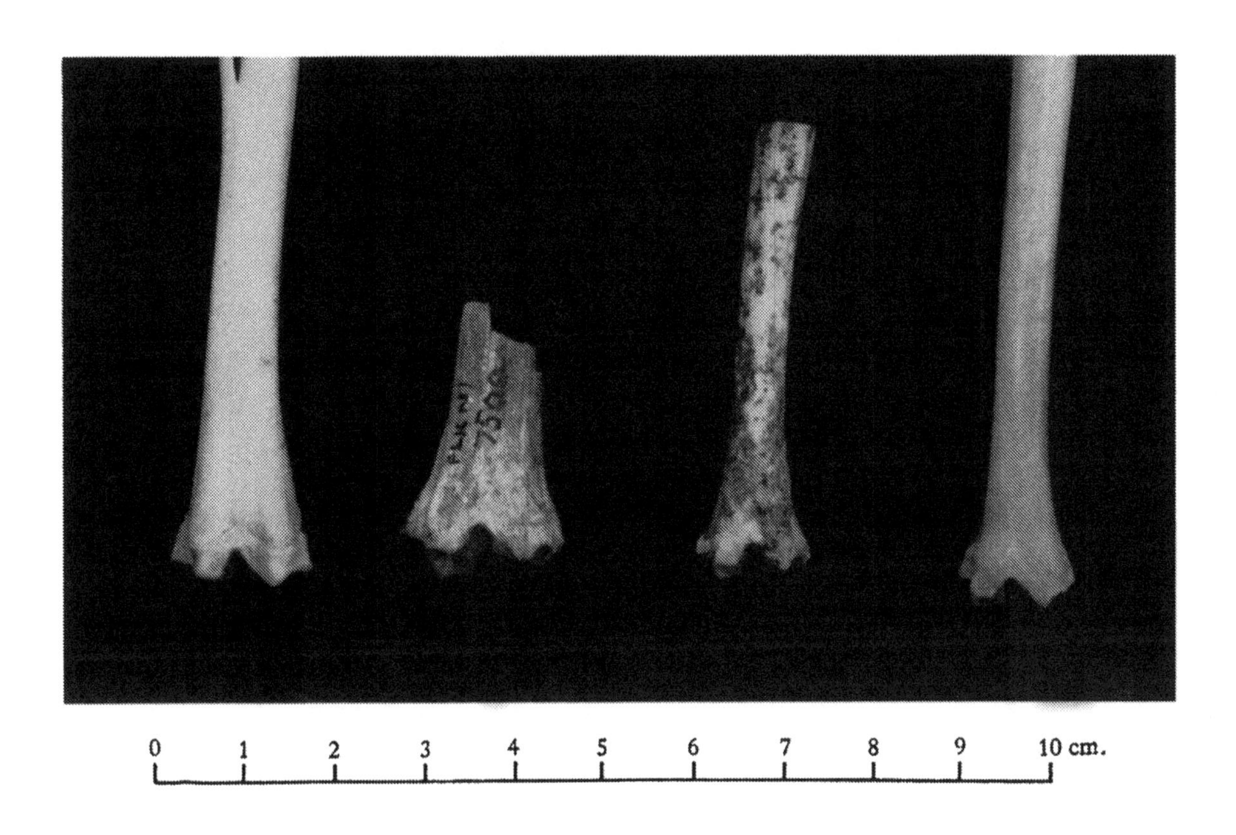

13. Lagomorph tibiae. From left to right: *Oryctolagus*; *Lepus* sp. from Olduvai; *Serengetilagus* from Laetolil; *Lepus capensis*

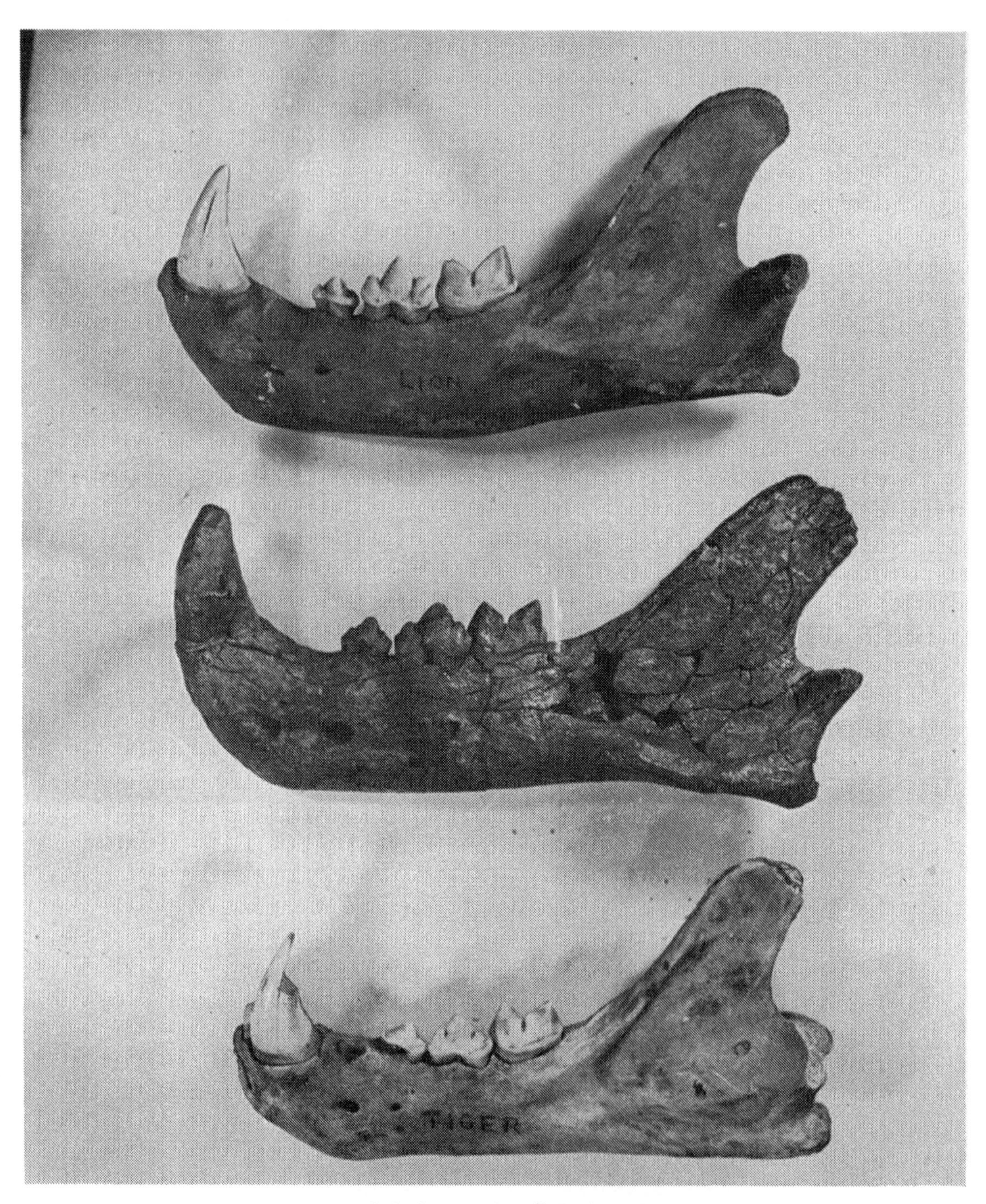

14. Felid mandibles. From top to bottom: modern lion; *Felis* cf. *tigris* from Olduvai; modern tiger

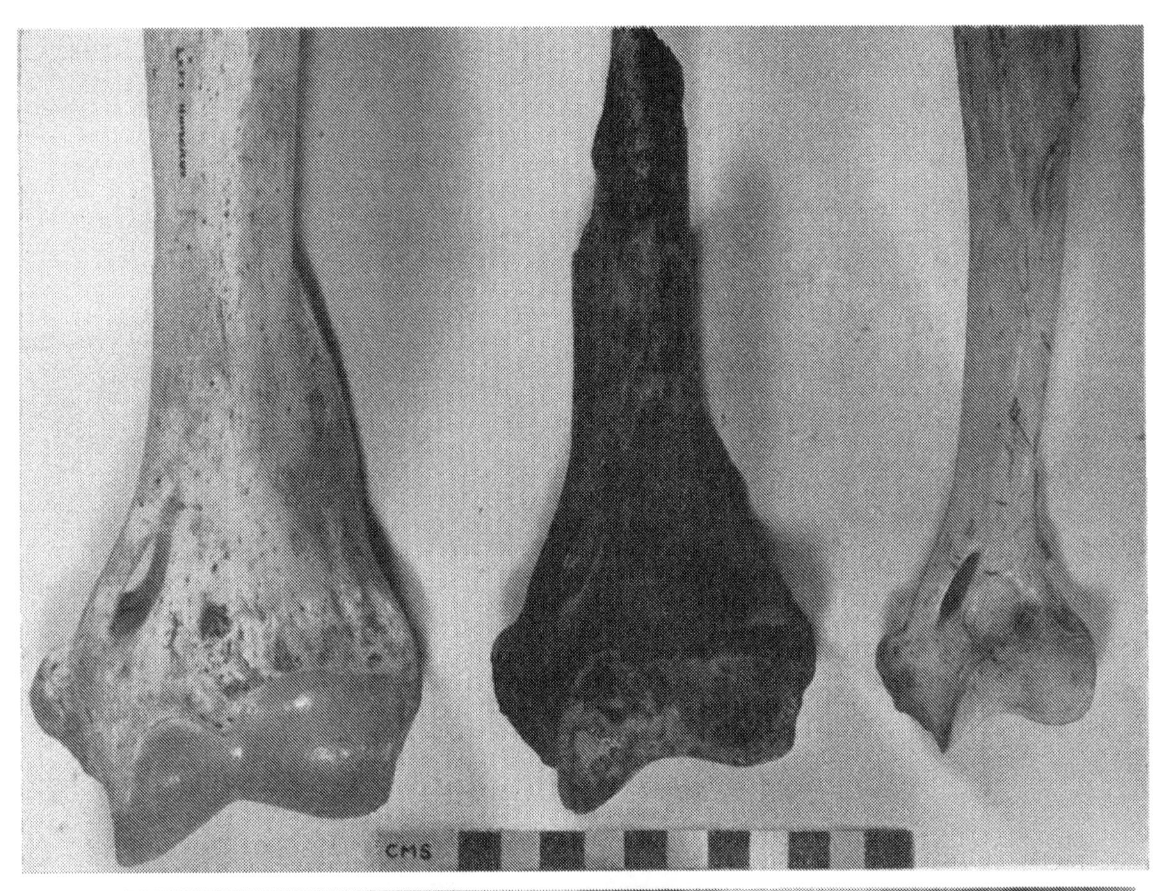

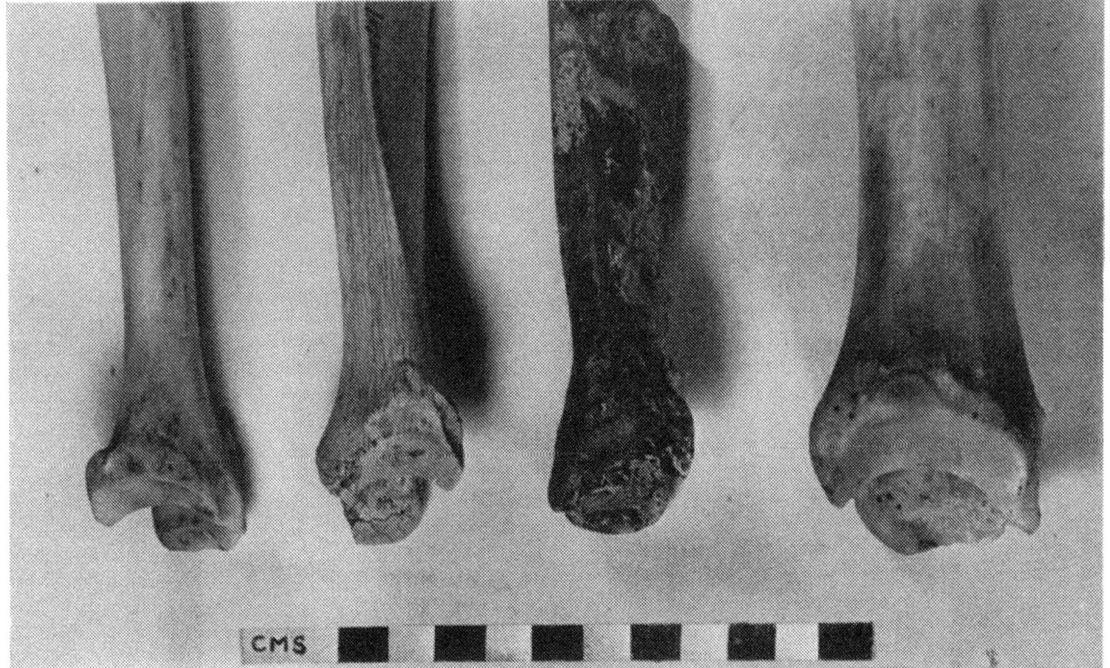

15. (Top) *Felis* sp. humerus, no. M. 14676 from Olduvai (centre) compared with: (left) modern lion; (right) modern leopard. (Bottom) *Felis* sp. tibiae (centre), nos. M. 20230 and M. 20231 compared with: (left) modern leopard; (right) modern lion

16. *Elephas* cf. *africanavus*. Broken molar

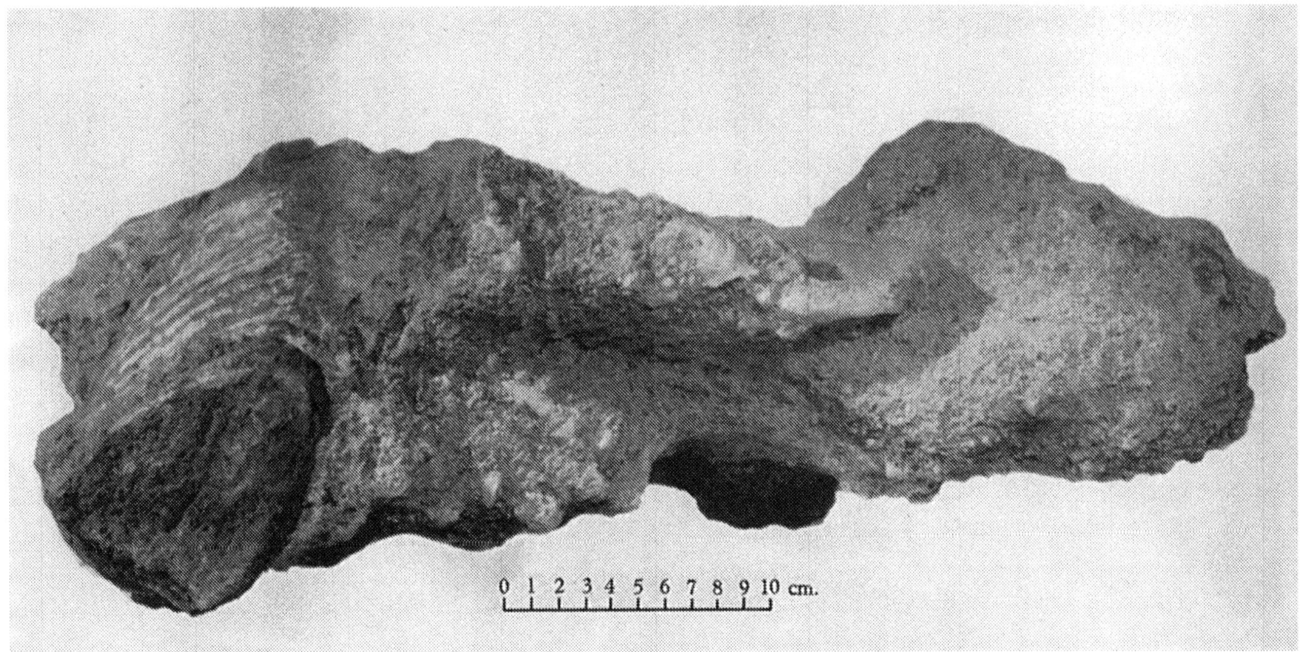

17. *Elephas* cf. *africanavus*. Part of mandible

18. *Elephas recki* (early form), no. M. 21465. Plate of molar

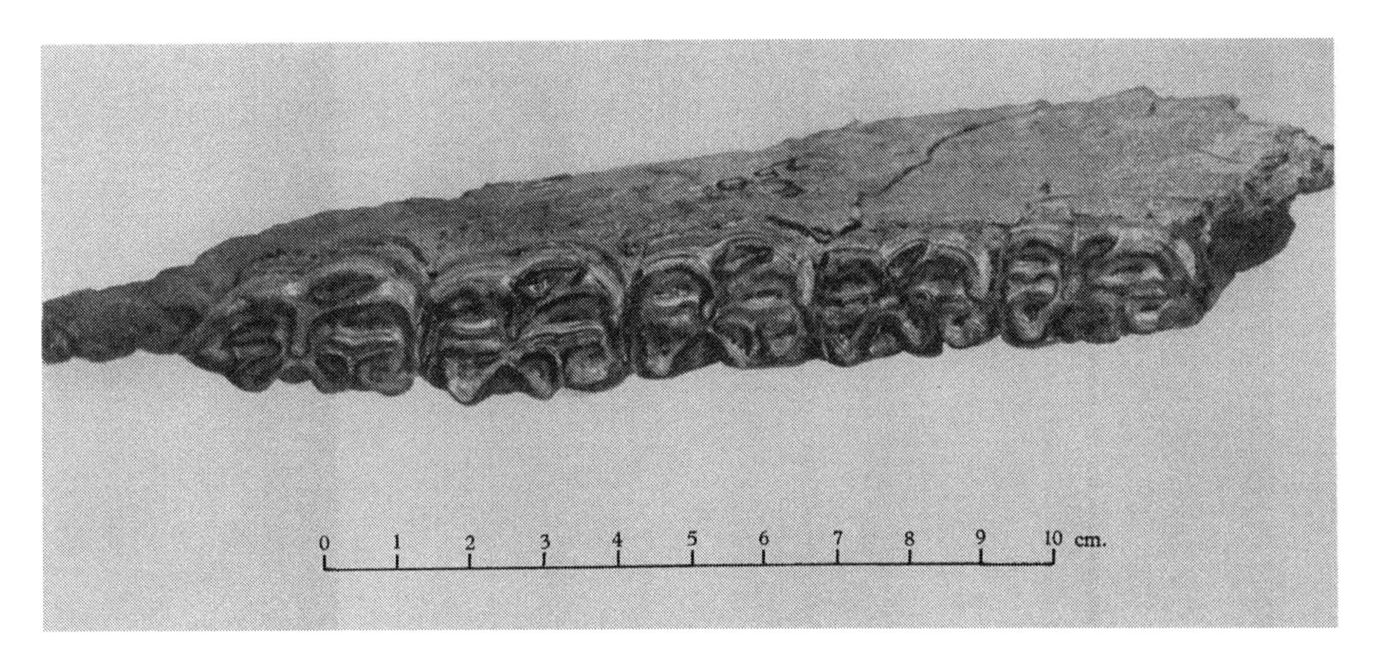

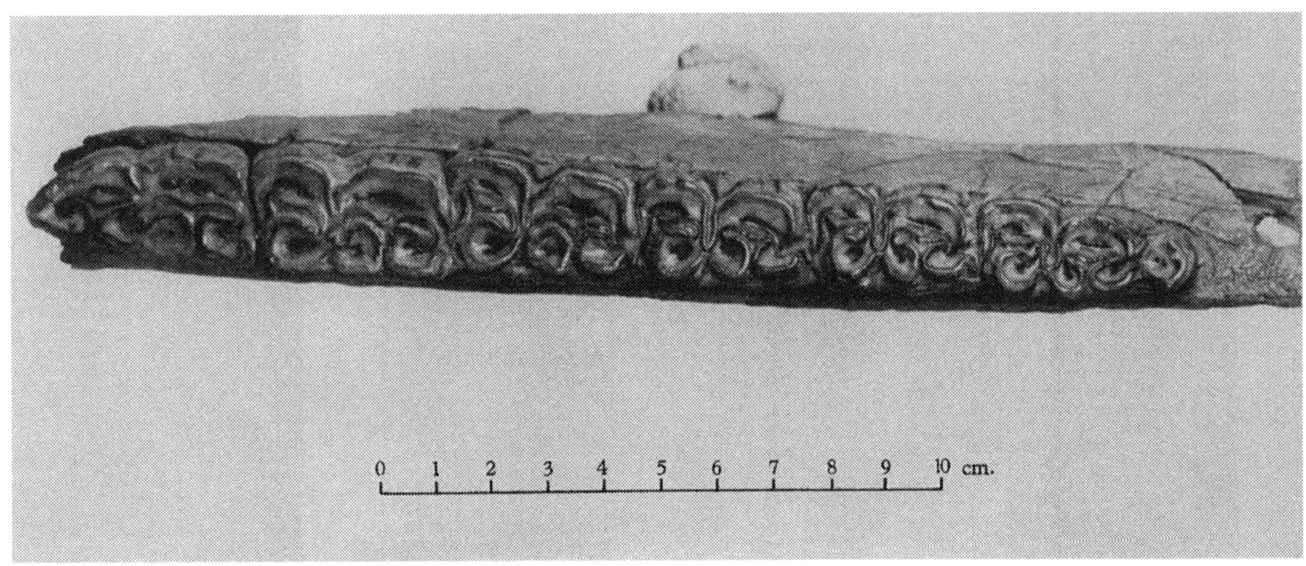

19. *Stylohipparion* (top) and *Equus* (bottom). Lower dentitions

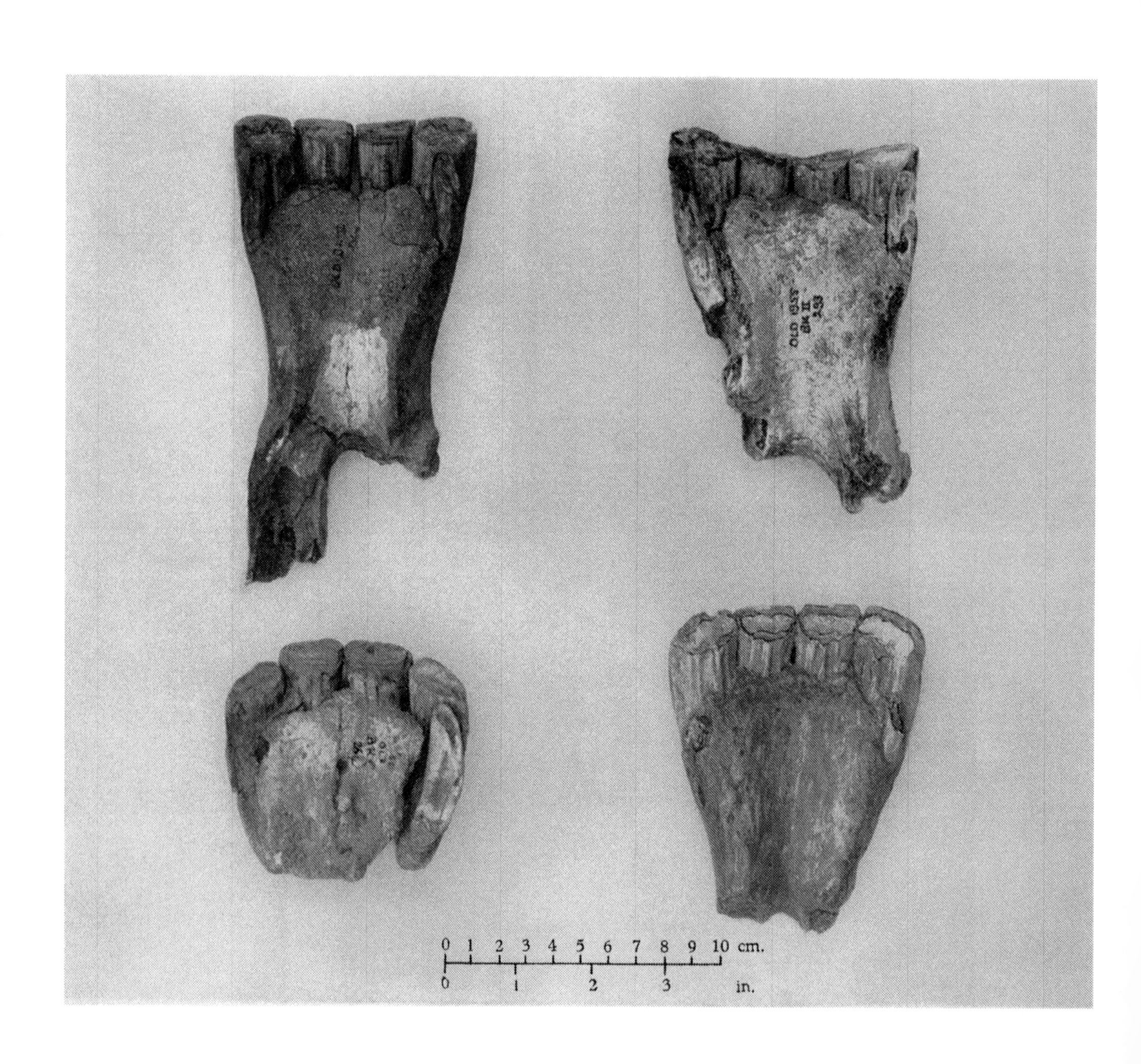

20. *Stylohipparion*. Anterior dentitions

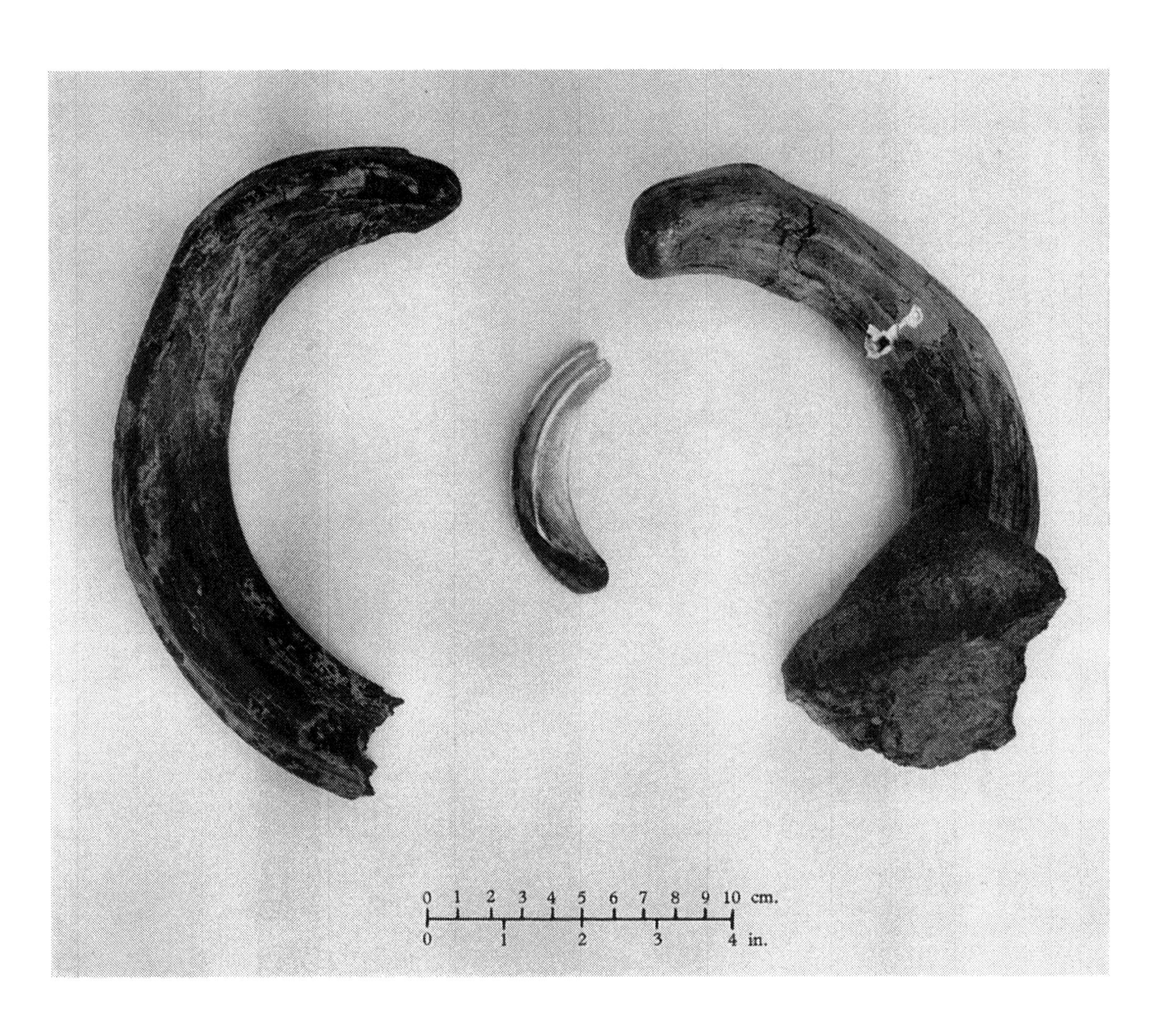

21. *Potamochoerus majus* and modern *Koiropotamus* (centre). Canines

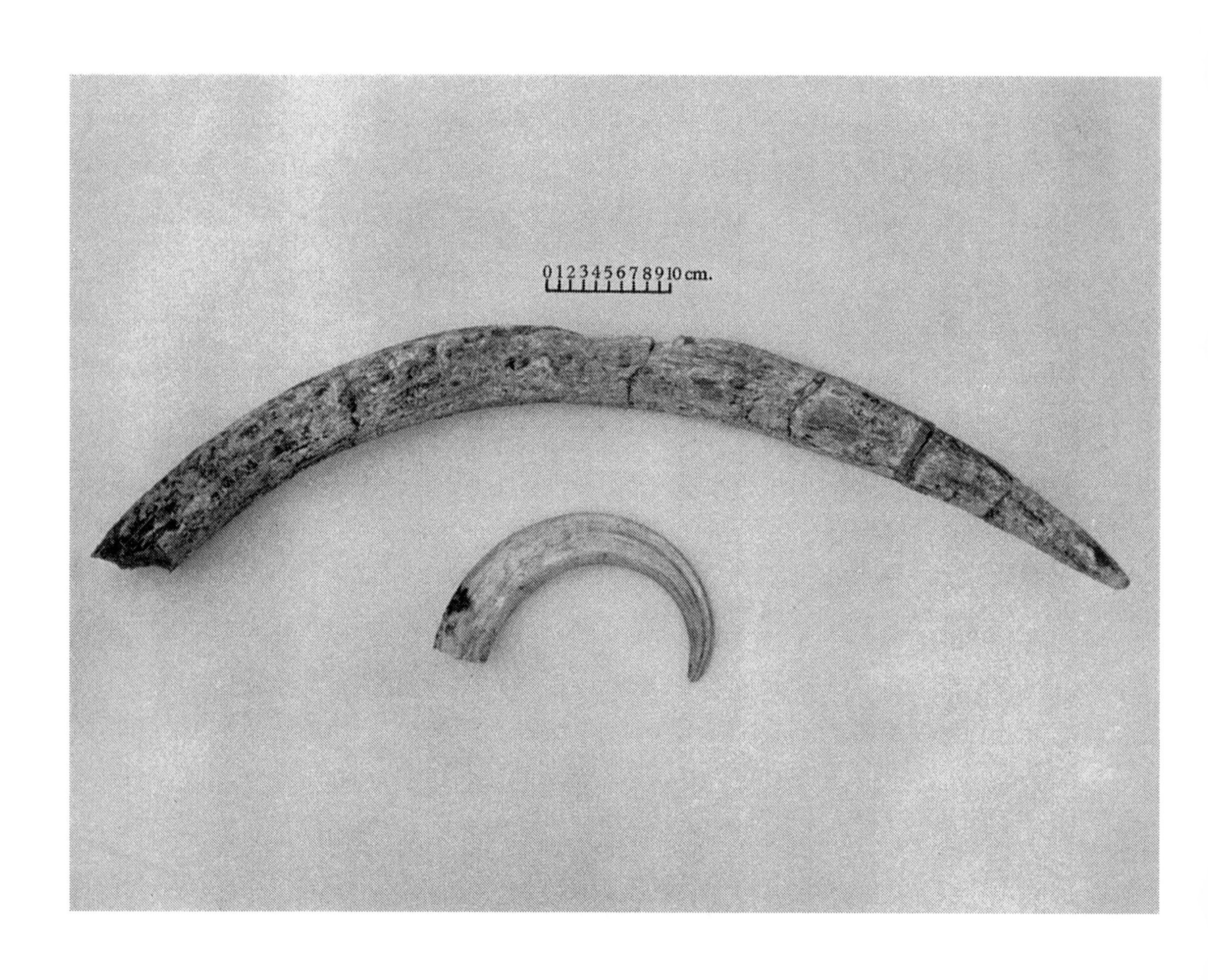

22. *Afrochoerus nicoli* (top) and modern *Phacochoerus* (bottom). Canines

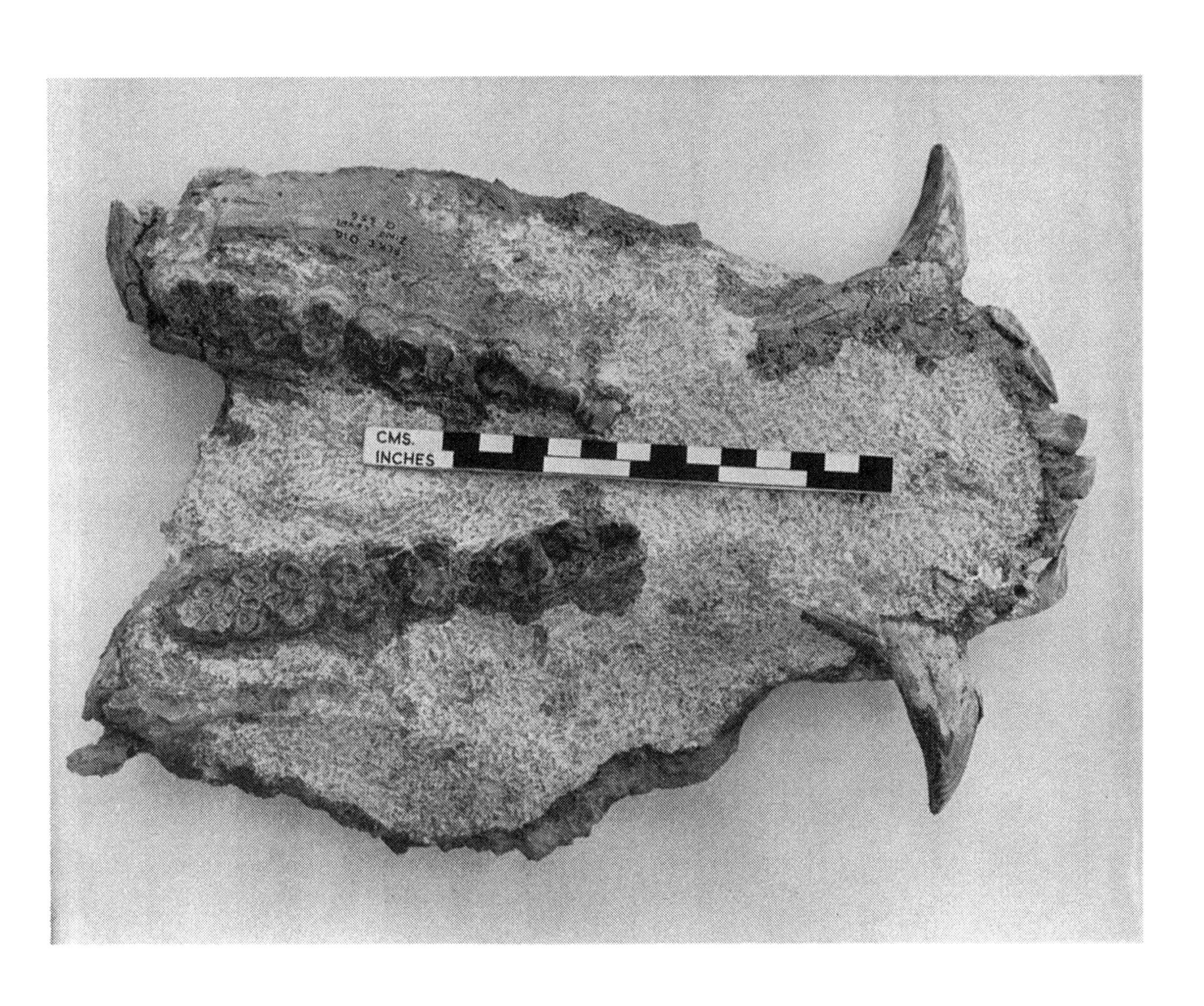

23. *Promesochoerus mukiri* mandible. Type, no. G. 356, 1960

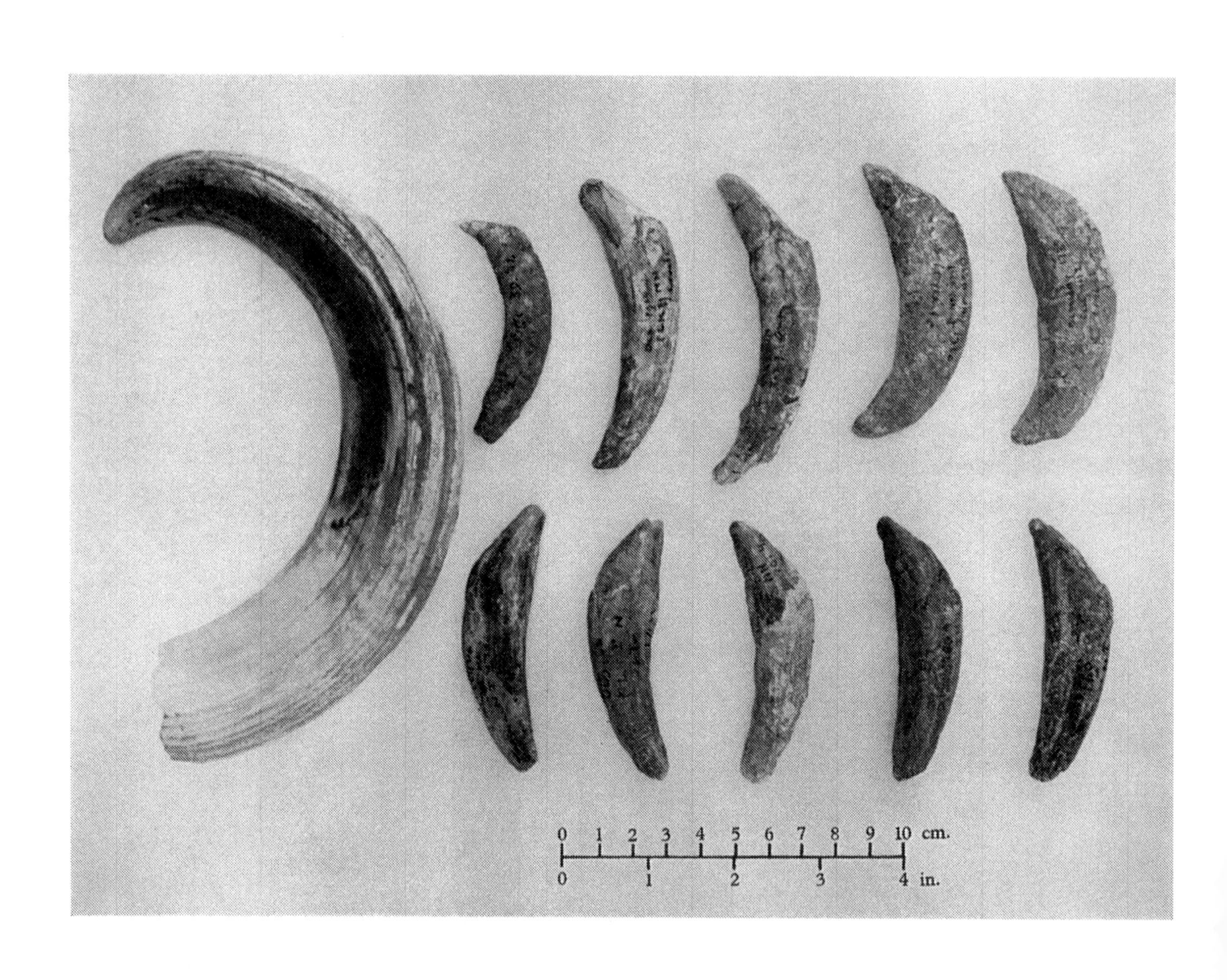

24. Canines of *Promesochoerus mukiri*, compared with modern *Phacochoerus* (left) and a Miocene suid (first left in top row)

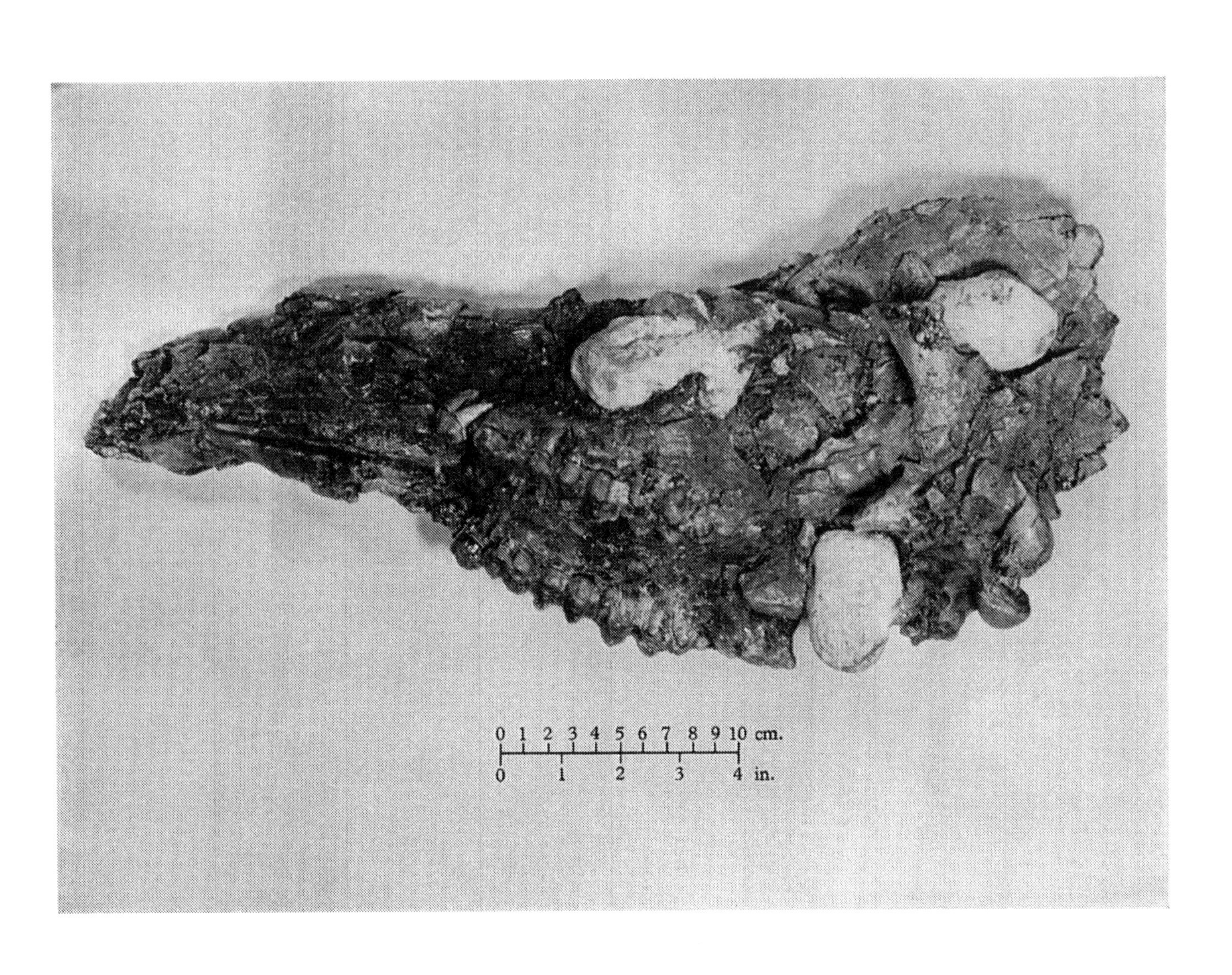

25. *Ectopotamochoerus dubius* skull. Type, no. FLK N I 1235

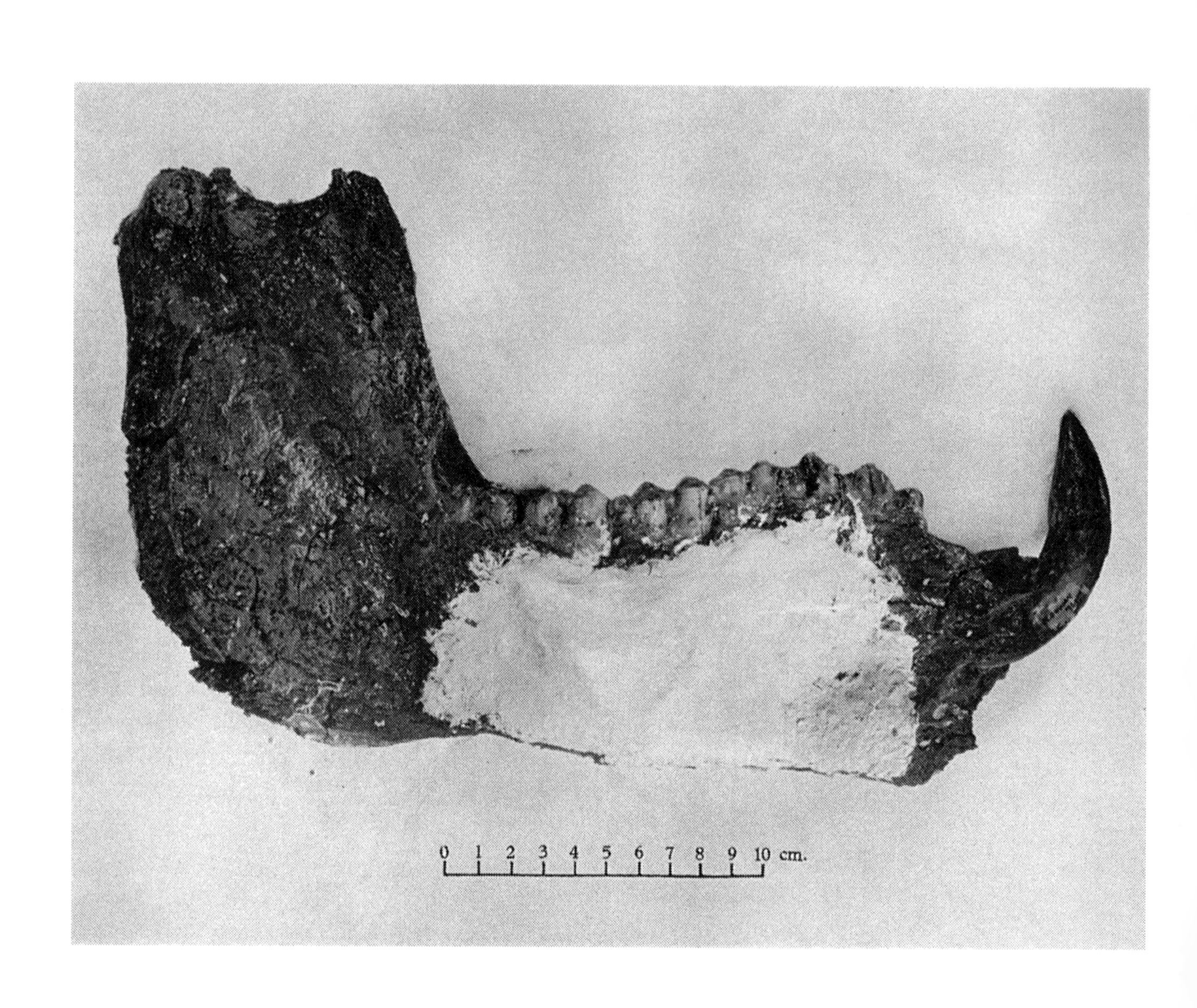

26. *Ectopotamochoerus dubius* mandible. Type, no. FLK N I 1236

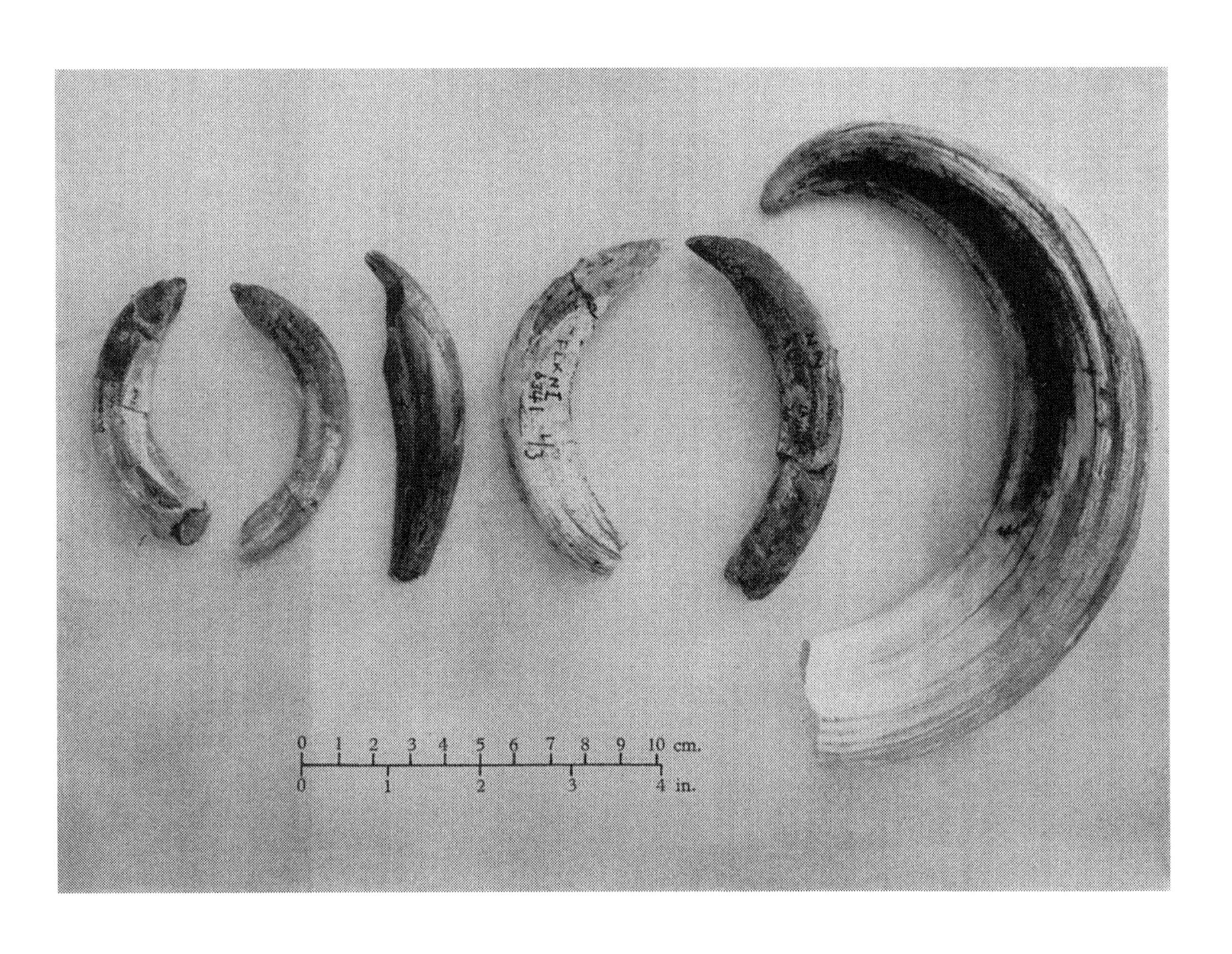

27. *Ectopotamochoerus dubius* canines, compared with modern *Phacochoerus* (right)

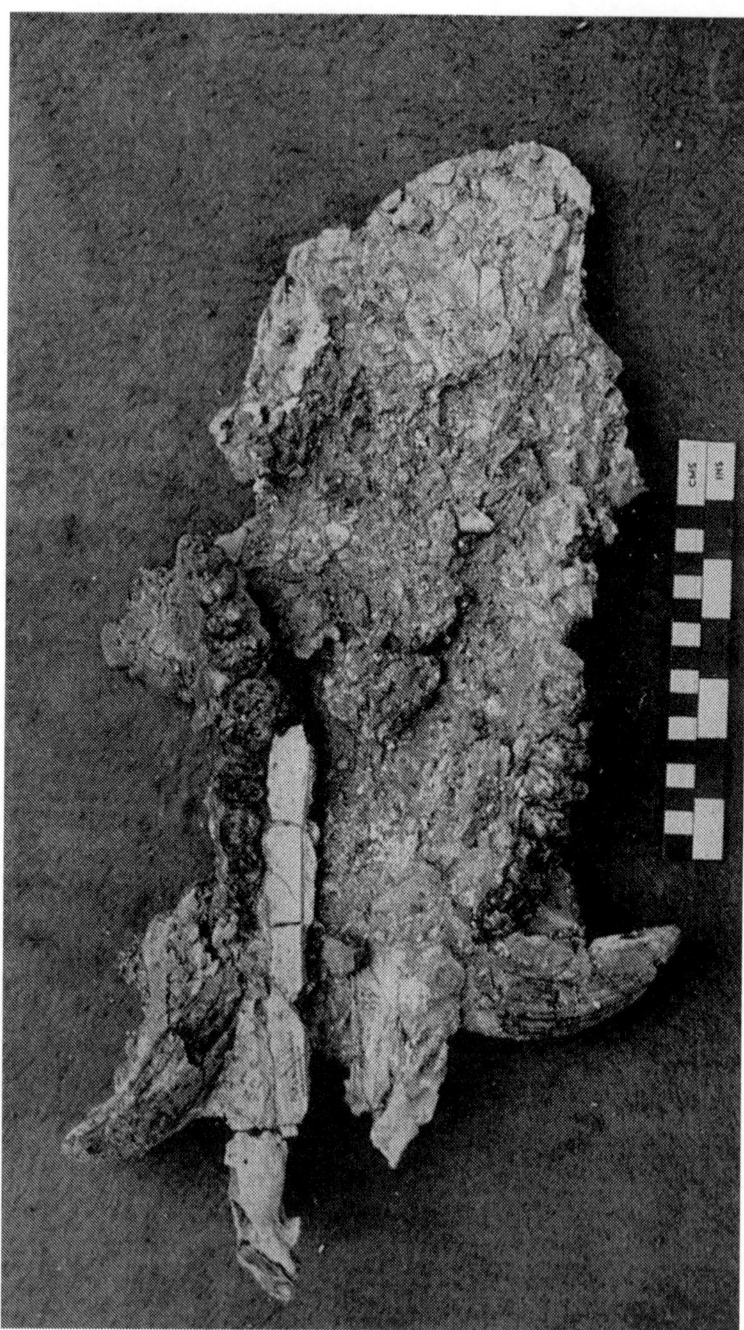

28. *Potamochoerus intermedius*. Type, no. FLK NN I 177. Upper dentition and crushed skull

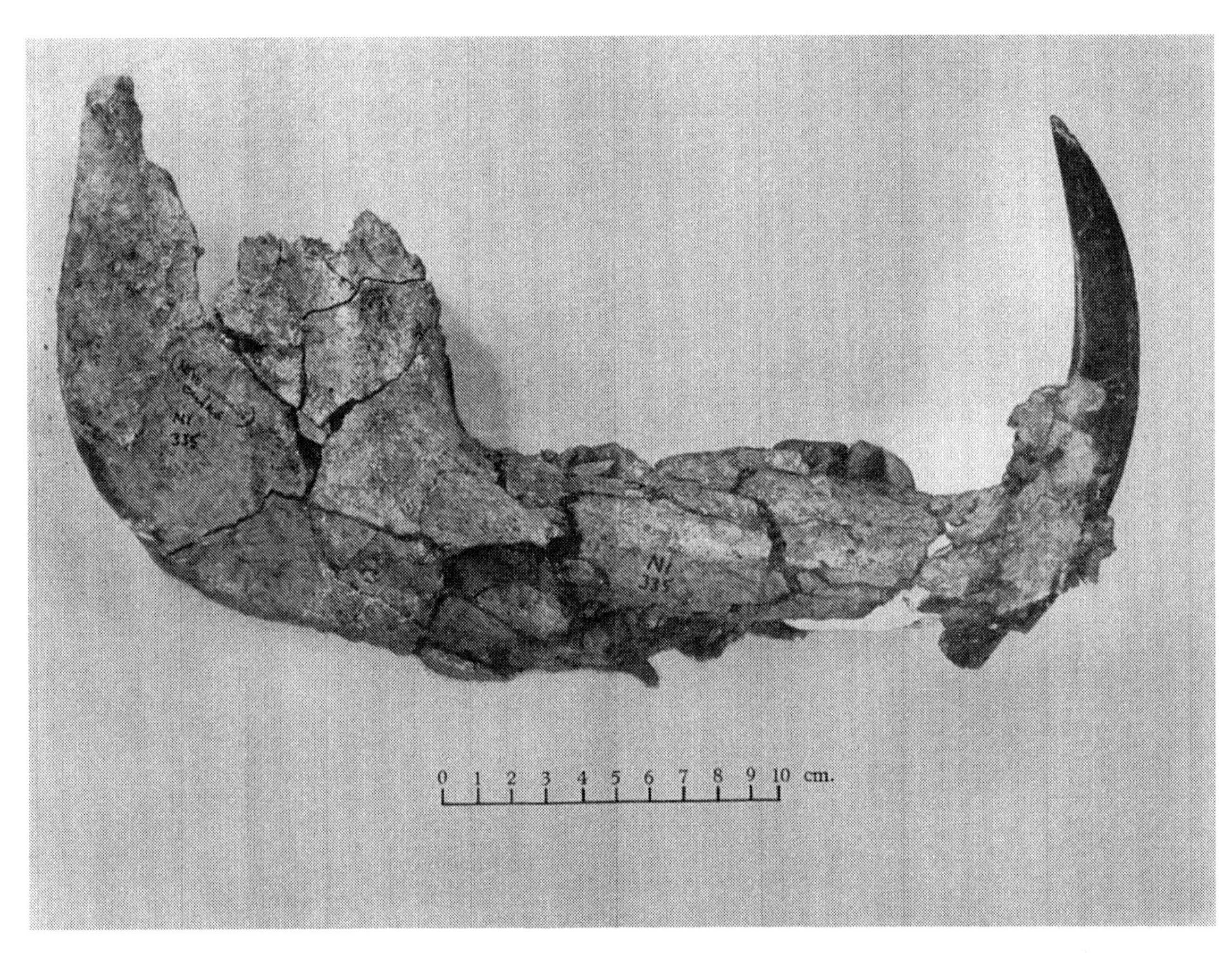

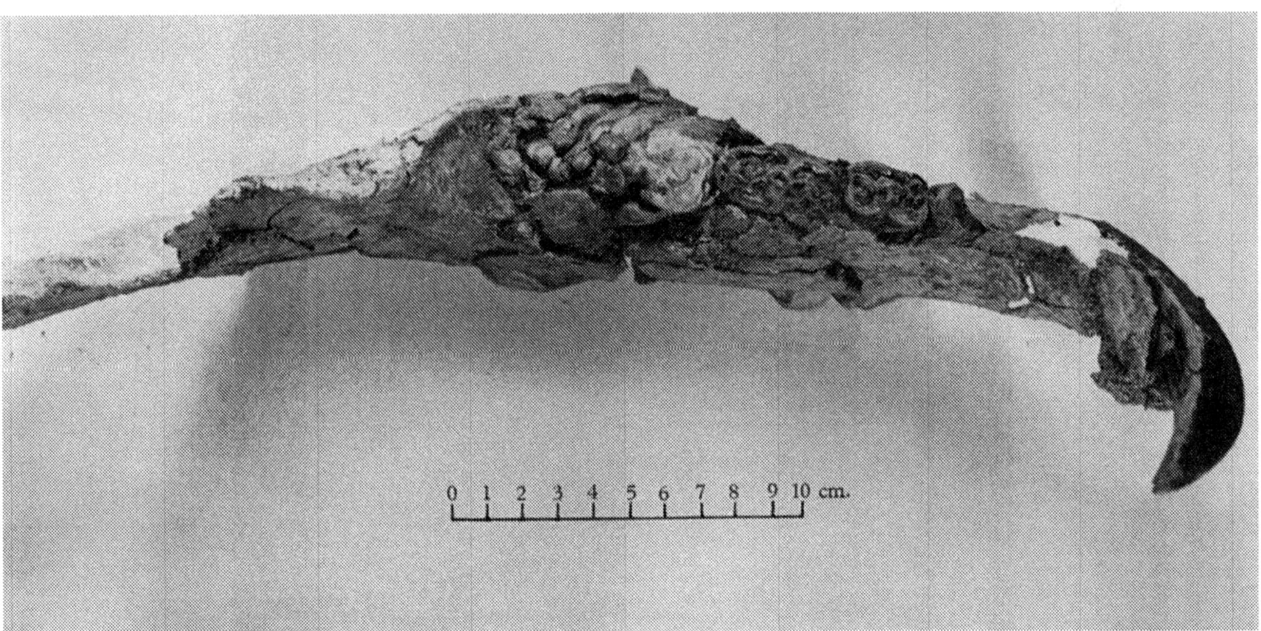

29. *Notochoerus* cf. *euilus*, no. FLK N I 335, 1960. Mandible

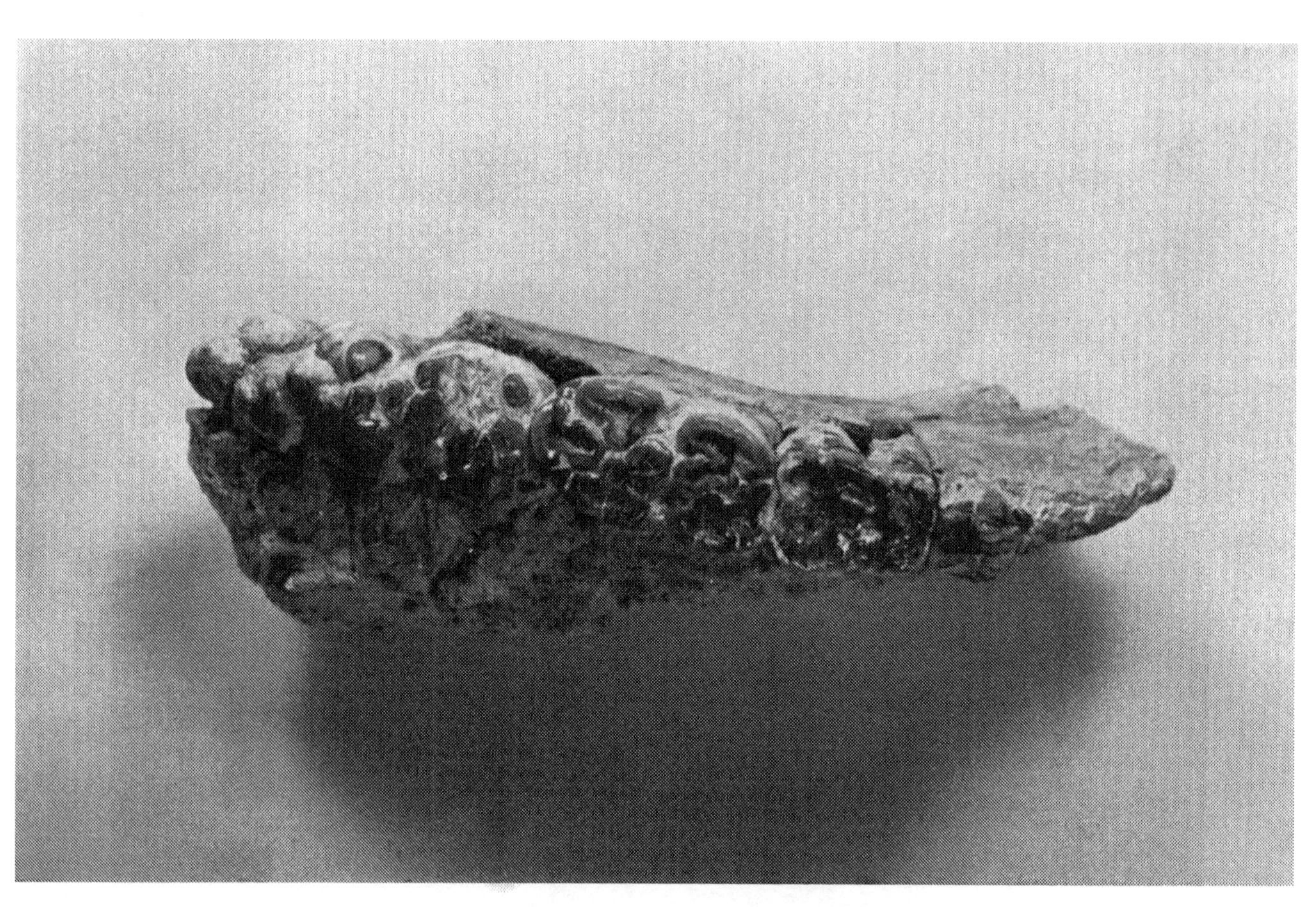

30. *Tapinochoerus* indet. Lower dentition

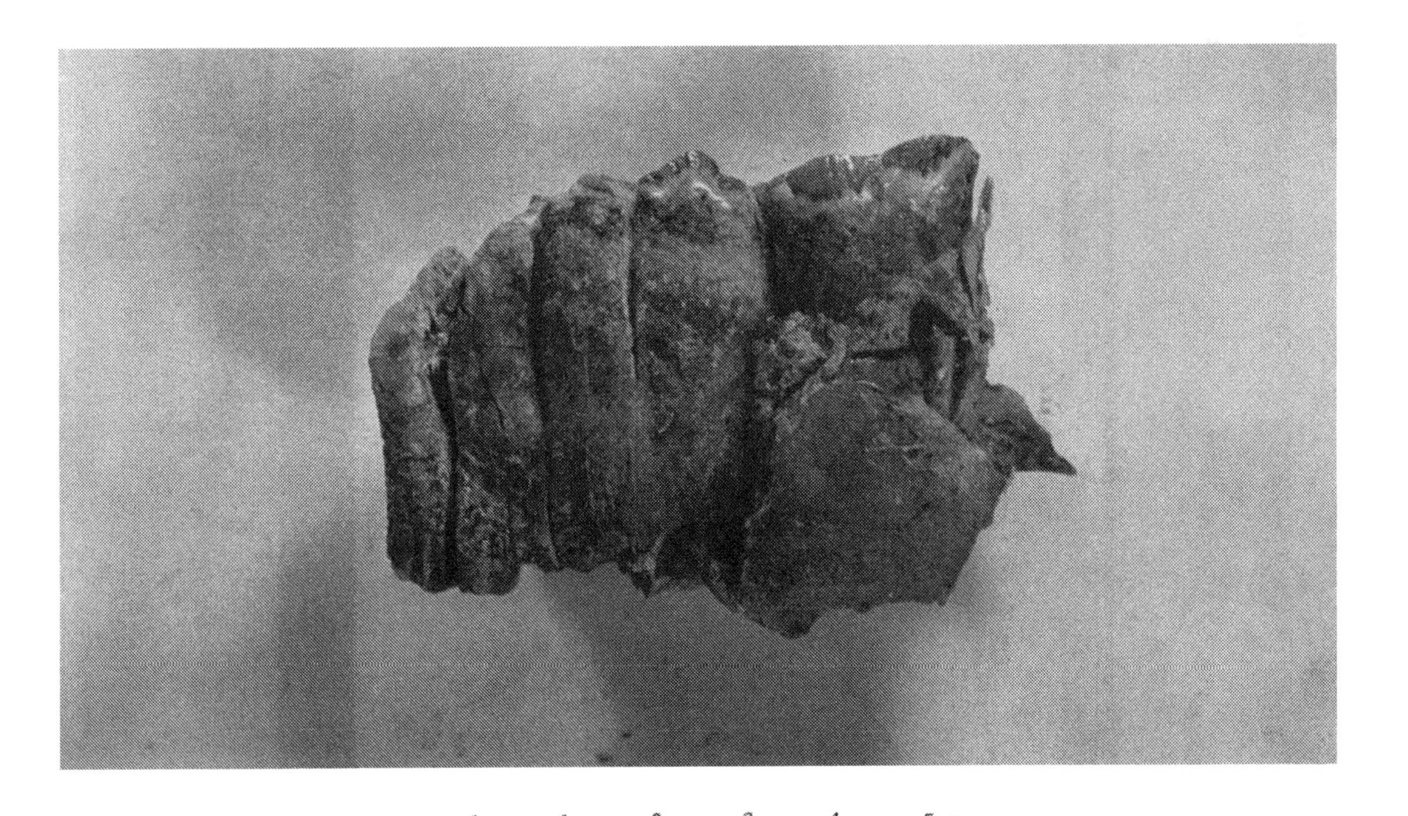

31. *Tapinochoerus* indet. 2nd and 3rd lower molars

32. *Tapinochoerus* indet. Lower 3rd molar

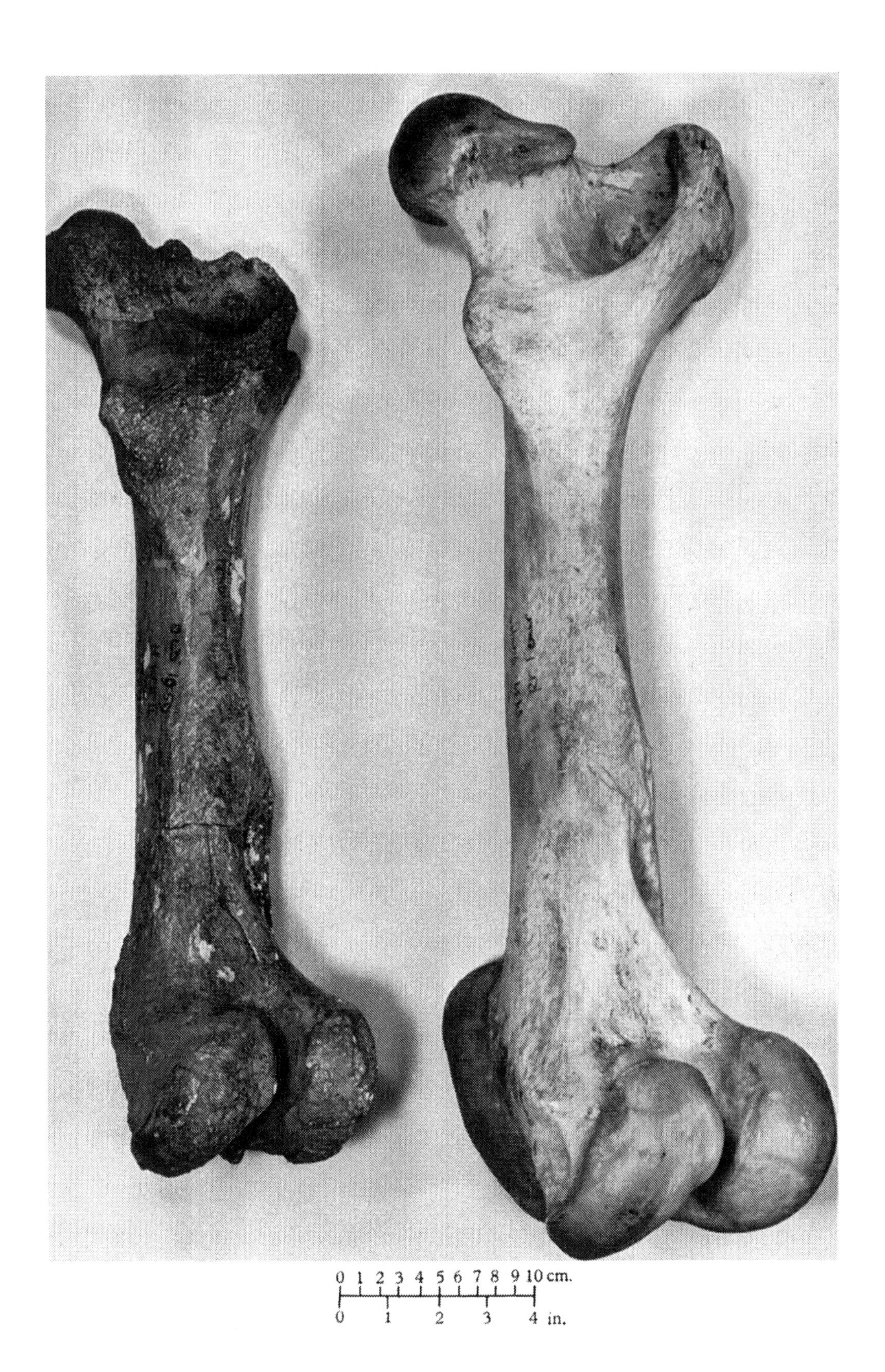

33. *Giraffa gracilis* femur (left), compared with that of *Giraffa camelopardalis* (right)

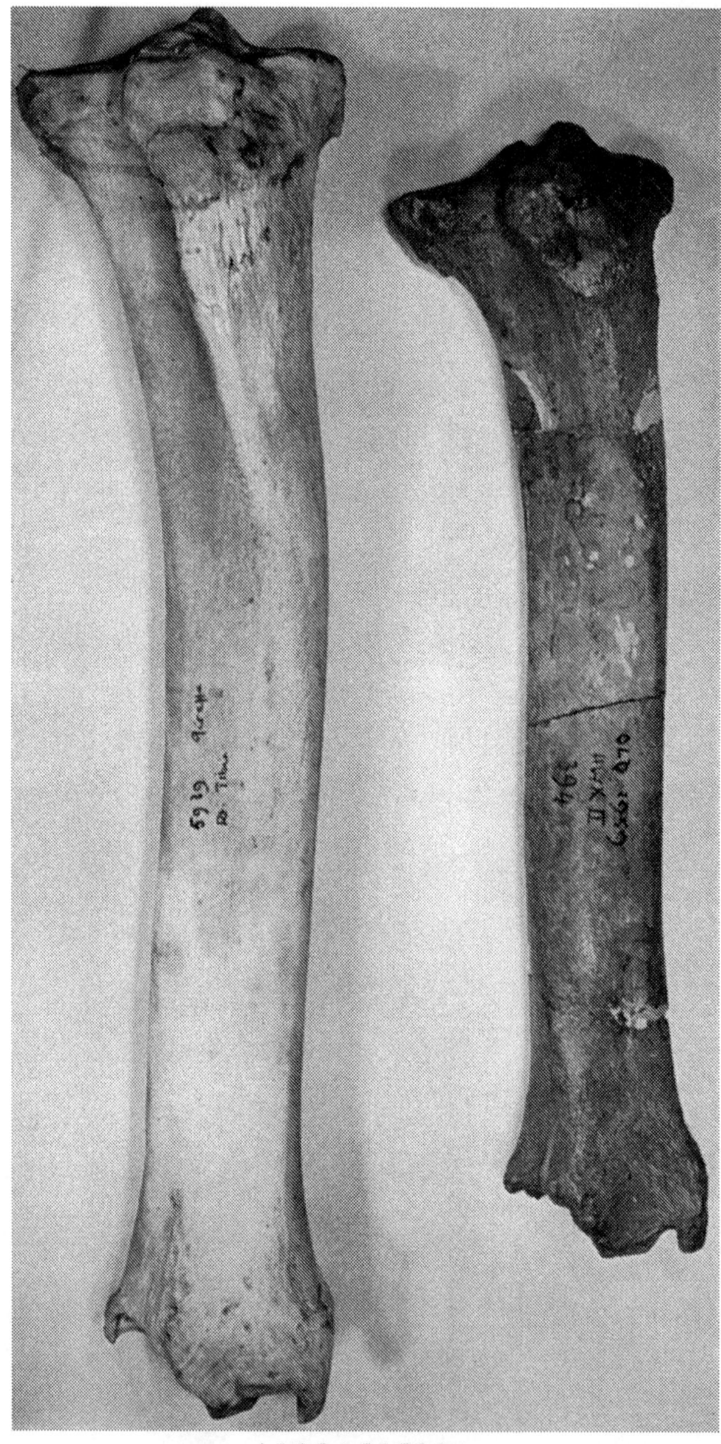

34. *Giraffa gracilis* tibia (right), compared with that of *Giraffa camelopardalis* (left)

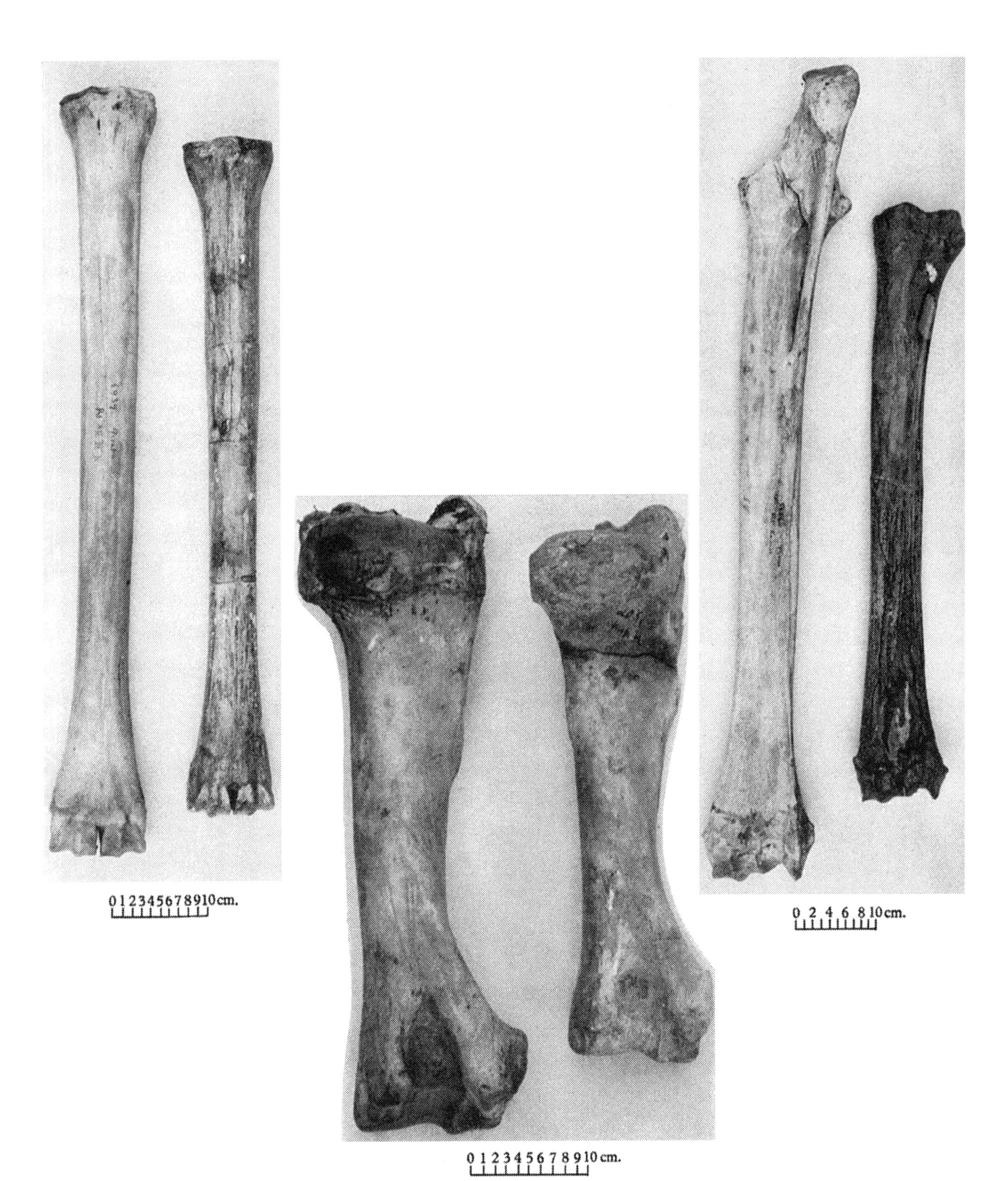

35. *Giraffa gracilis* metacarpal, radio-ulna and humerus, compared with those of *Giraffa camelopardalis*. The fossil is on the right in each case

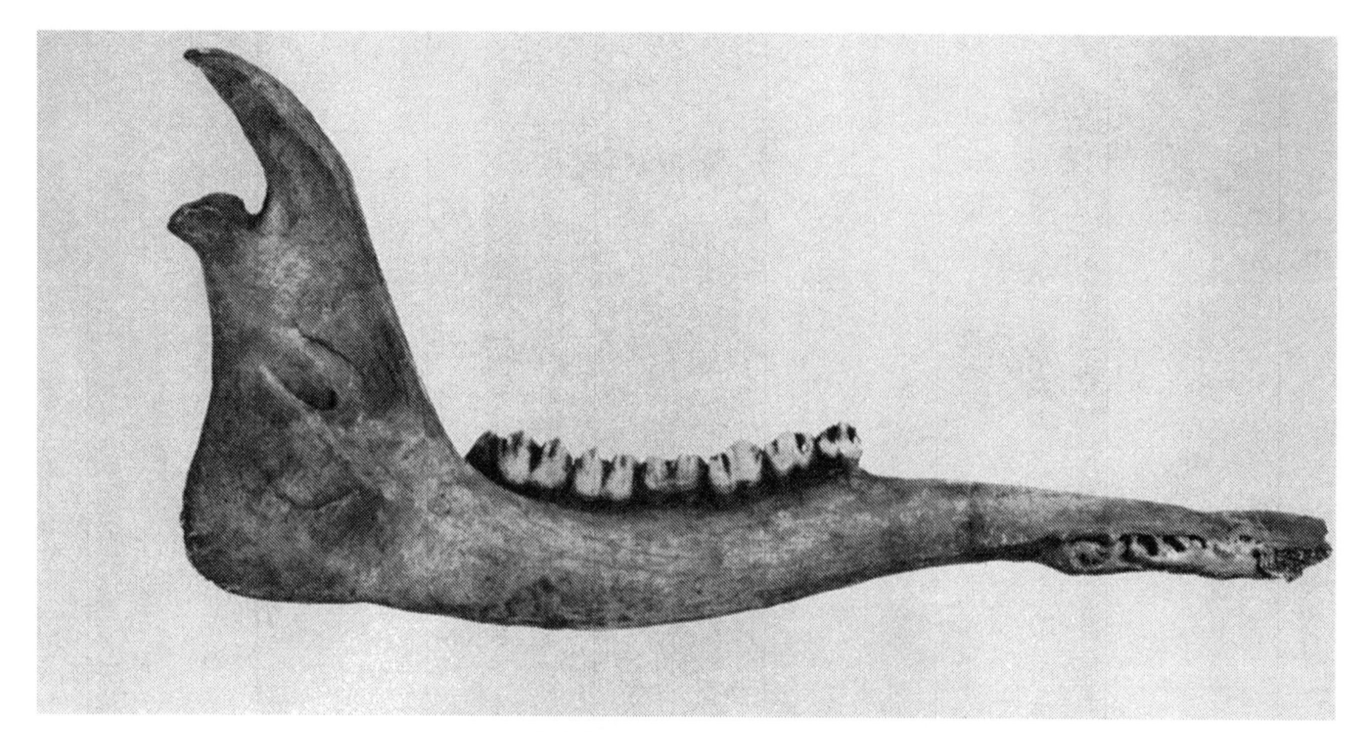

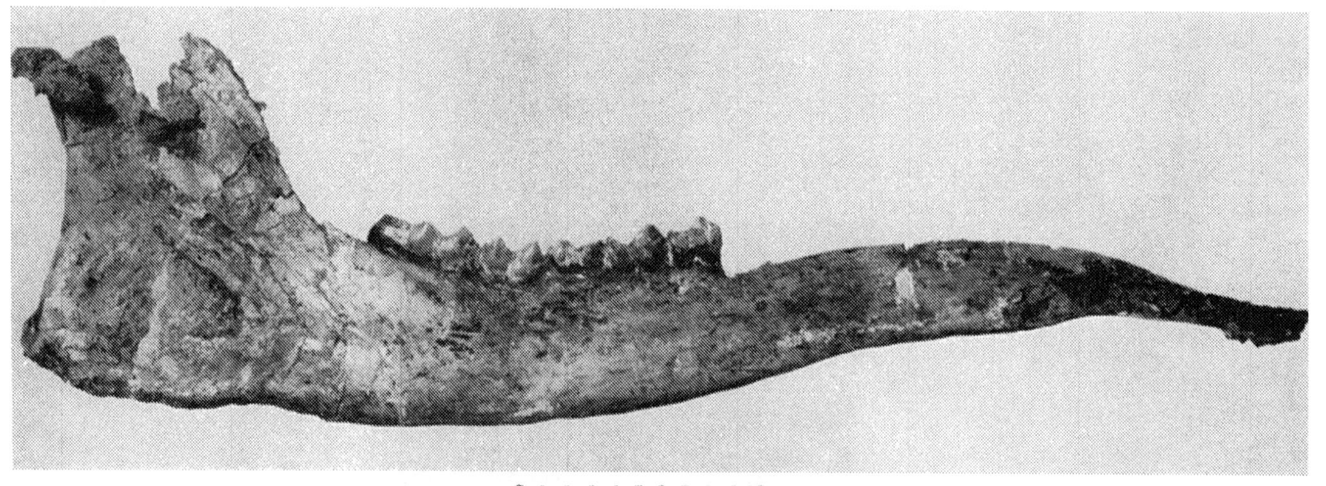

36. *Giraffa camelopardalis* mandible (top), compared with Type of *Giraffa jumae* (bottom), no. M. 21466

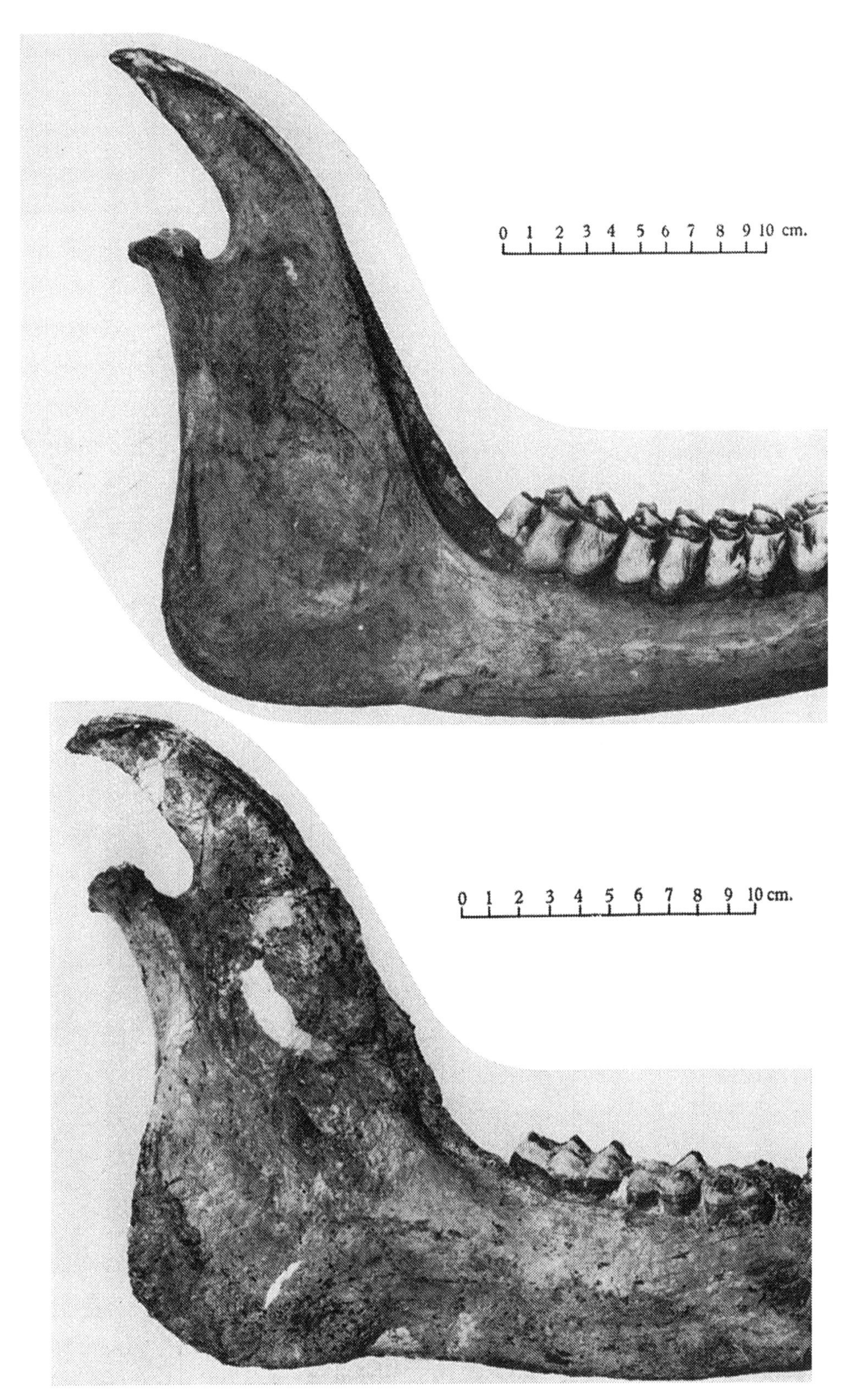

37. Enlargement of the posterior part of the mandibles of *Giraffa camelopardalis* (top) and *Giraffa jumae* (bottom)

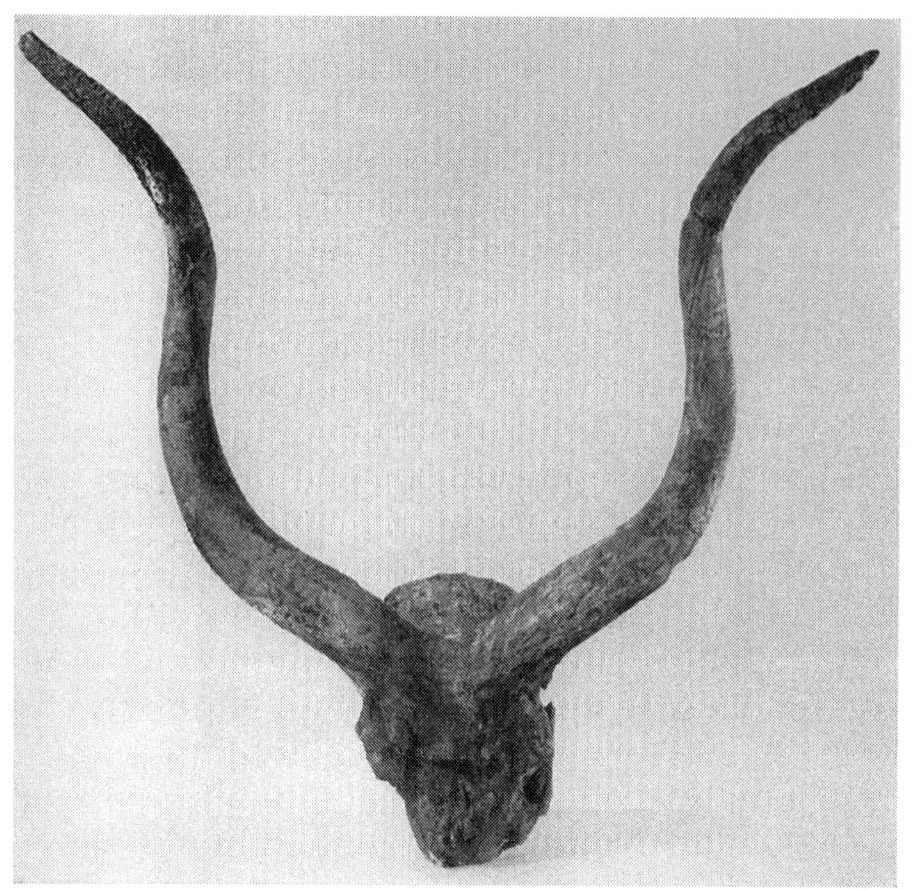

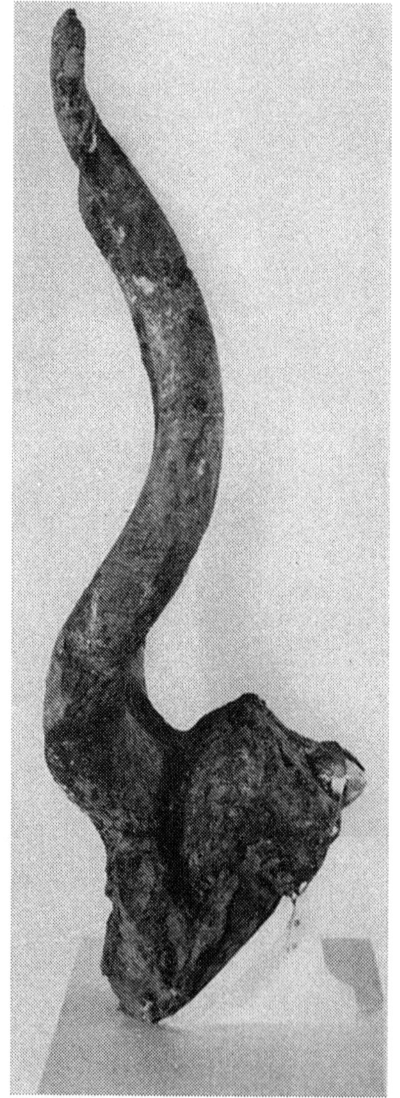

38. *Strepsiceros grandis*. Type, no. M. 21461. Cranium and horn cores

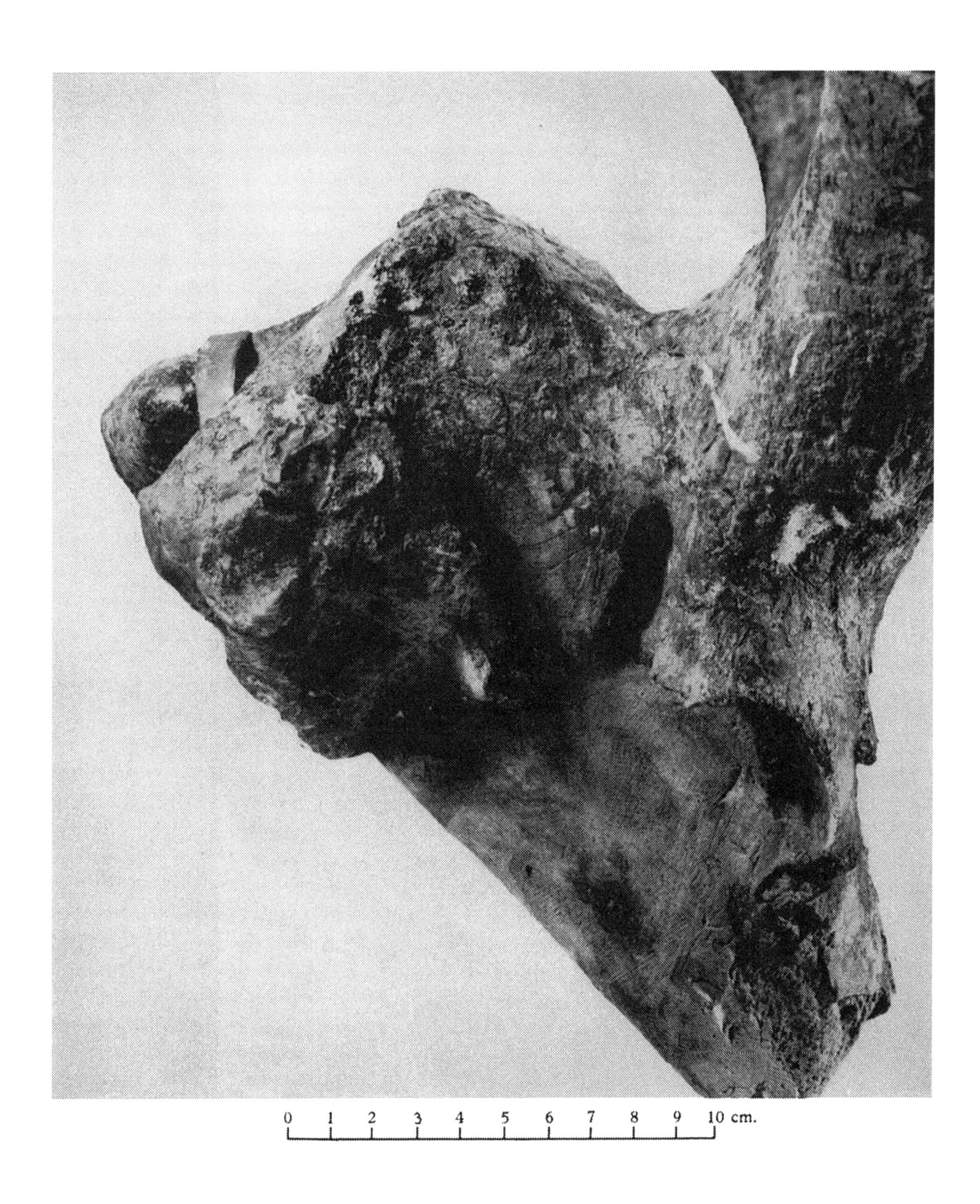

39. *Strepsiceros grandis*. Type. Enlargement of right side of skull

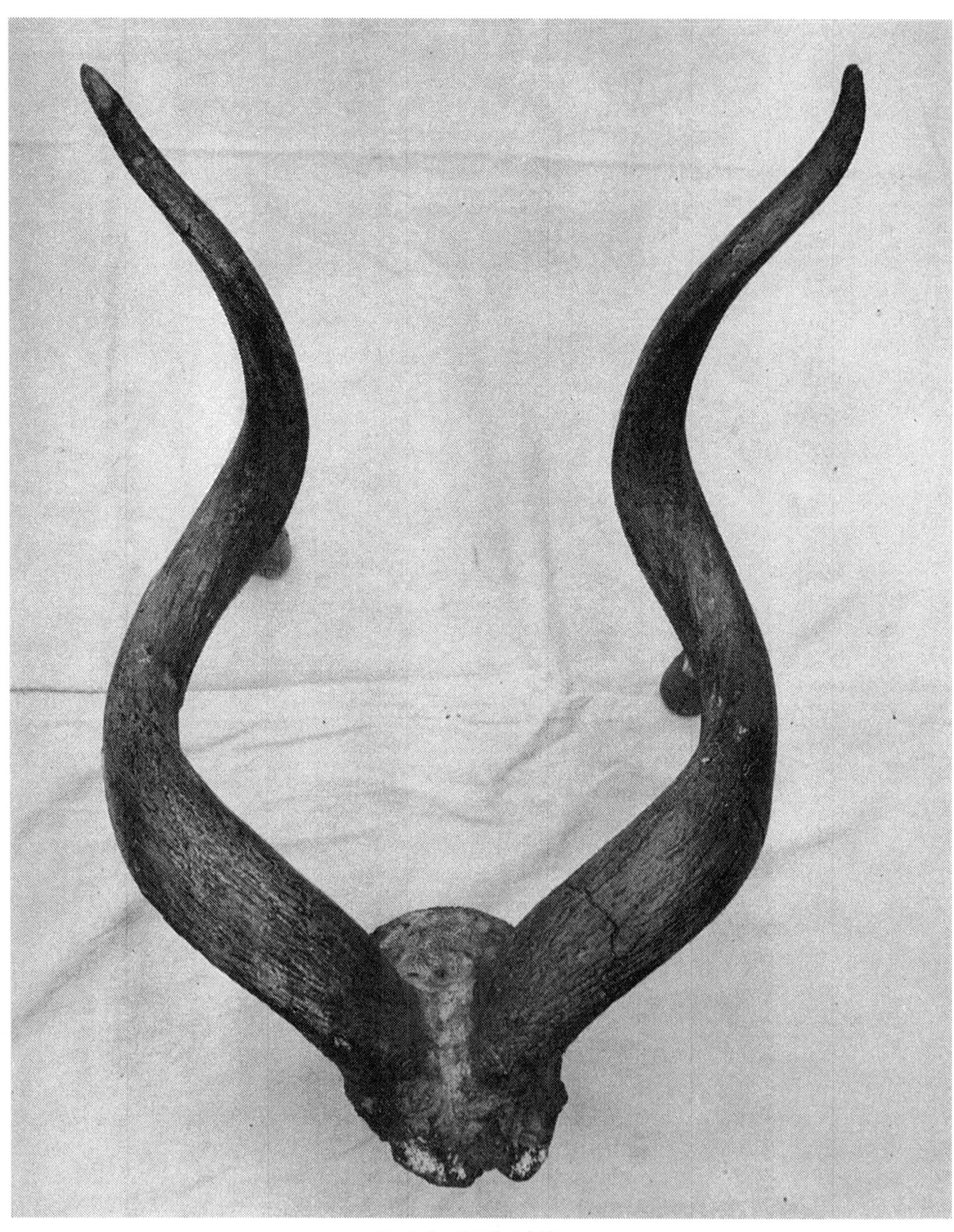

40. *Strepsiceros maryanus*. Type. Cranium and horn cores

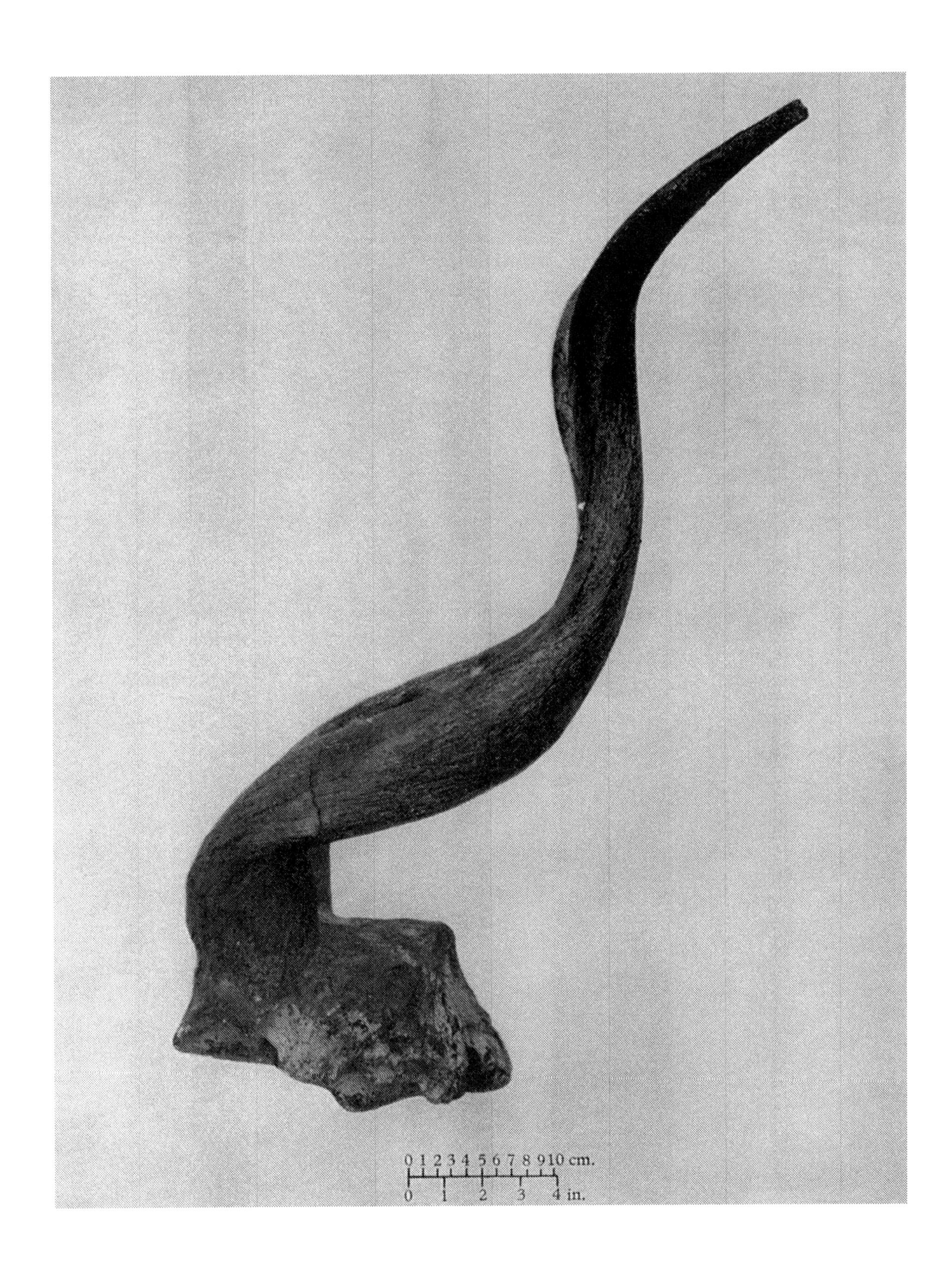

41. *Strepsiceros maryanus*. Type. Side view

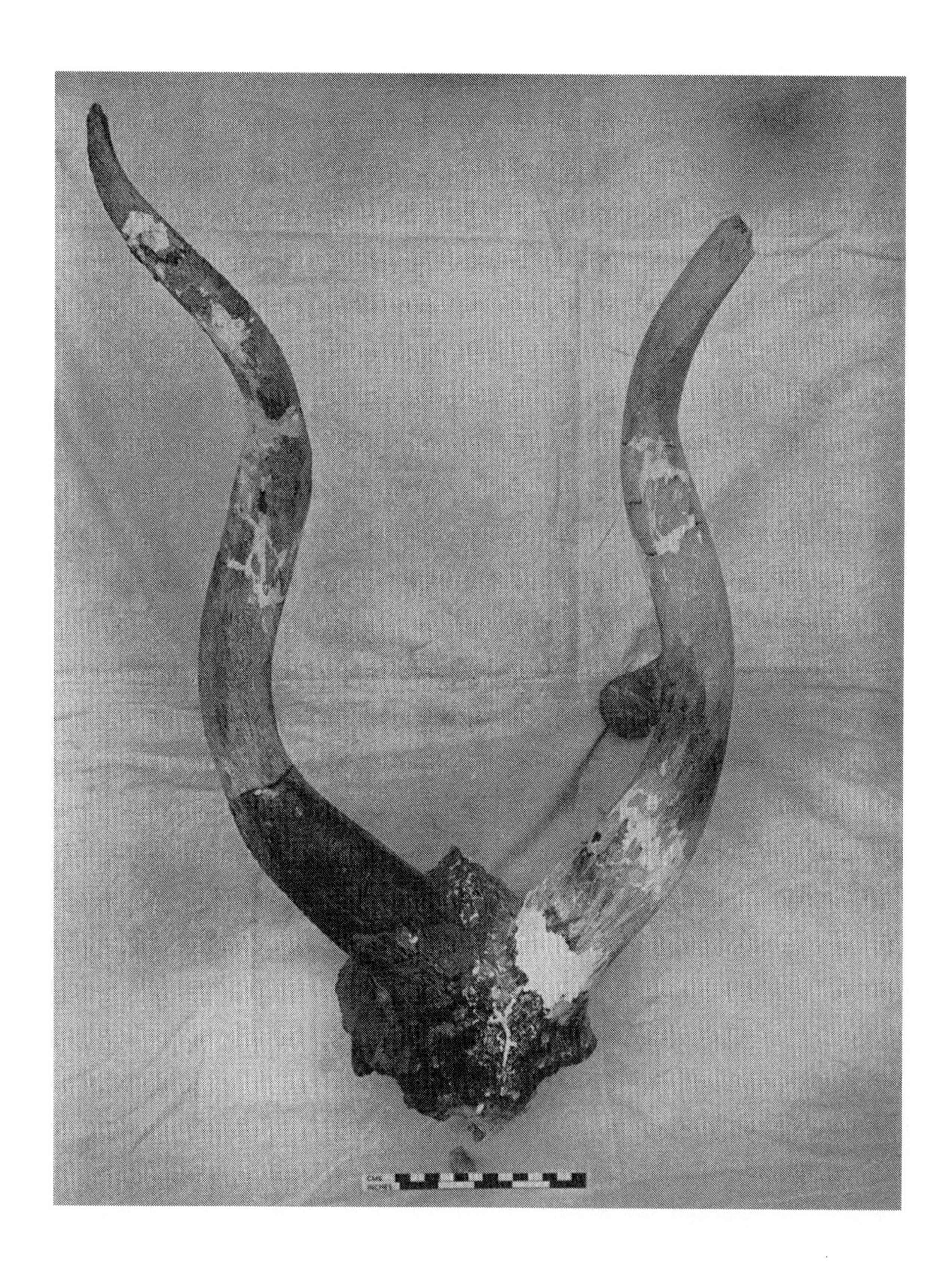

42. *Strepsiceros maryanus*. A second specimen

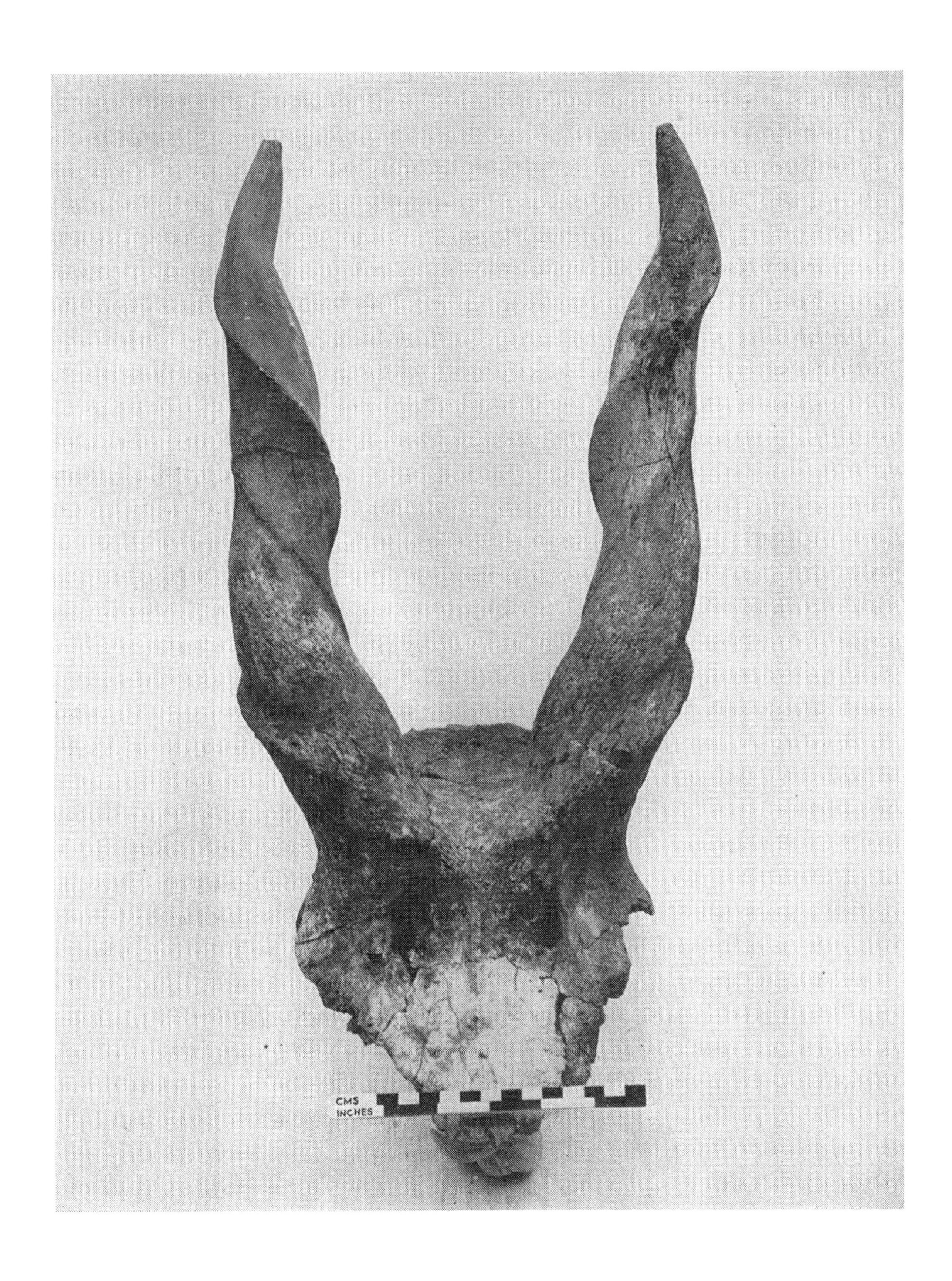

43. *Taurotragus arkelli*. Type, no. F. 3665. Cranium and horn cores

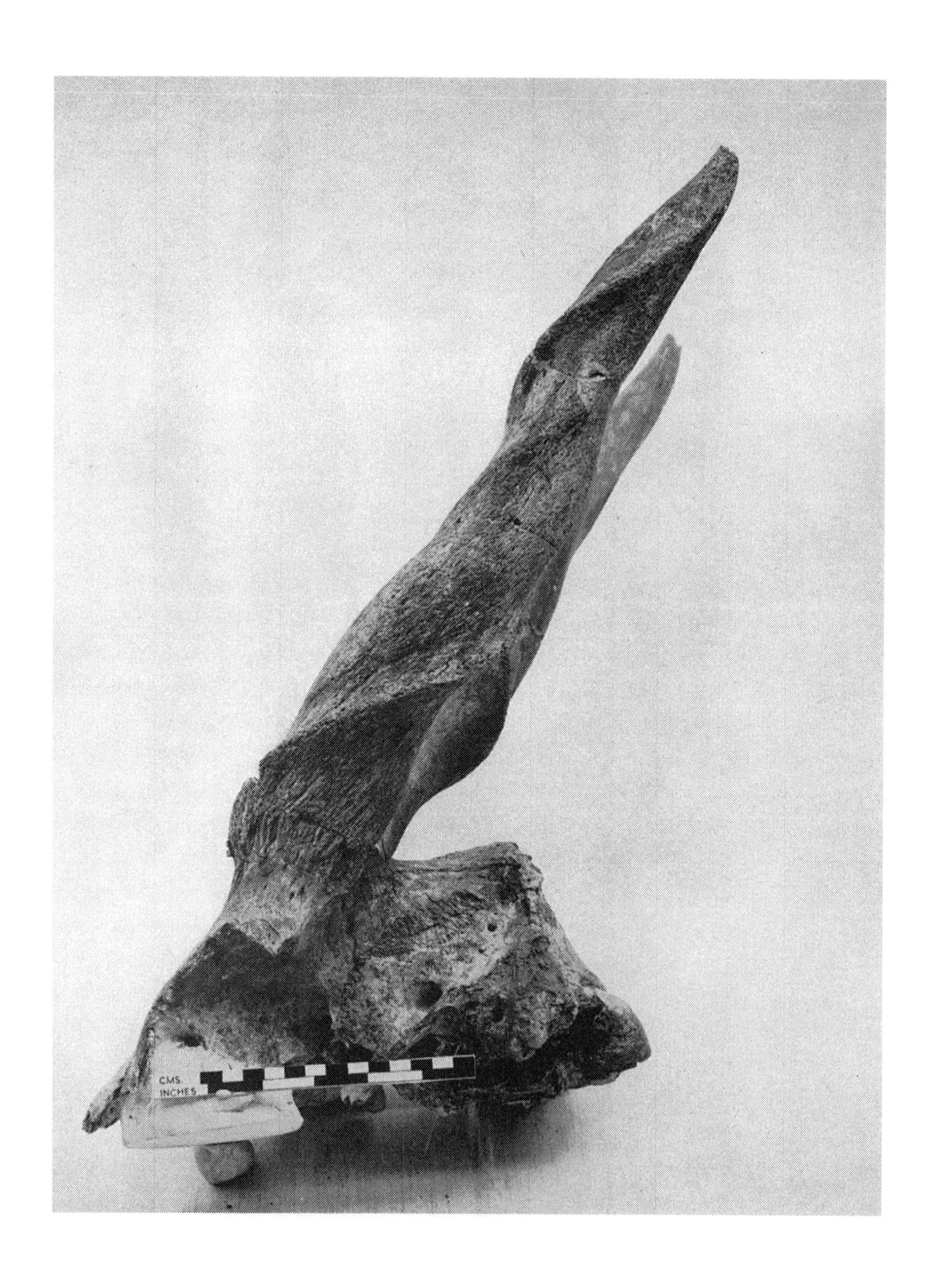

44. *Taurotragus arkelli*. Type. Cranium and horn cores, side view

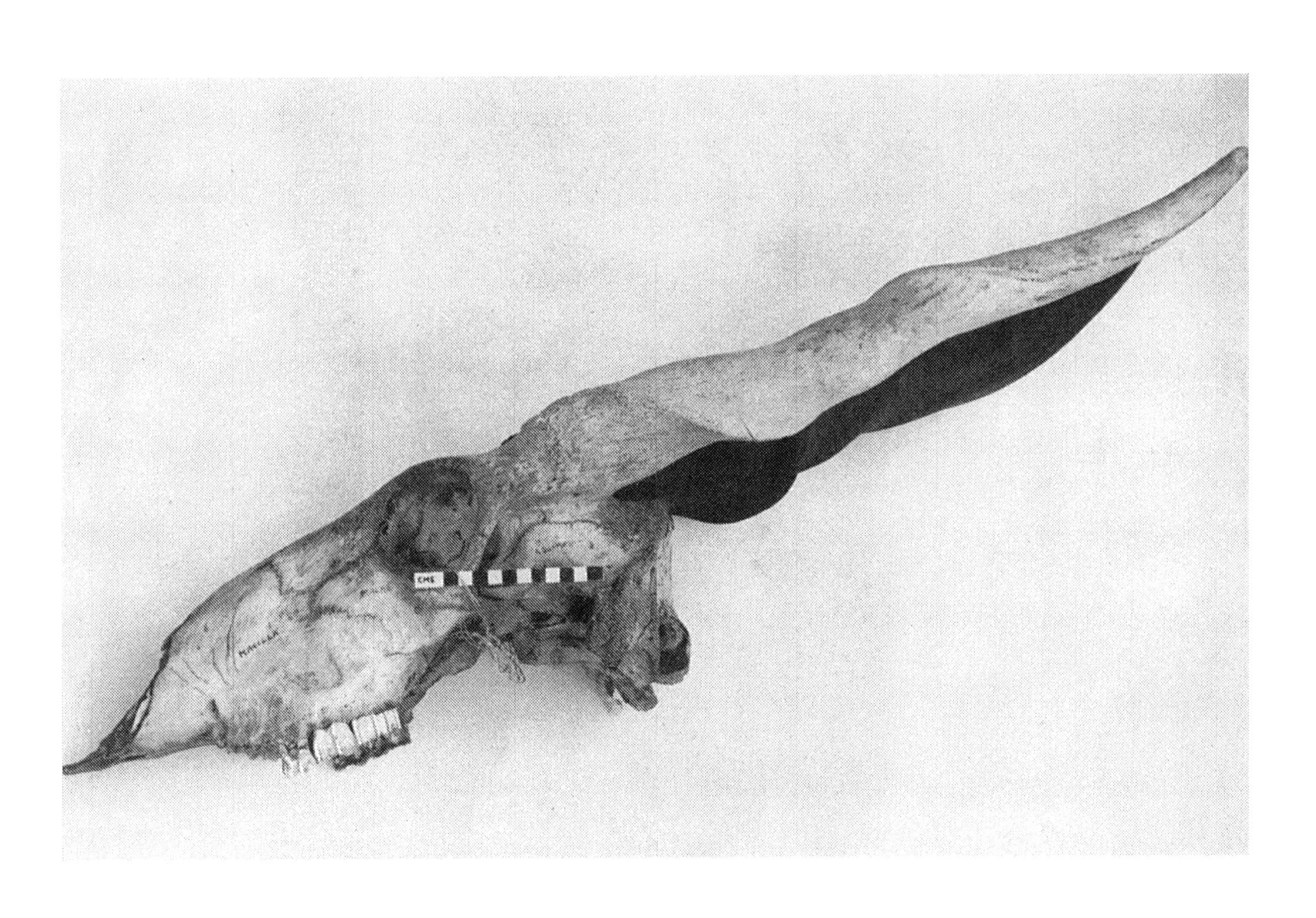

45. *Taurotragus oryx*. Skull and horn cores of modern eland

46. *Bularchus arok*. Type, no. M. 14947. Horn cores

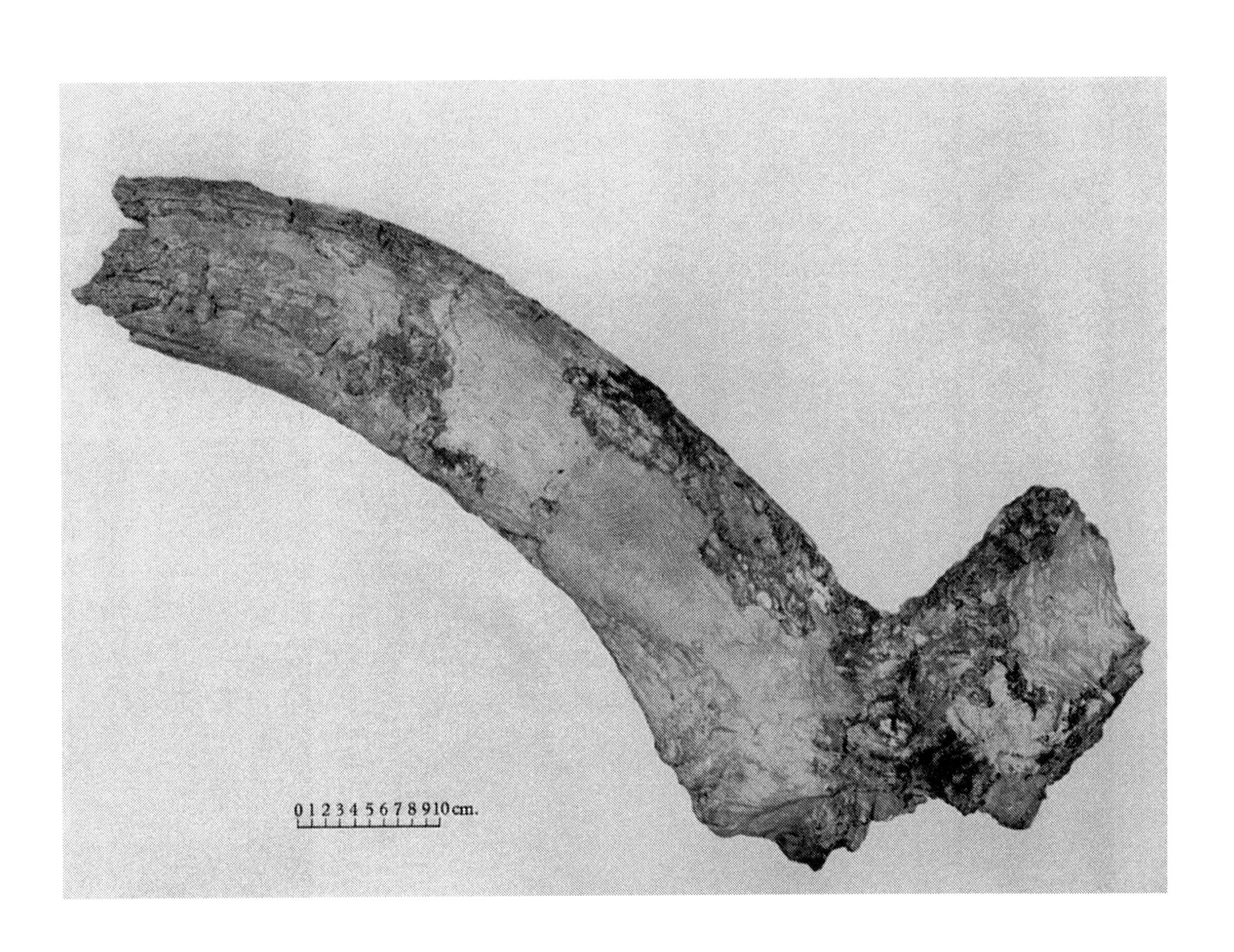

47. *Bularchus arok*. Paratype, no. M. 14948. Horn core

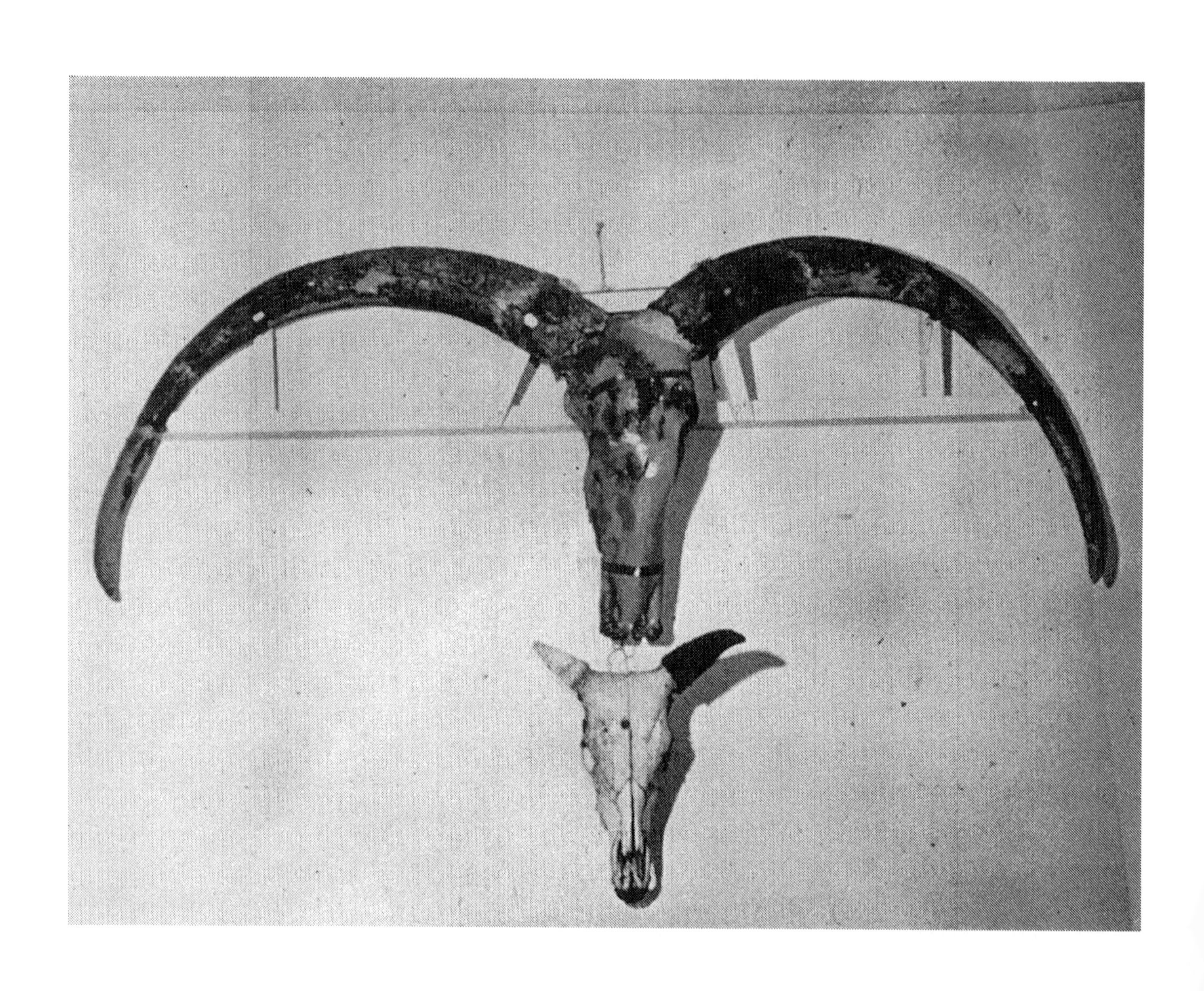

48. *Bularchus arok*. Skull and horn cores, compared with modern ox (in Coryndon Memorial Museum)

49. *Gorgon olduvaiensis*. Type, no. M. 21451. Side view

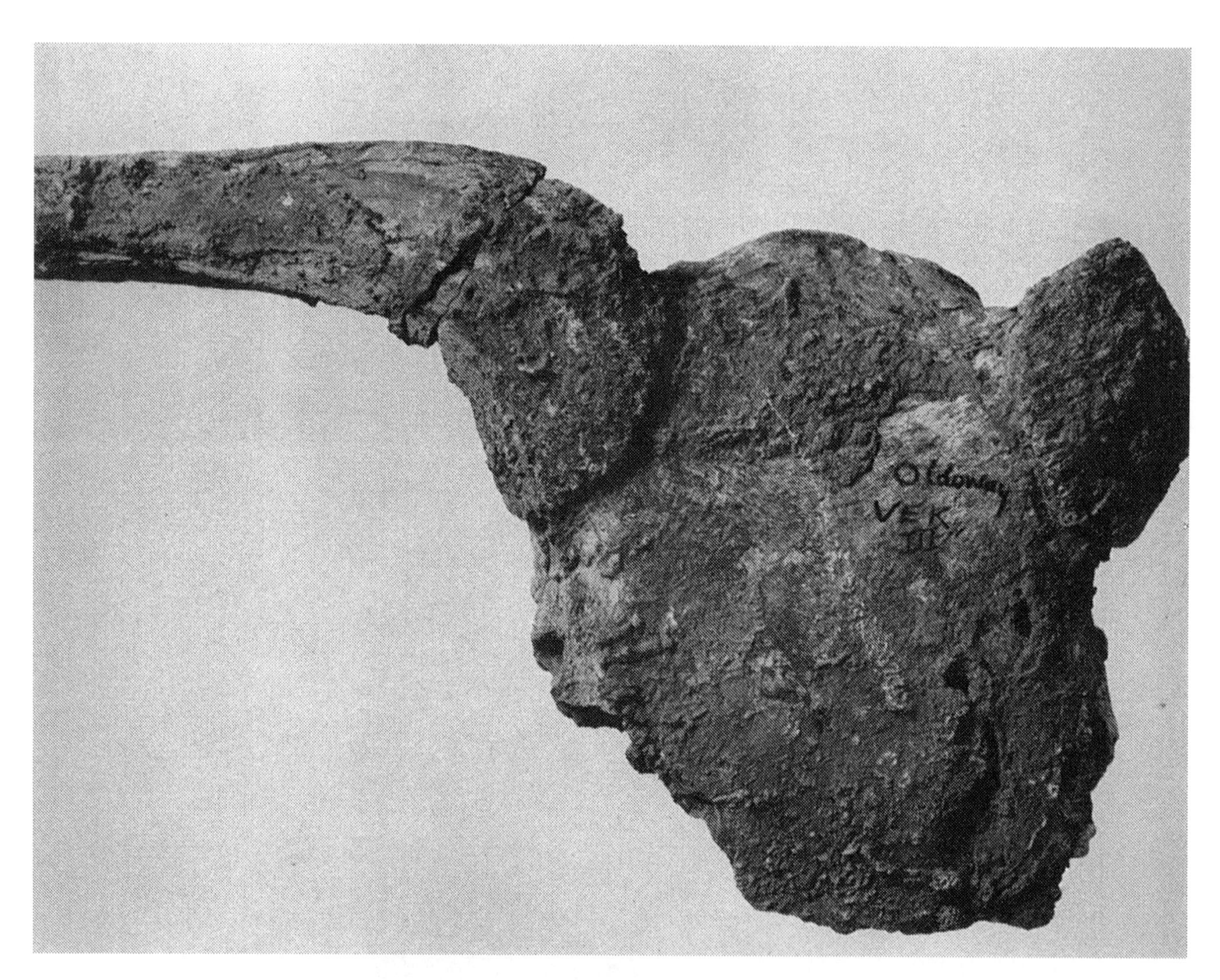

50. *Gorgon olduvaiensis*. Type, no. M. 21451. Frontlet and horn core

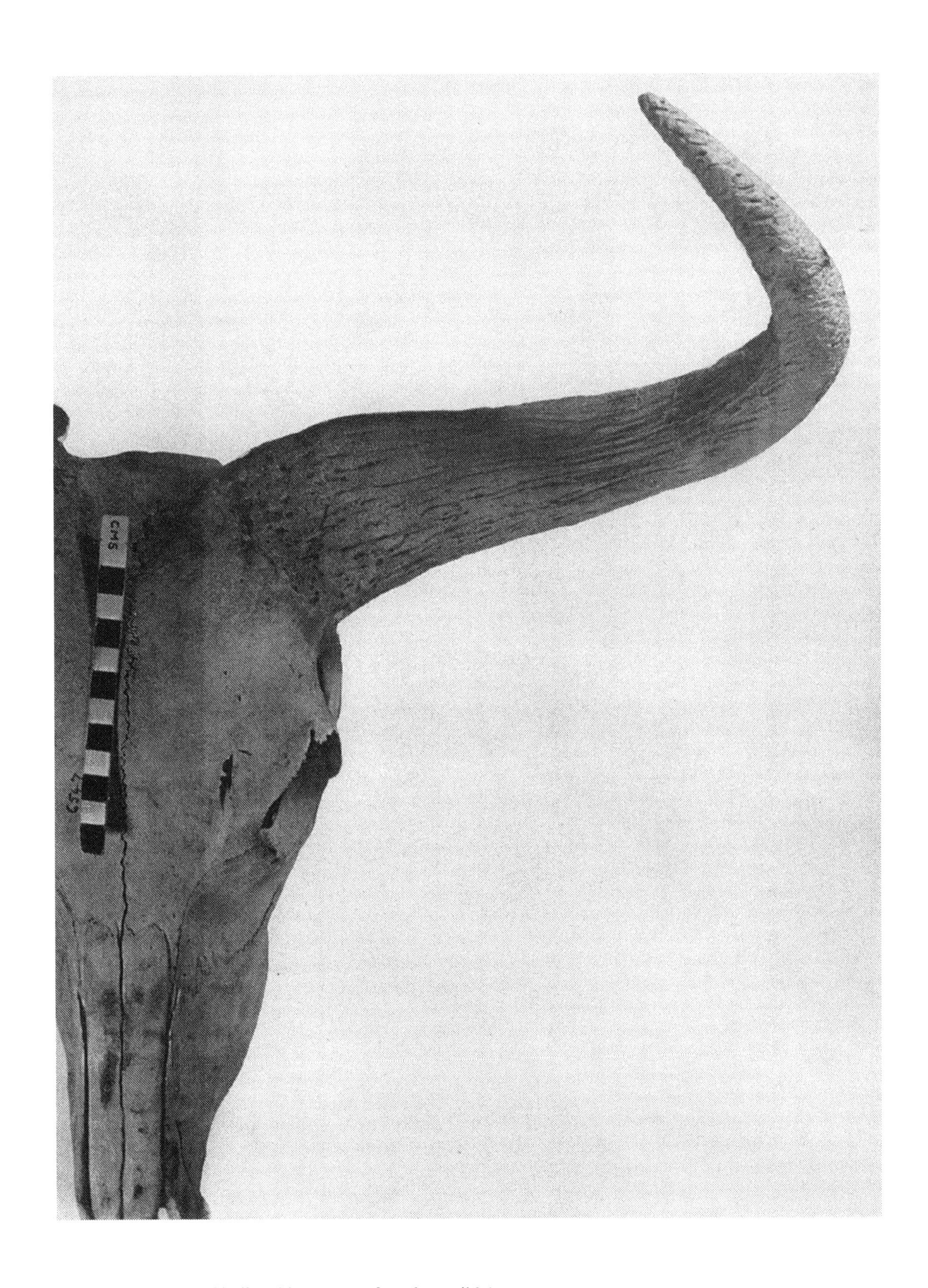

51. *Gorgon taurinus*. Skull and horn core of modern wildebeest

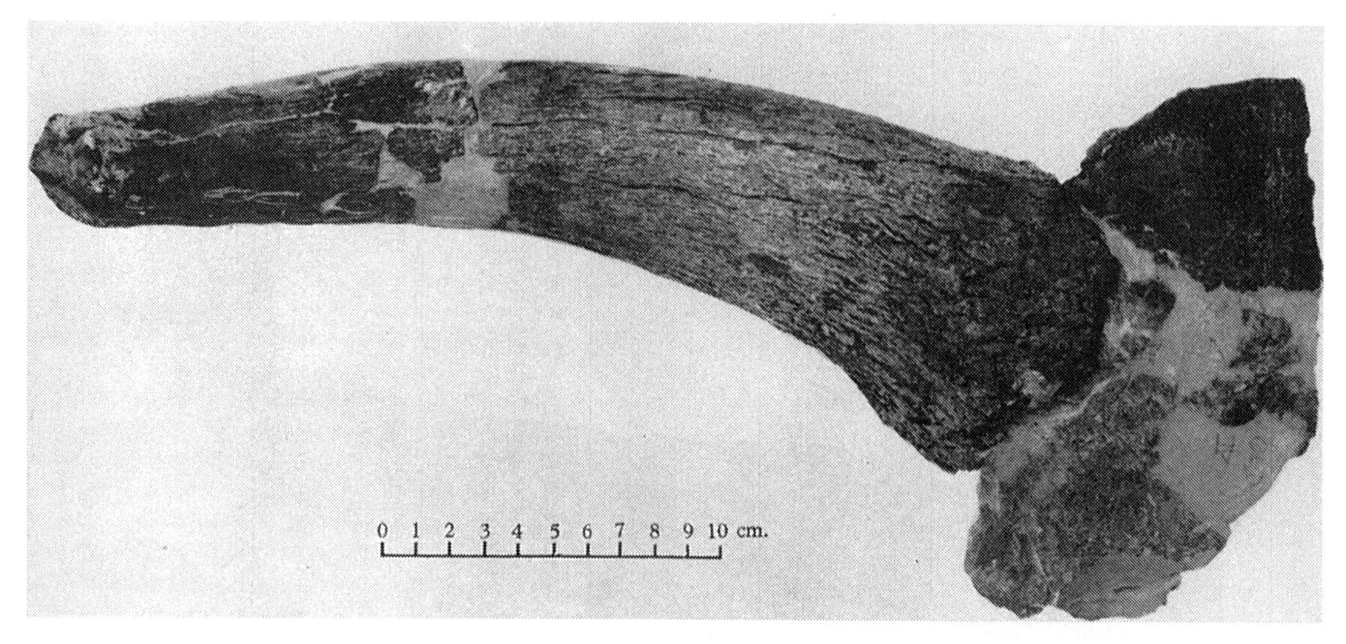

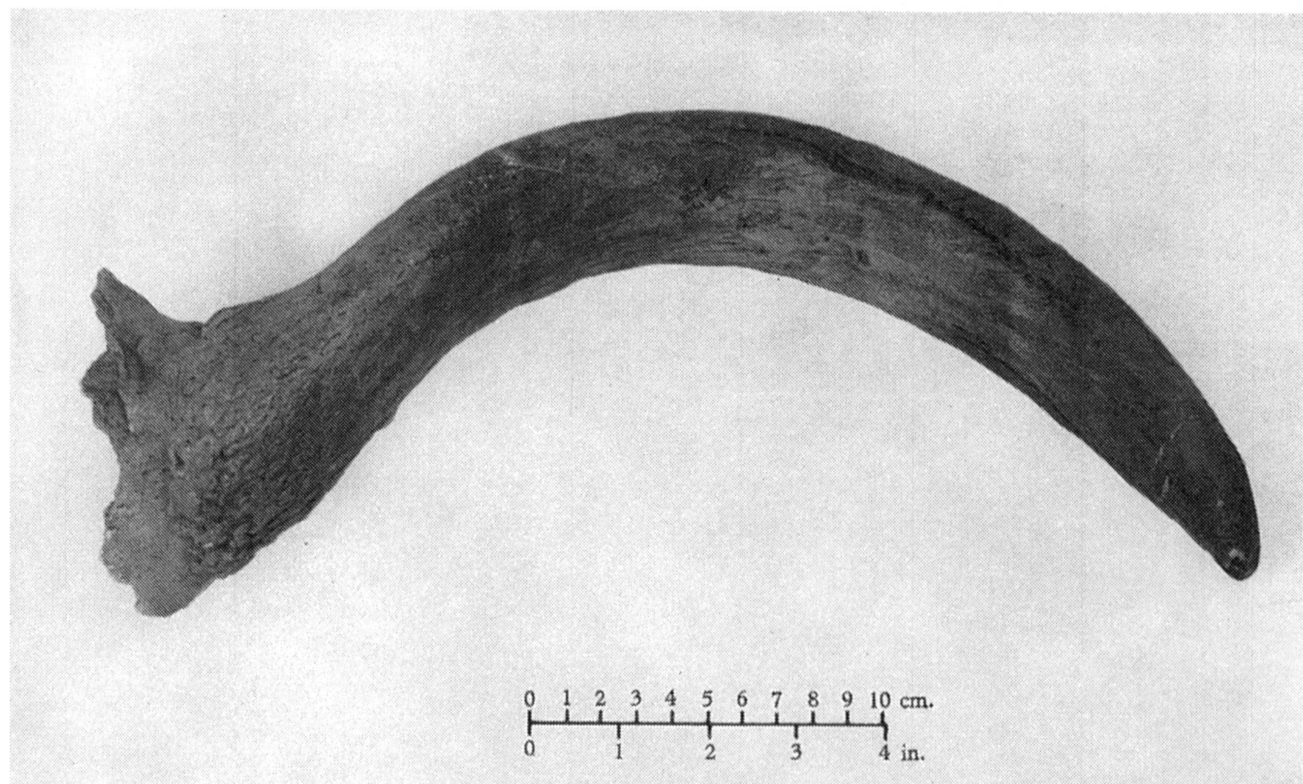

52. *Gorgon olduvaiensis*. Horn core: no. M. 21452 (top) from the front and (bottom) vertically from above

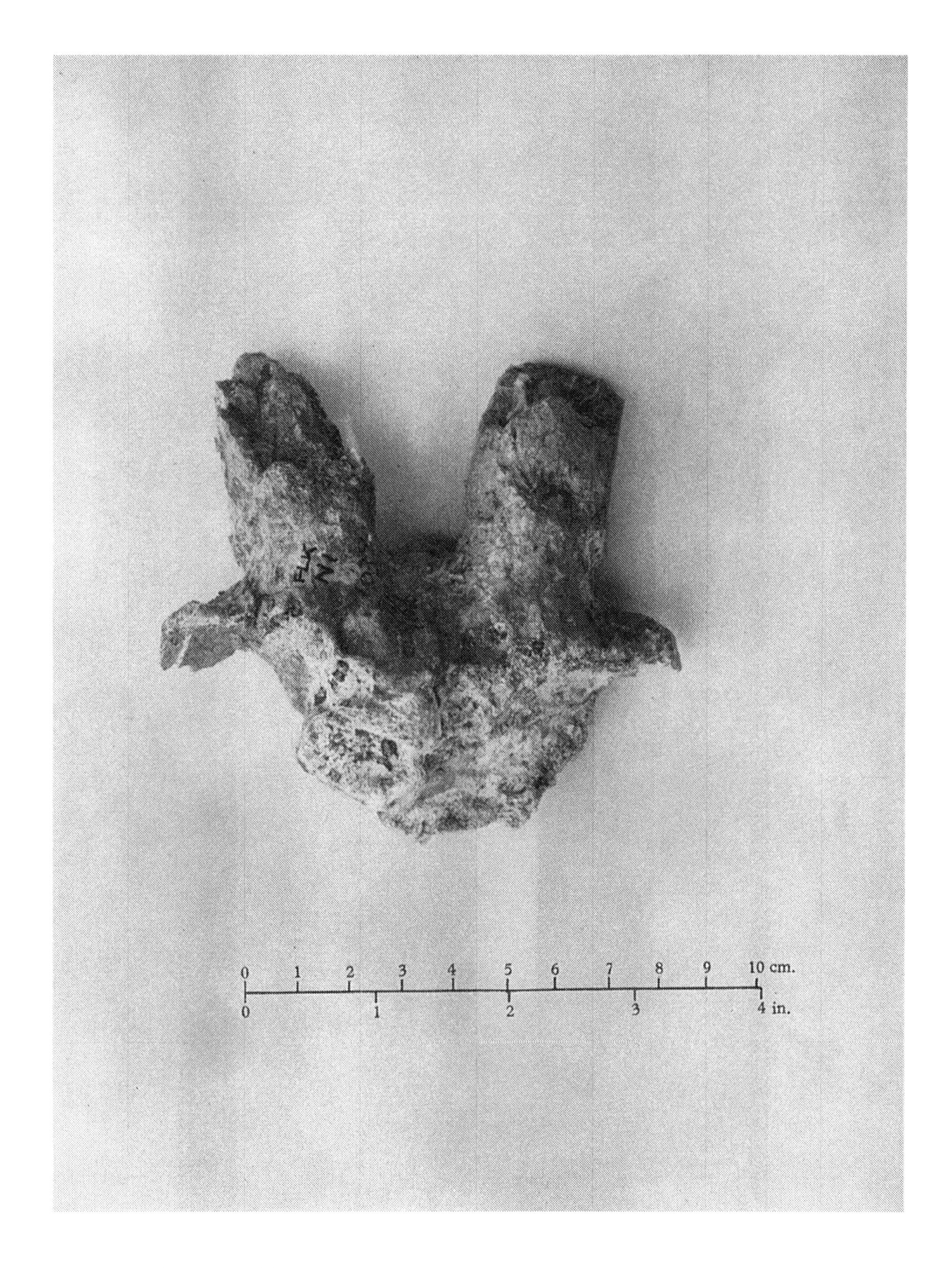

53. Reduncini indet. Cranial fragment

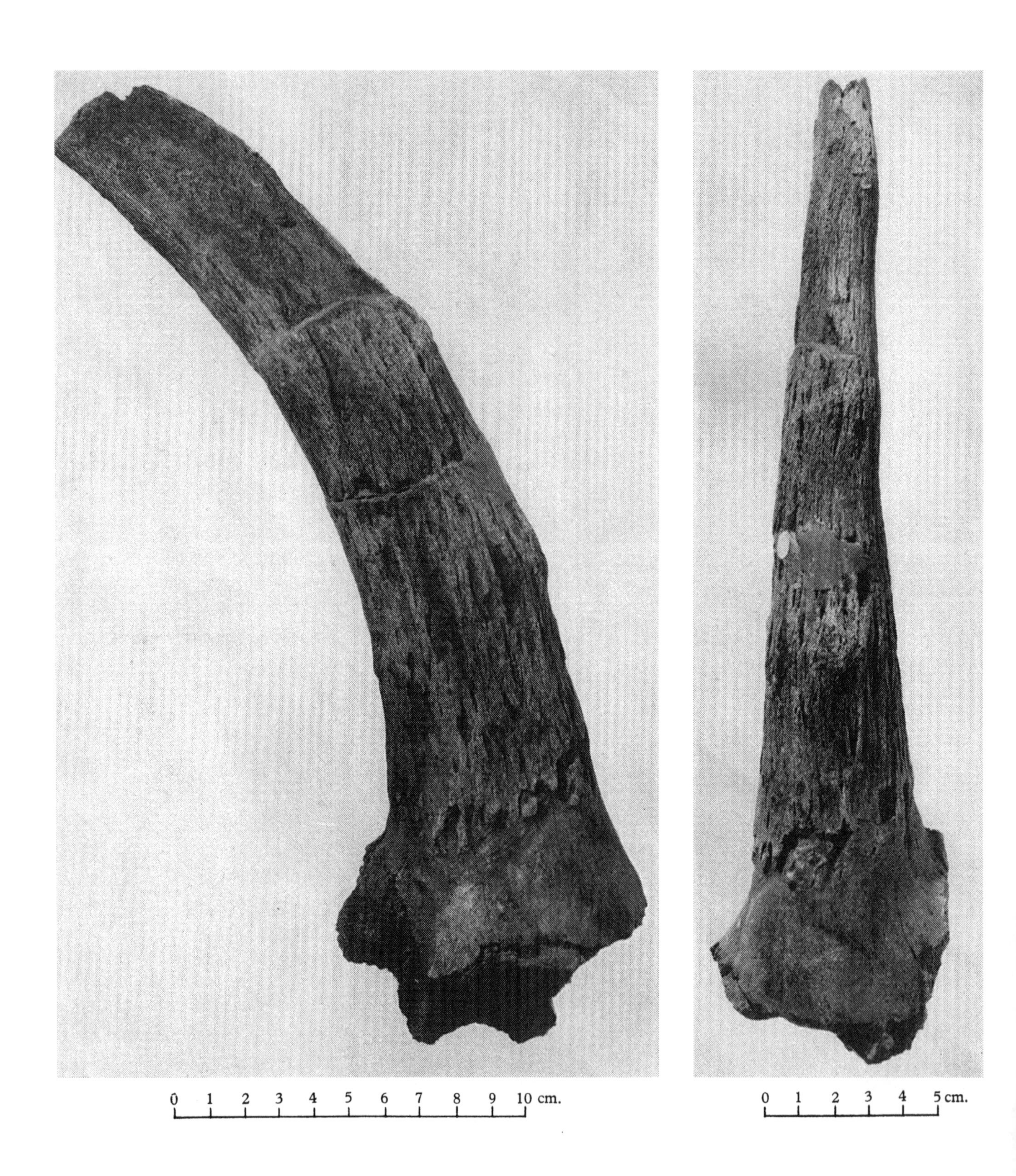

54. *Hippotragus niro*. Type, no. M. 14561. Right horn core, profile and front views

55. *Hippotragus niro*. Additional horn cores

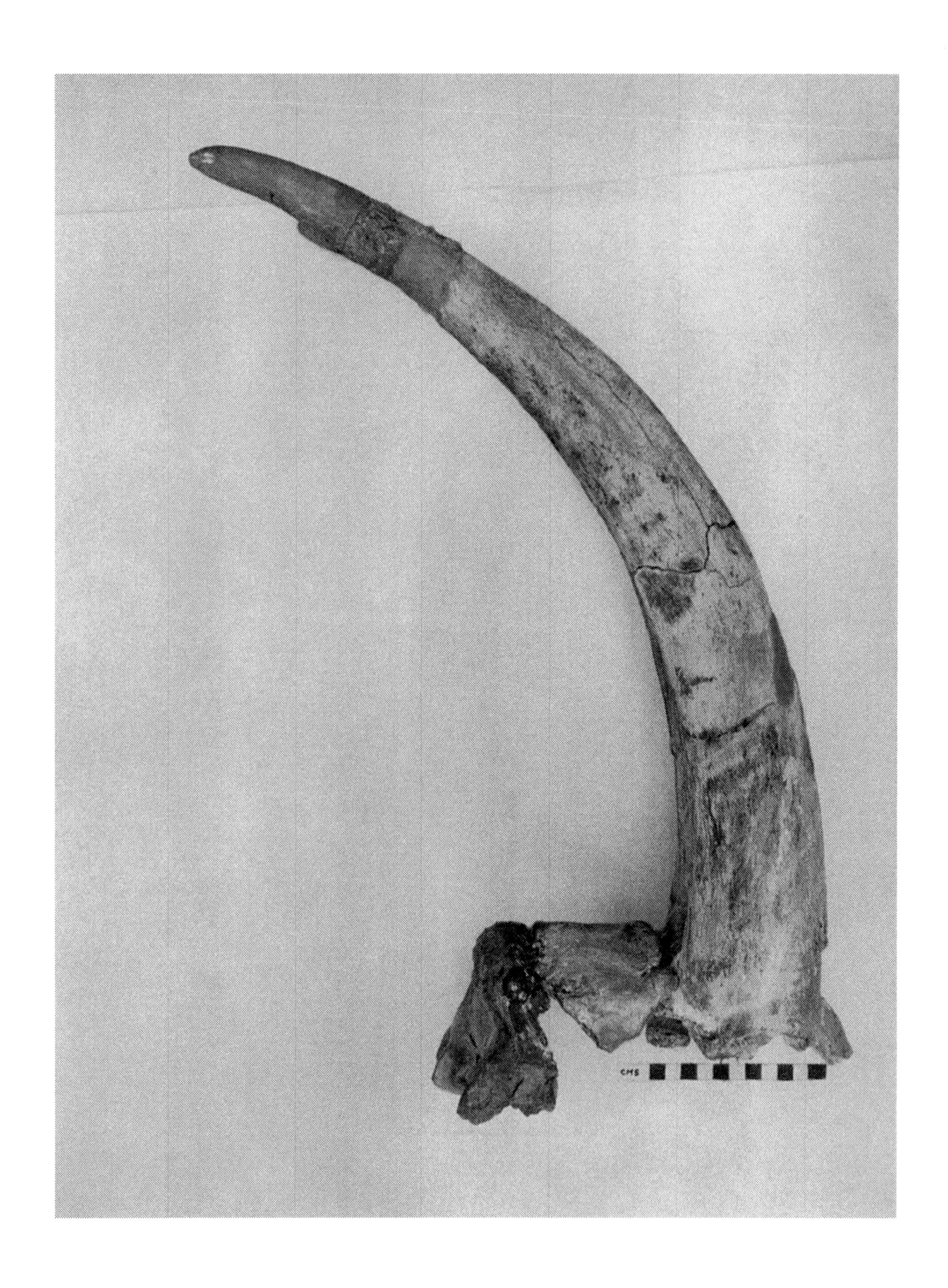

56. *Hippotragus gigas* (male). Type, no. P.P.T. 2. Cranium and horn cores, side view

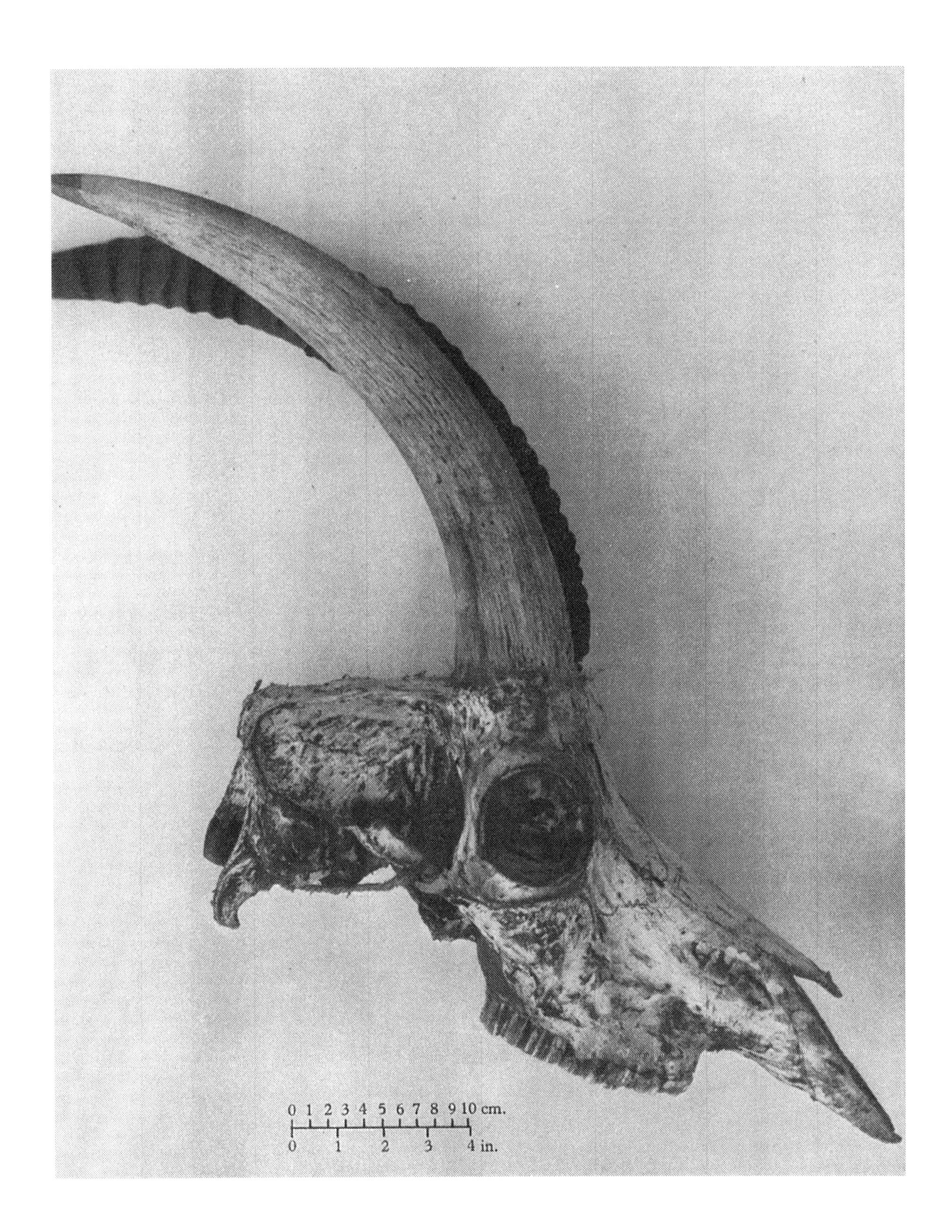

57. *Hippotragus equinus*. Skull and horn cores of modern roan antelope

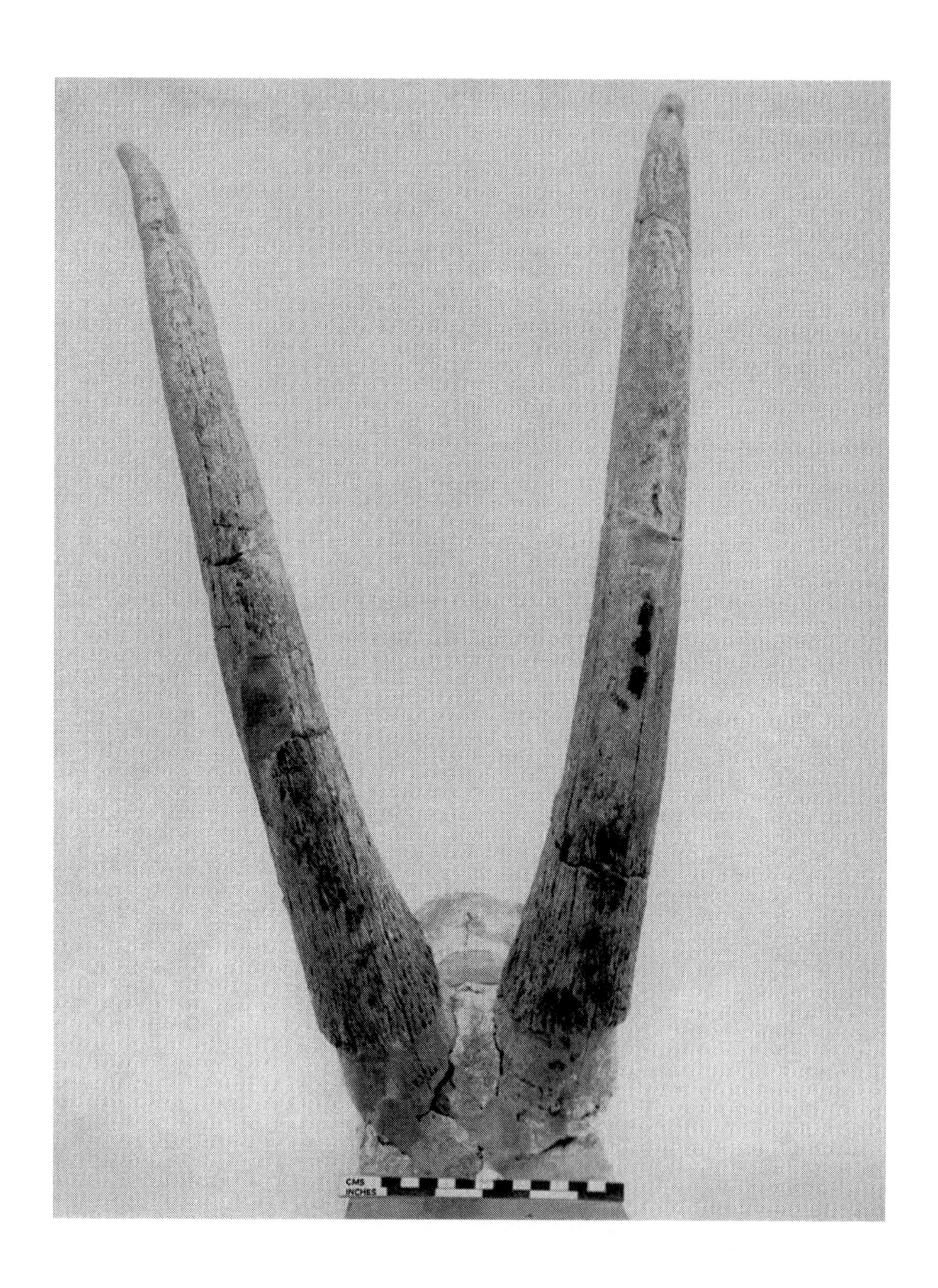

58. *Hippotragus gigas* (male). Type. Cranium and horn cores, front view

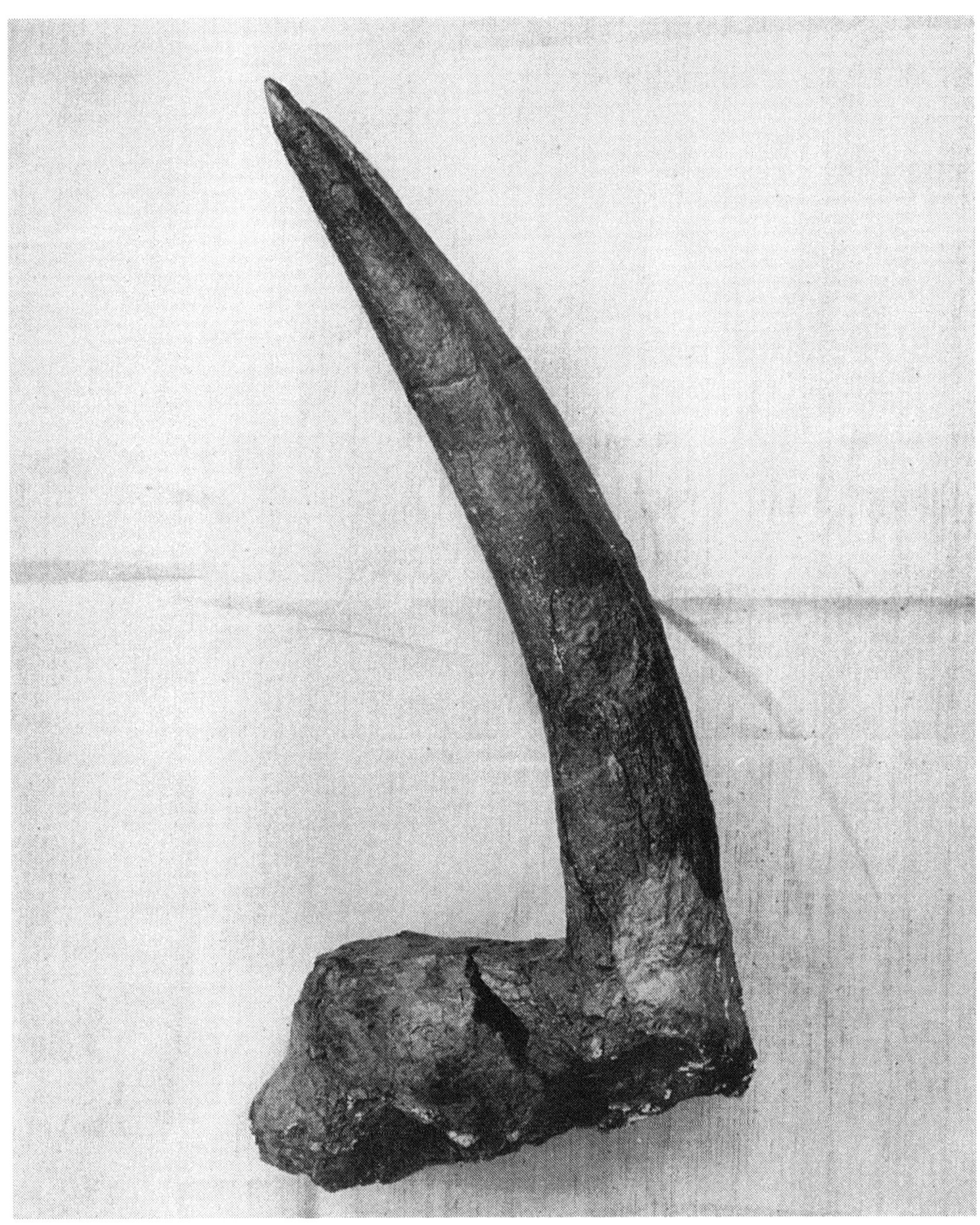

59. *Hippotragus gigas* (female). Paratype. Cranium and horn cores, side view

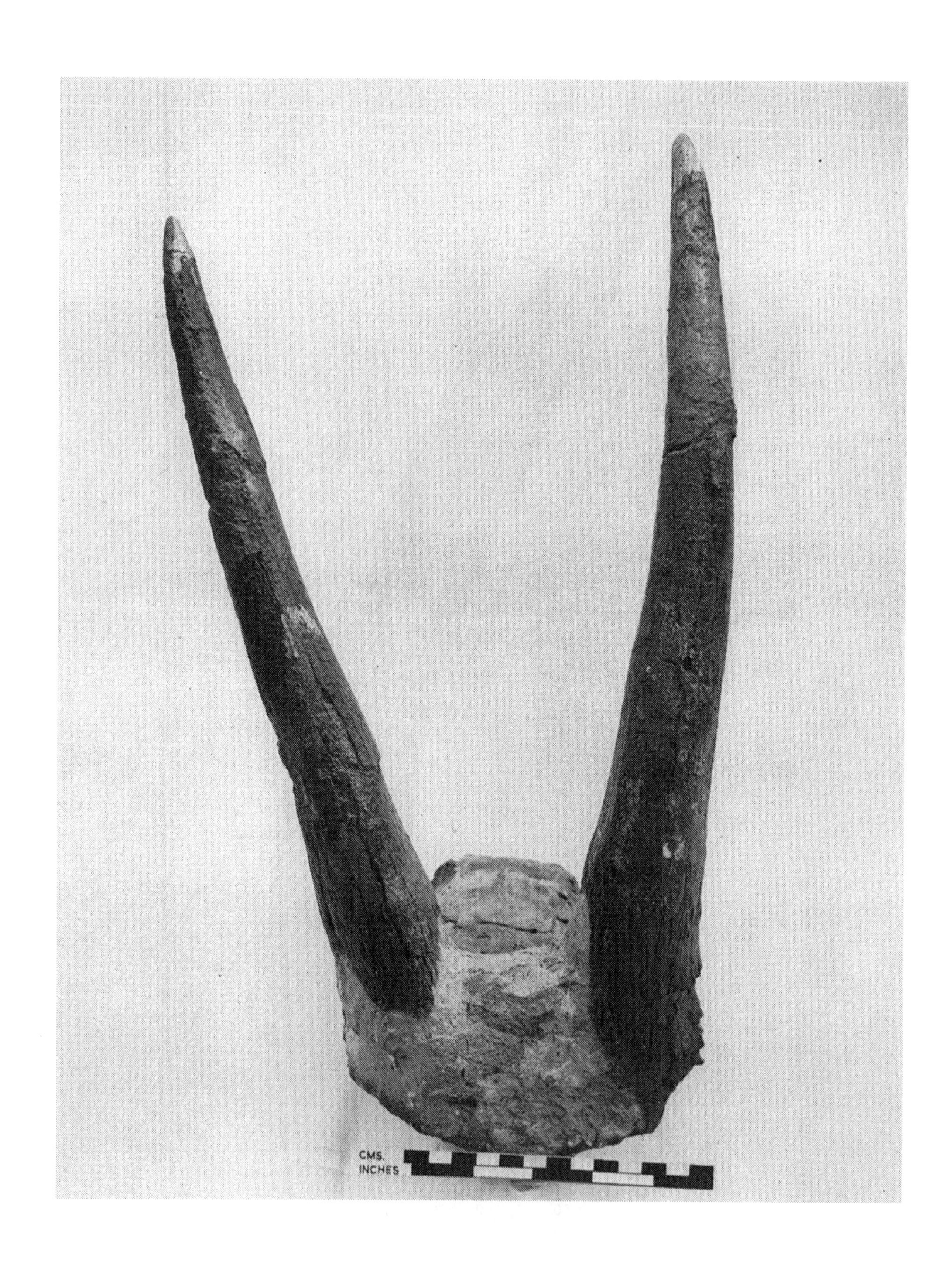

60. *Hippotragus gigas* (female). Paratype. Cranium and horn cores, front view

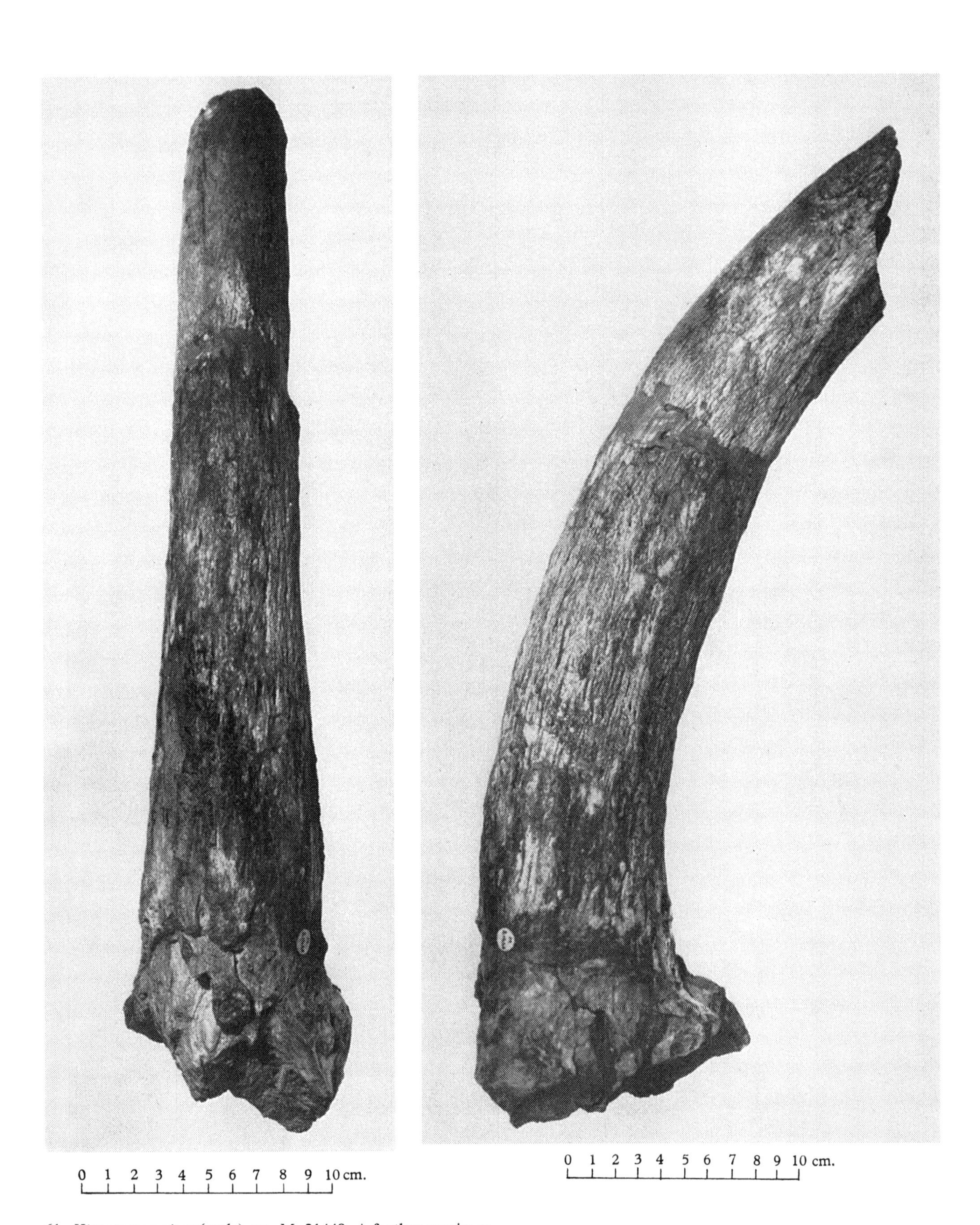

61. *Hippotragus gigas* (male), no. M. 21448. A further specimen

62. *Oryx* sp. indet., no. FLK I G. 390, 1960. Left horn core

63. *Damaliscus angusticornis*, no. M. 14553. Right horn core. (The original paratype described by Dr Schwarz.)

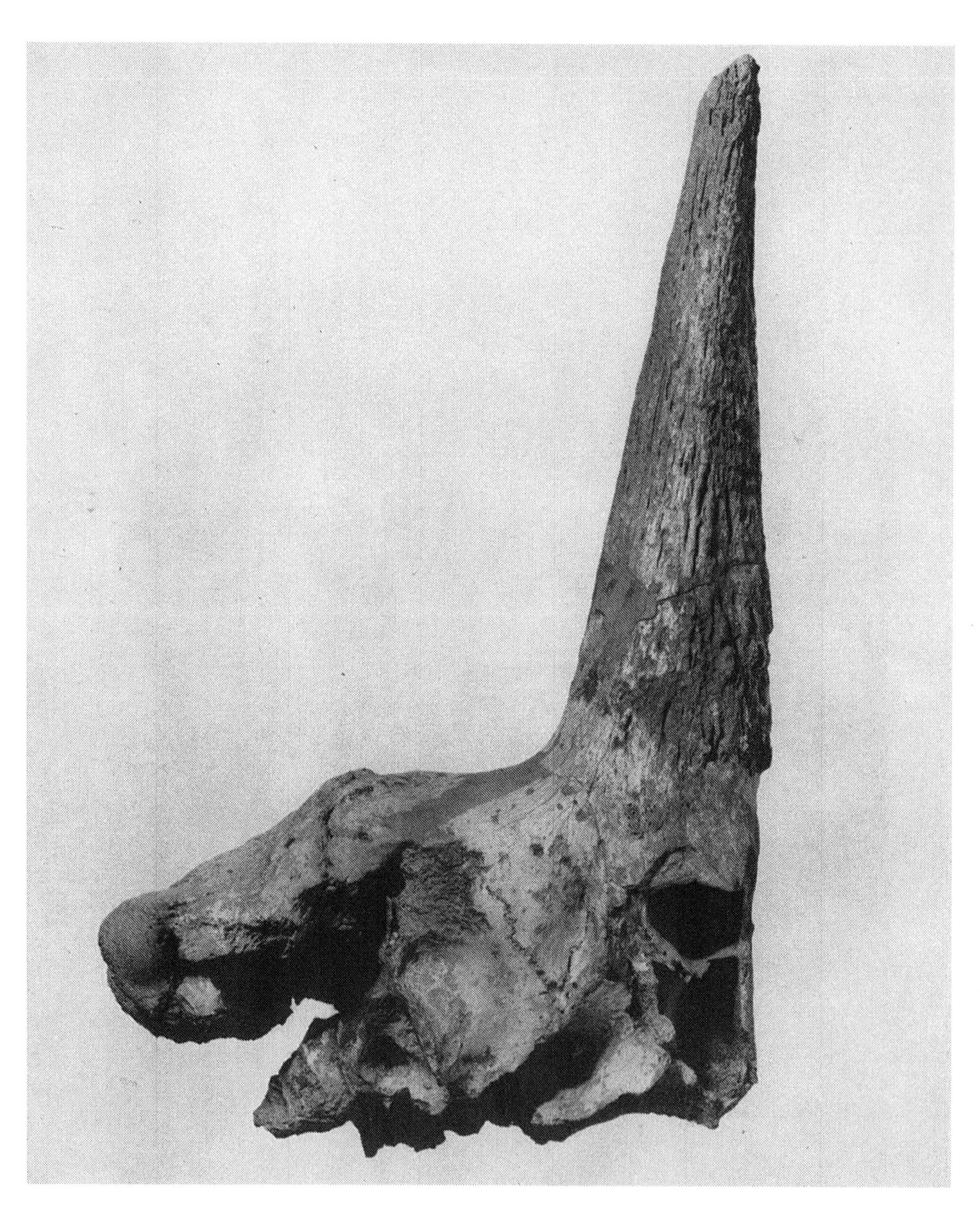

64. *Damaliscus angusticornis*, no. M. 21425. Cranium and horn cores, side view

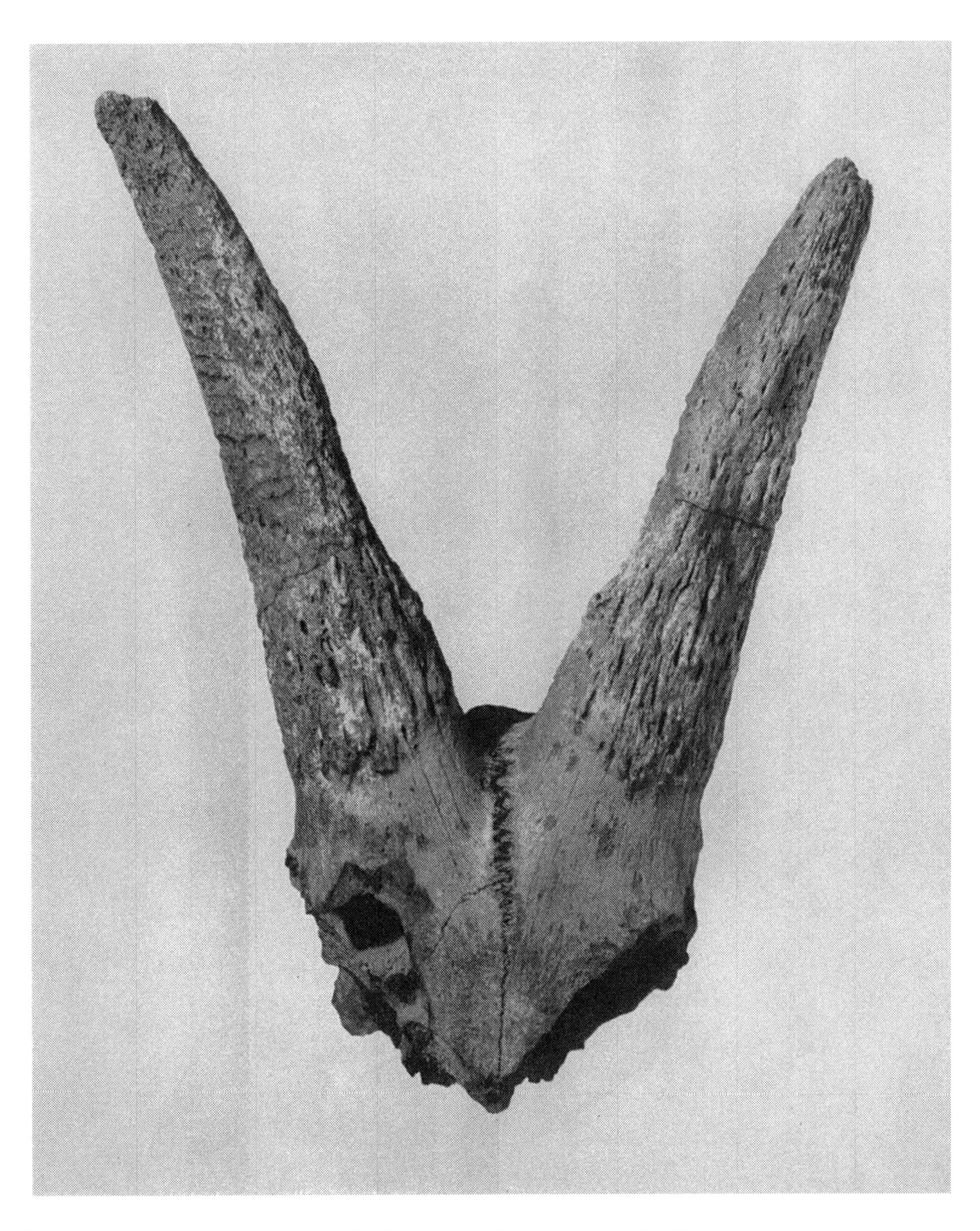

65. *Damaliscus angusticornis*, no. M. 21425, front view

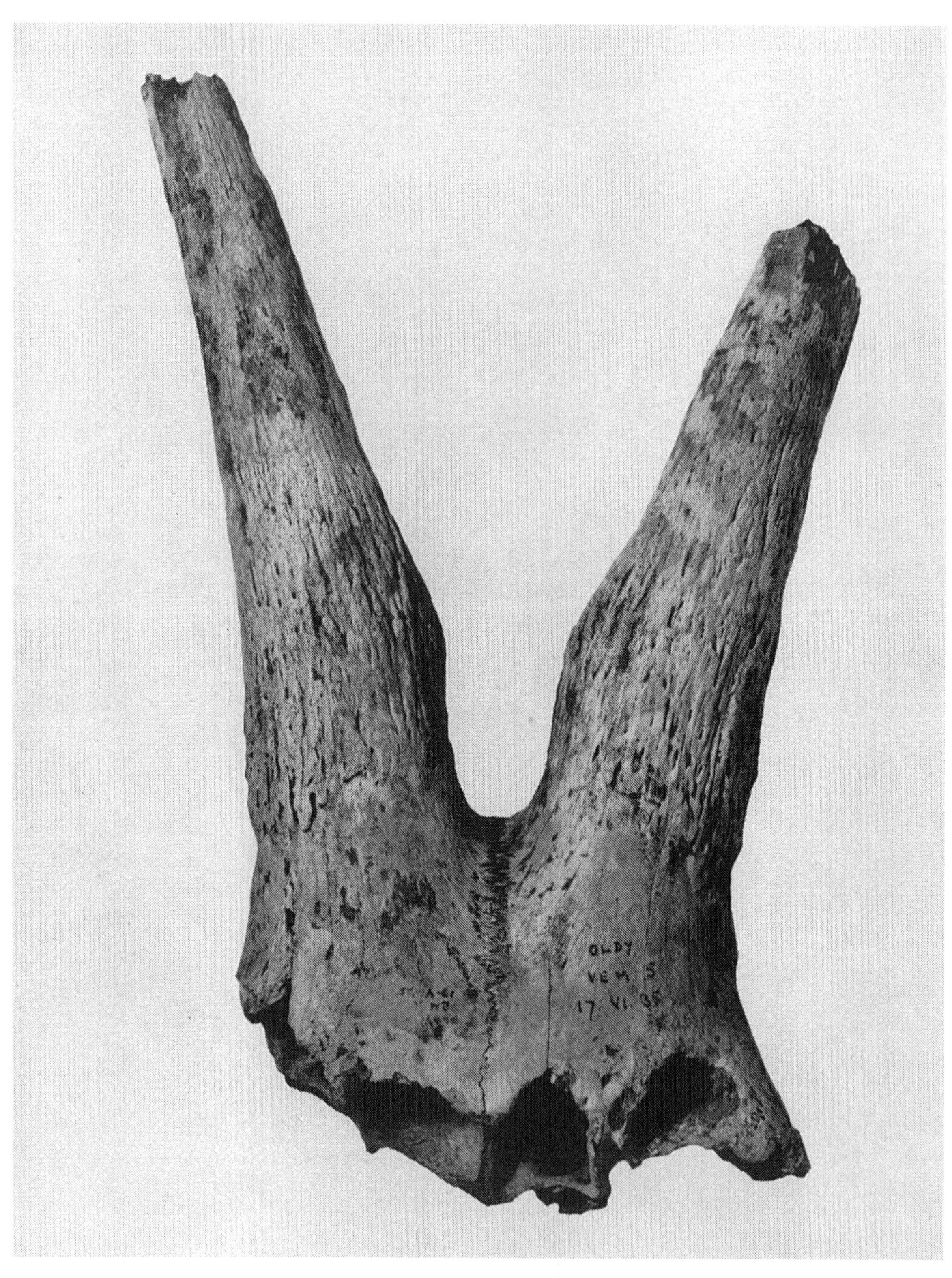

0 1 2 3 4 5 6 7 8 9 10 cm.

66. *Damaliscus angusticornis*, no. M. 21422. Additional specimen

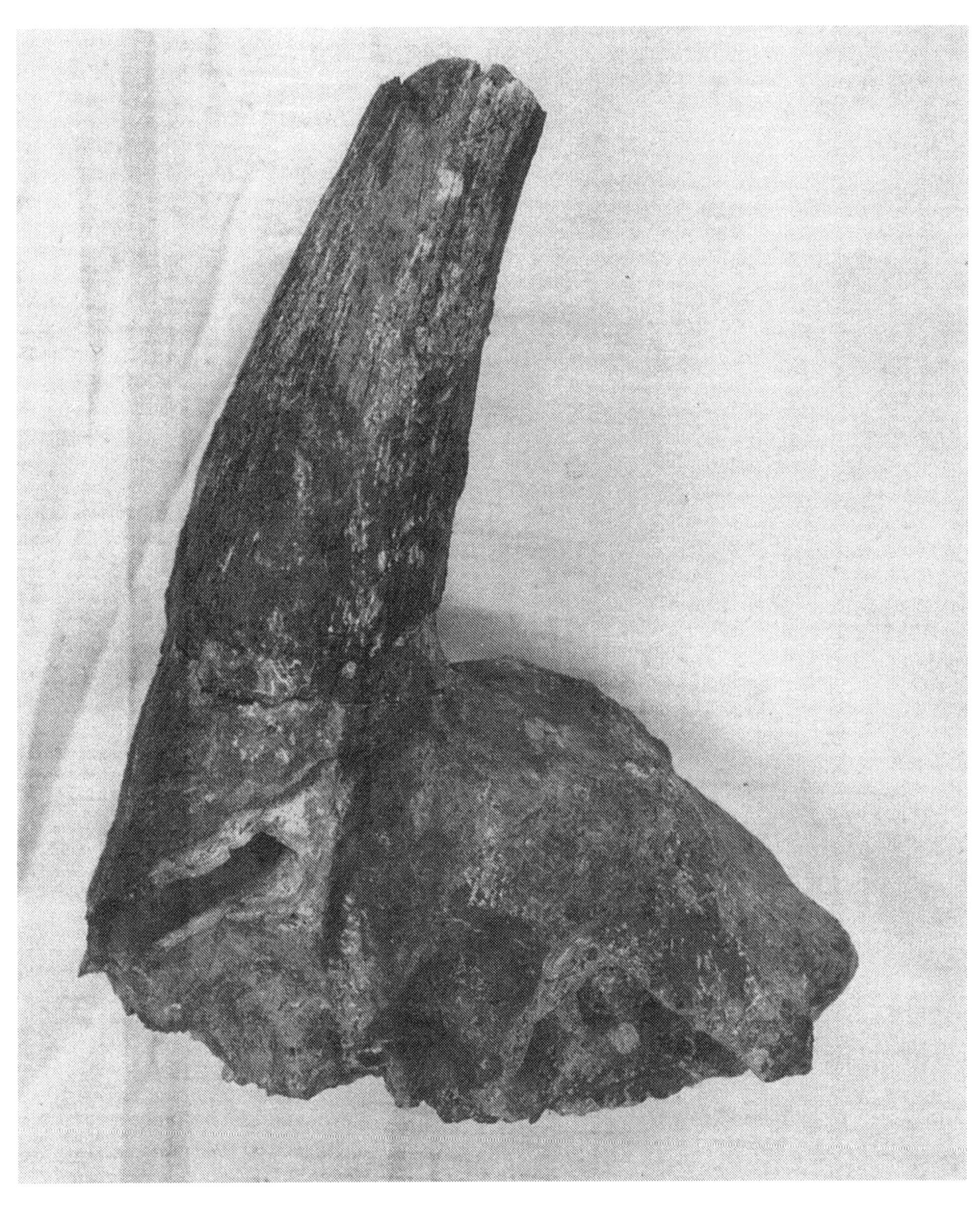

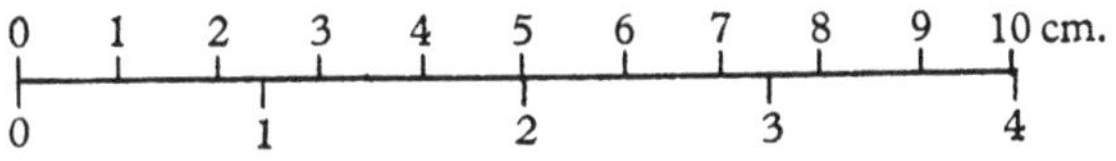

67. *Damaliscus antiquus*. Type, no. P.P.T. 3. Cranium and horn cores, side view

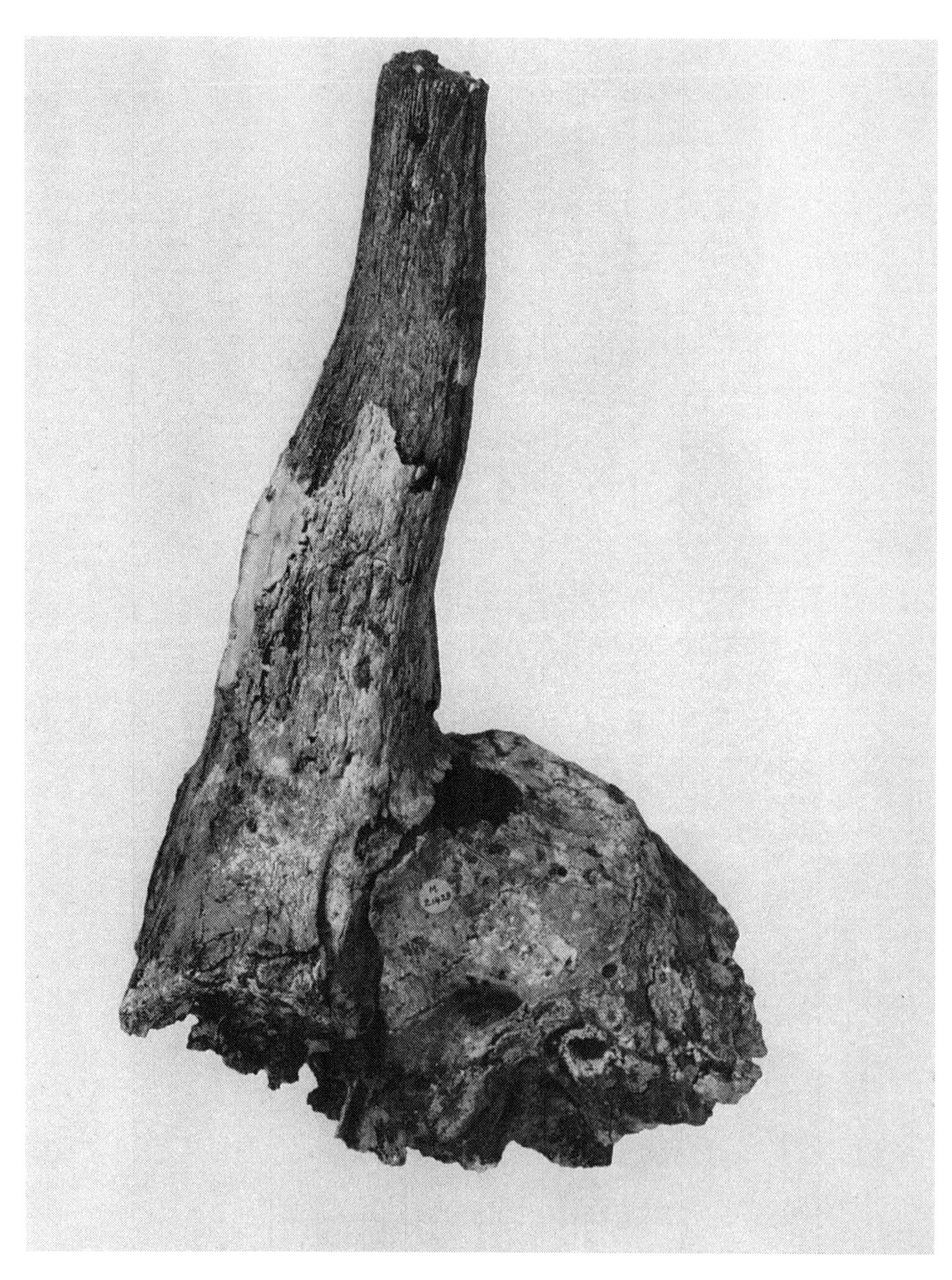

68. *Damaliscus antiquus*. Paratype, no. M. 21428. Cranium and horn cores, side view

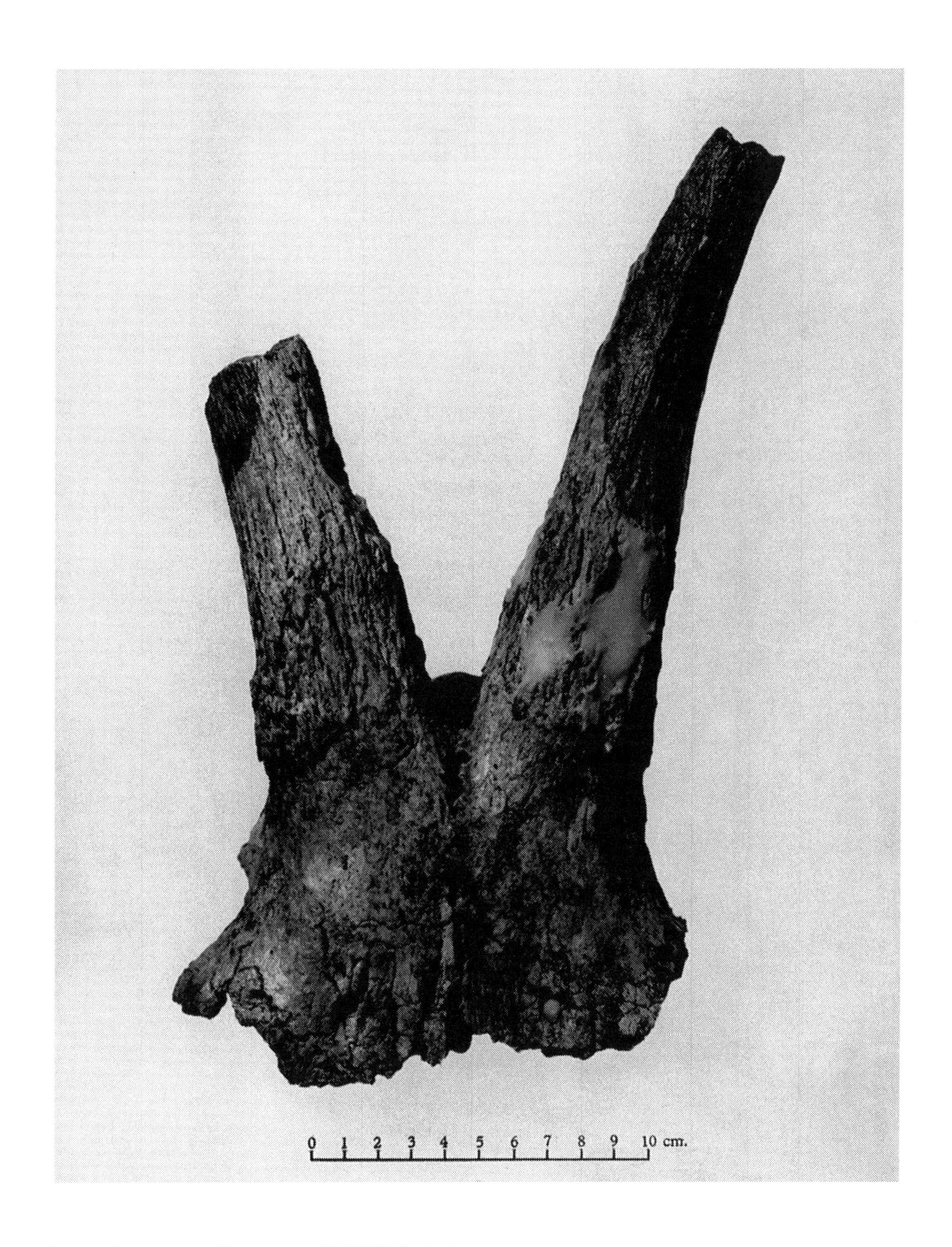

69. *Damaliscus antiquus*. Paratype, no. M. 21428, front view

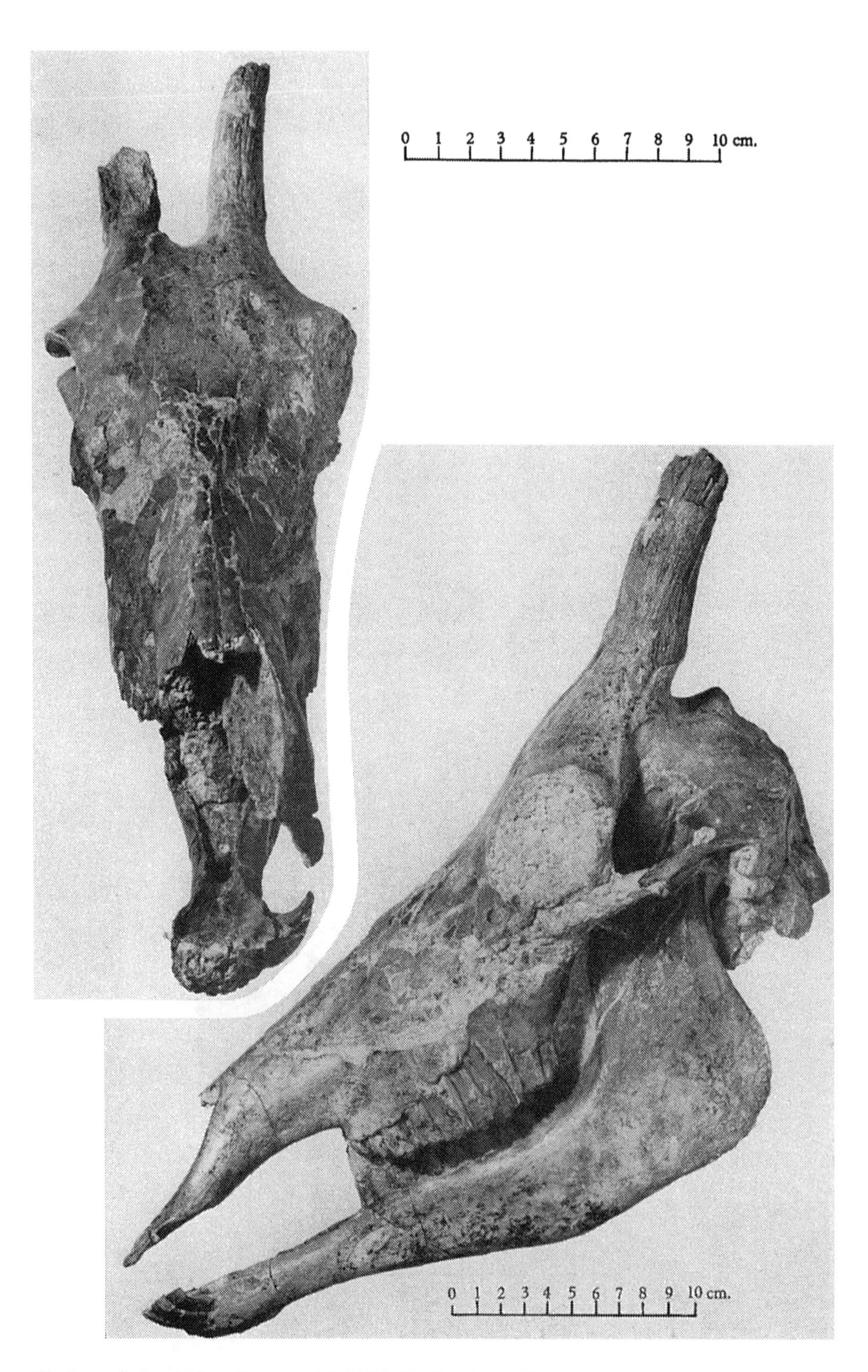

70. *Parmularius altidens*. Type, no. M. 14689. Skull and mandible

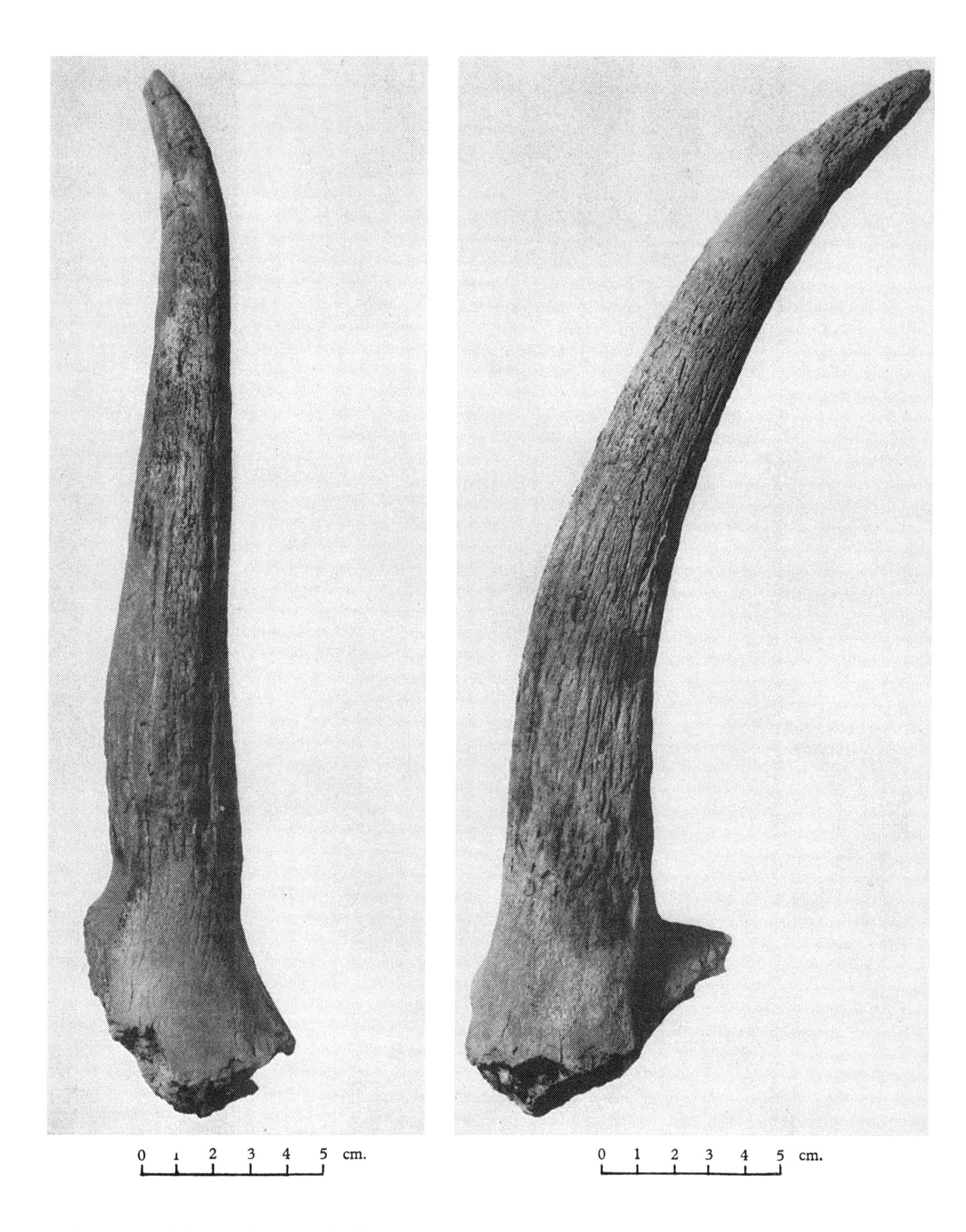

71. *Parmularius altidens*, no. M. 14514. Left horn core

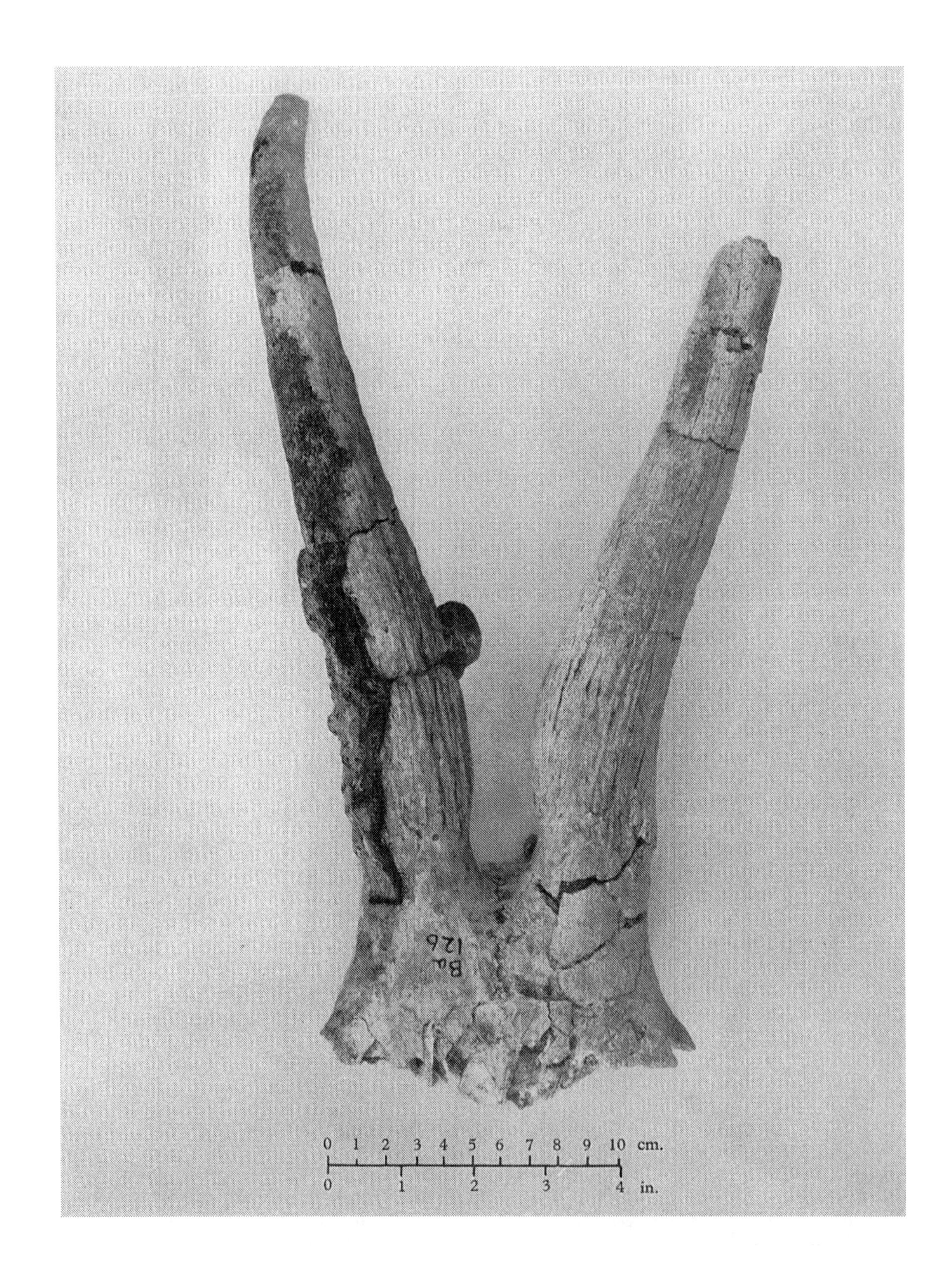

72. *Parmularius altidens*, no. FLK Ba. 126, 1960. Frontlet and horn cores

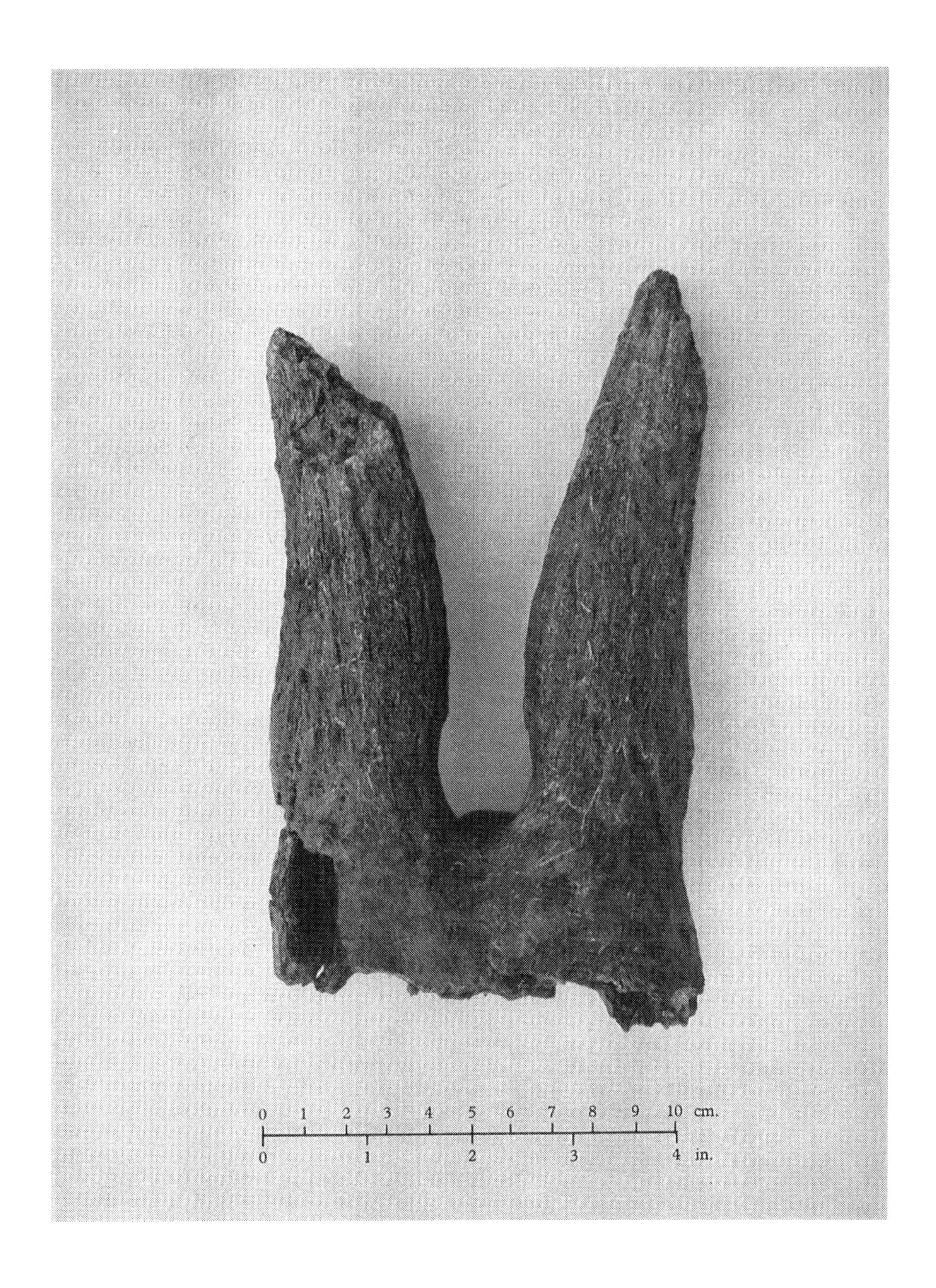

73. *Parmularius altidens*, FLK N I, no. 1315, 1960. Frontlet and horn cores

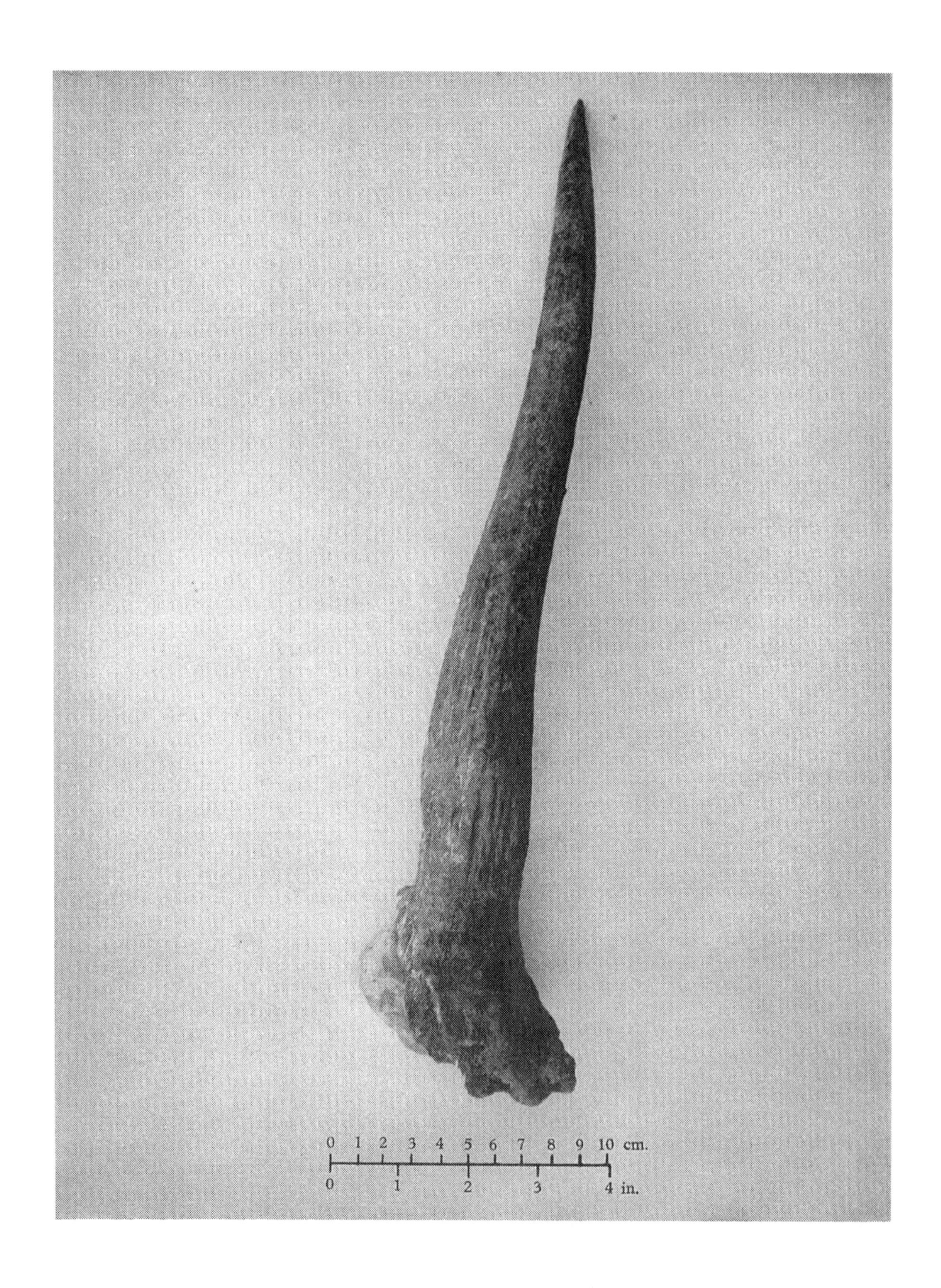

74. *Parmularius altidens*, FLK I, no. F. 206, 1960. Left horn core

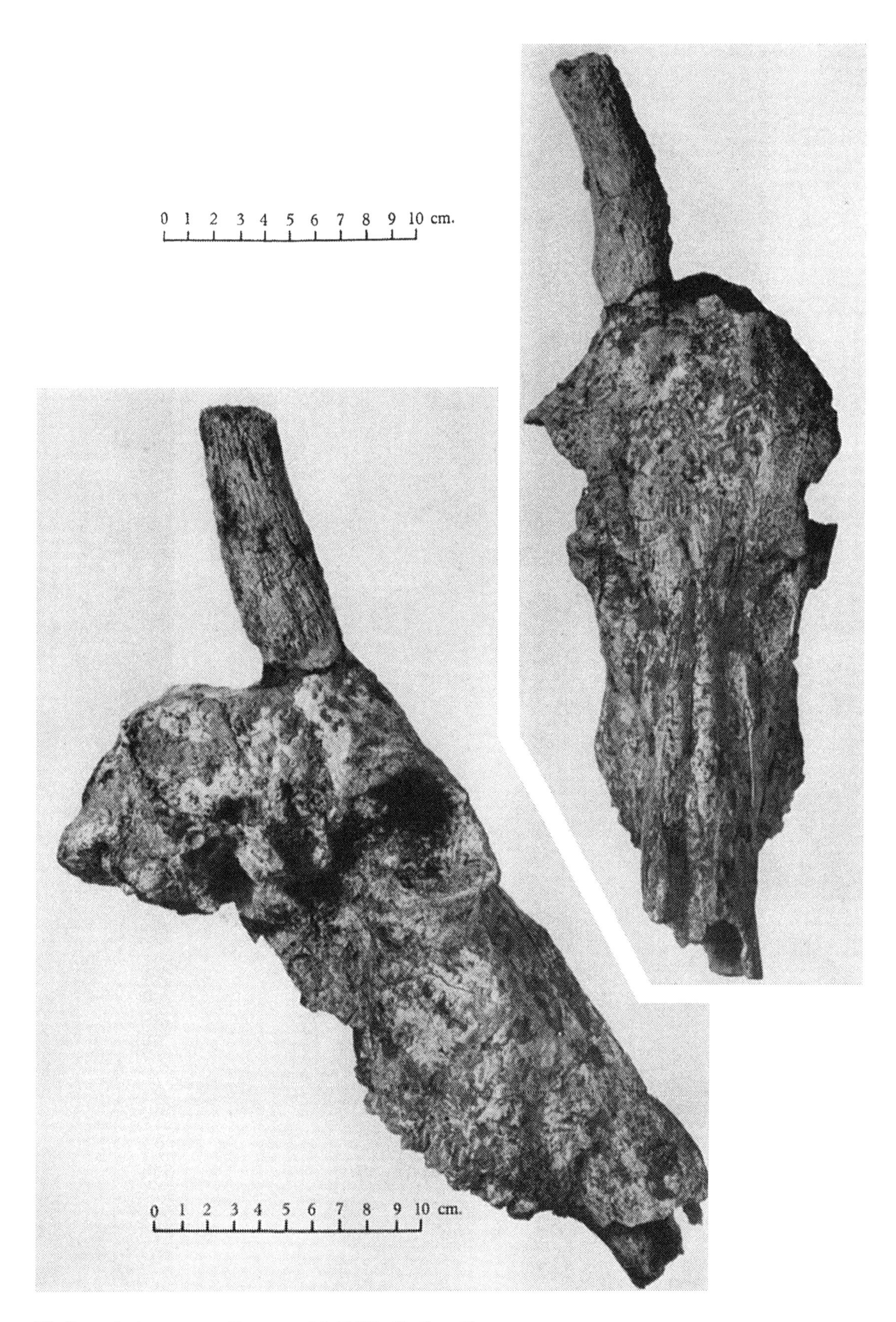

75. *Parmularius rugosus*. Type, no. M. 21430. Skull and horn core

76. *Parmularius rugosus*. Type, no. M. 21430. Palate

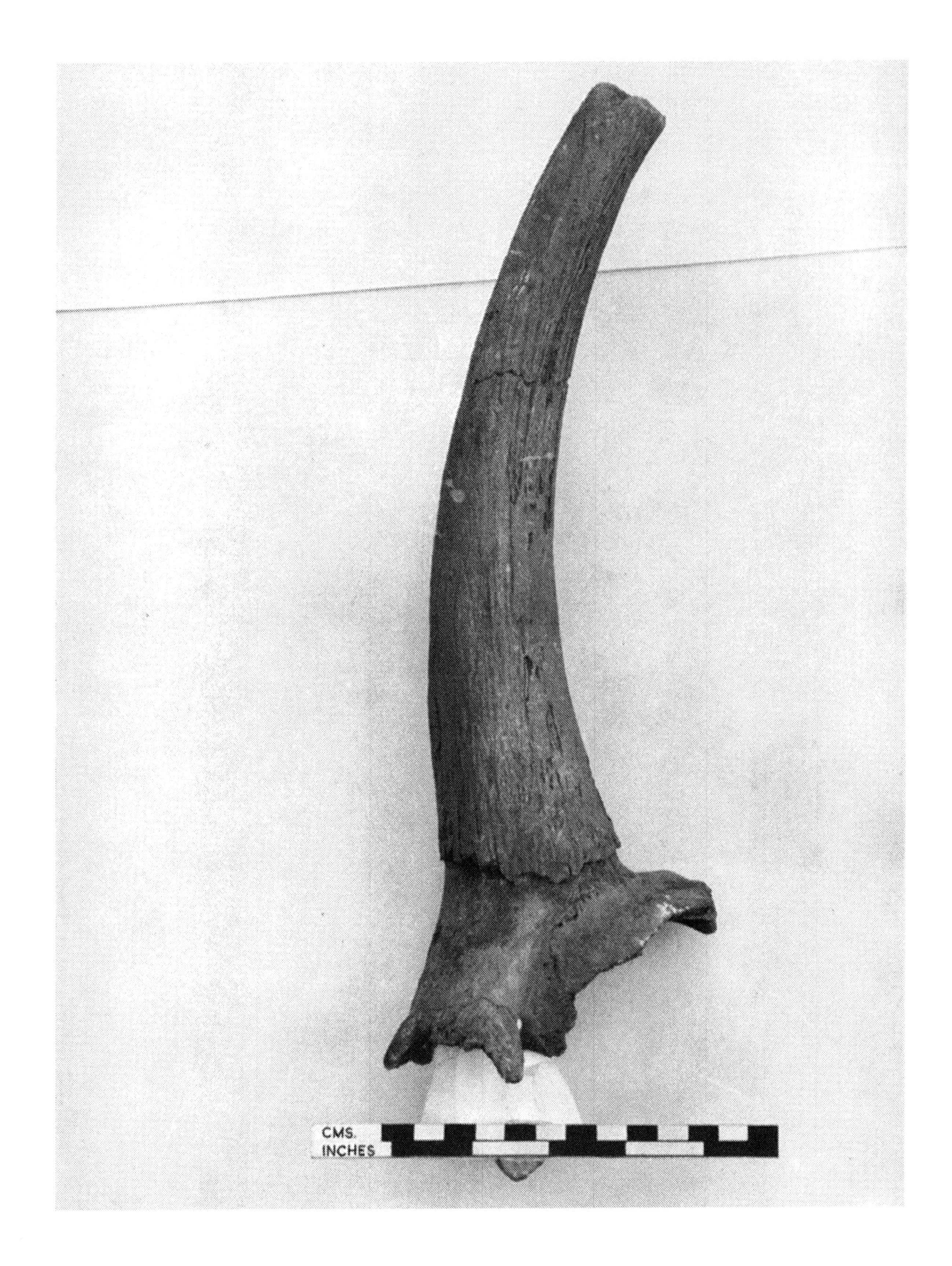

77. *Parmularius* indet., no. F. 3001. Left horn core

78. *Alcelaphus* cf. *kattwinkeli*, no. F 3013. Frontlet and horn cores

79. *Alcelaphus howardi*. Type, no. M. 14950. Frontlet and horn cores, side and front views

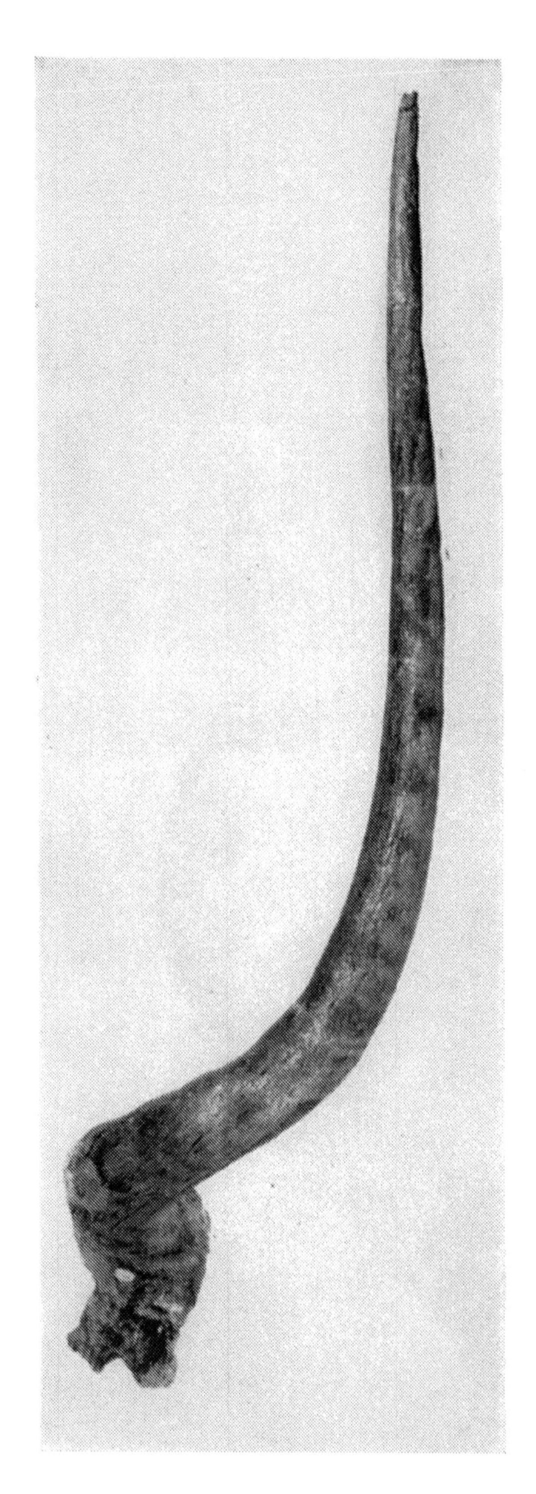

80. *Beatragus antiquus*. Paratype, no. M. 21446. Front and side views of left horn core

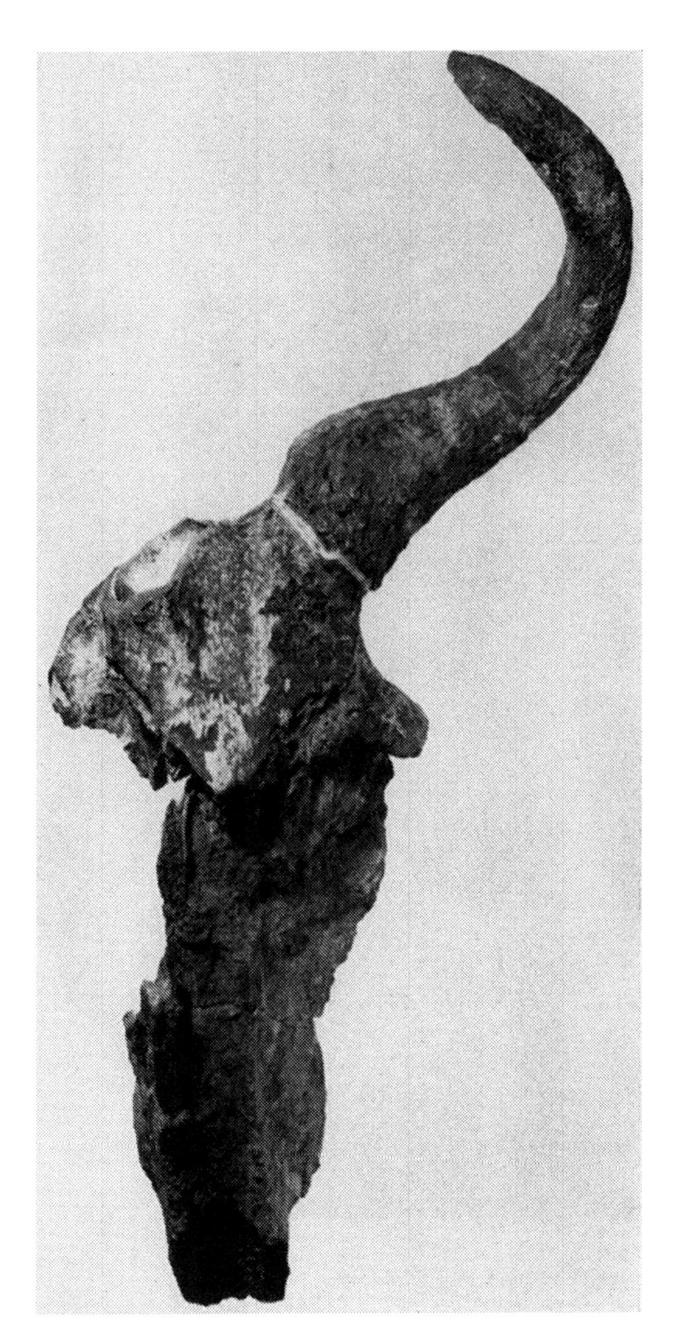

81. *Xenocephalus robustus*. Type, no. M. 21447. Skull and horn core

82. *Xenocephalus robustus*. Palate of Type (enlarged)

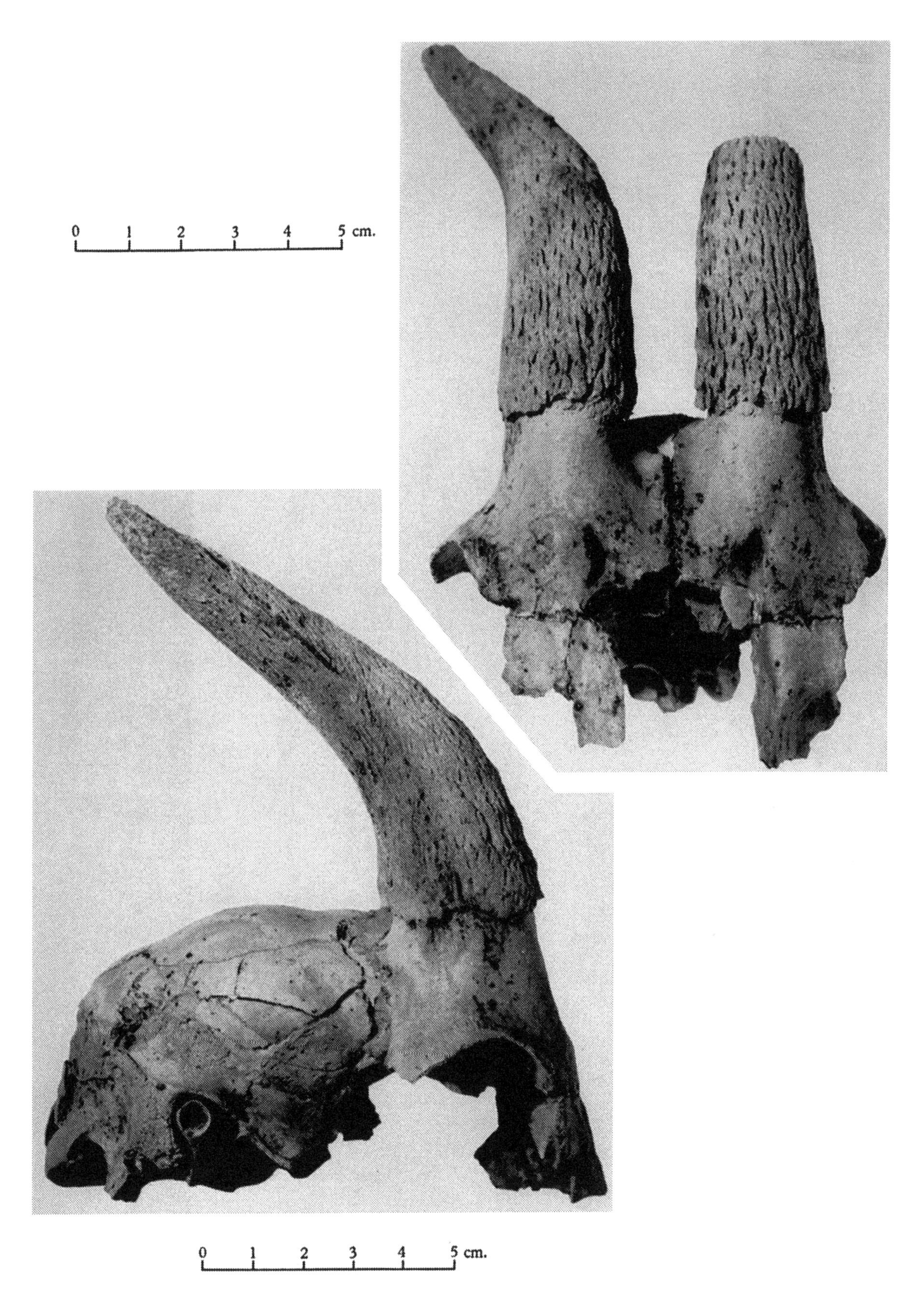

83. *Gazella* sp. (*a*) no. M. 21464. Cranium and horn cores

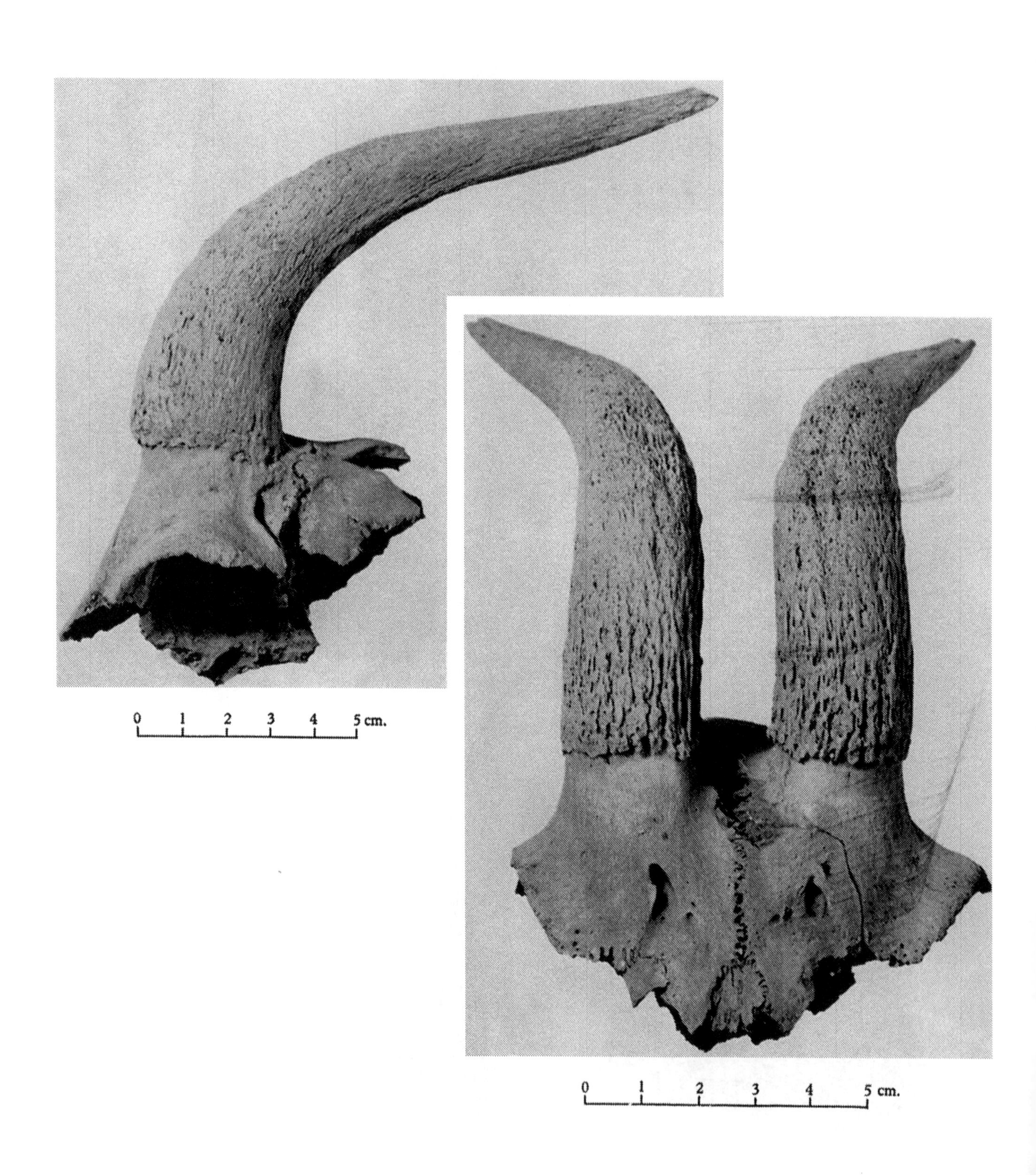

84. *Gazella* sp. (*a*) no. M. 21463. Second specimen

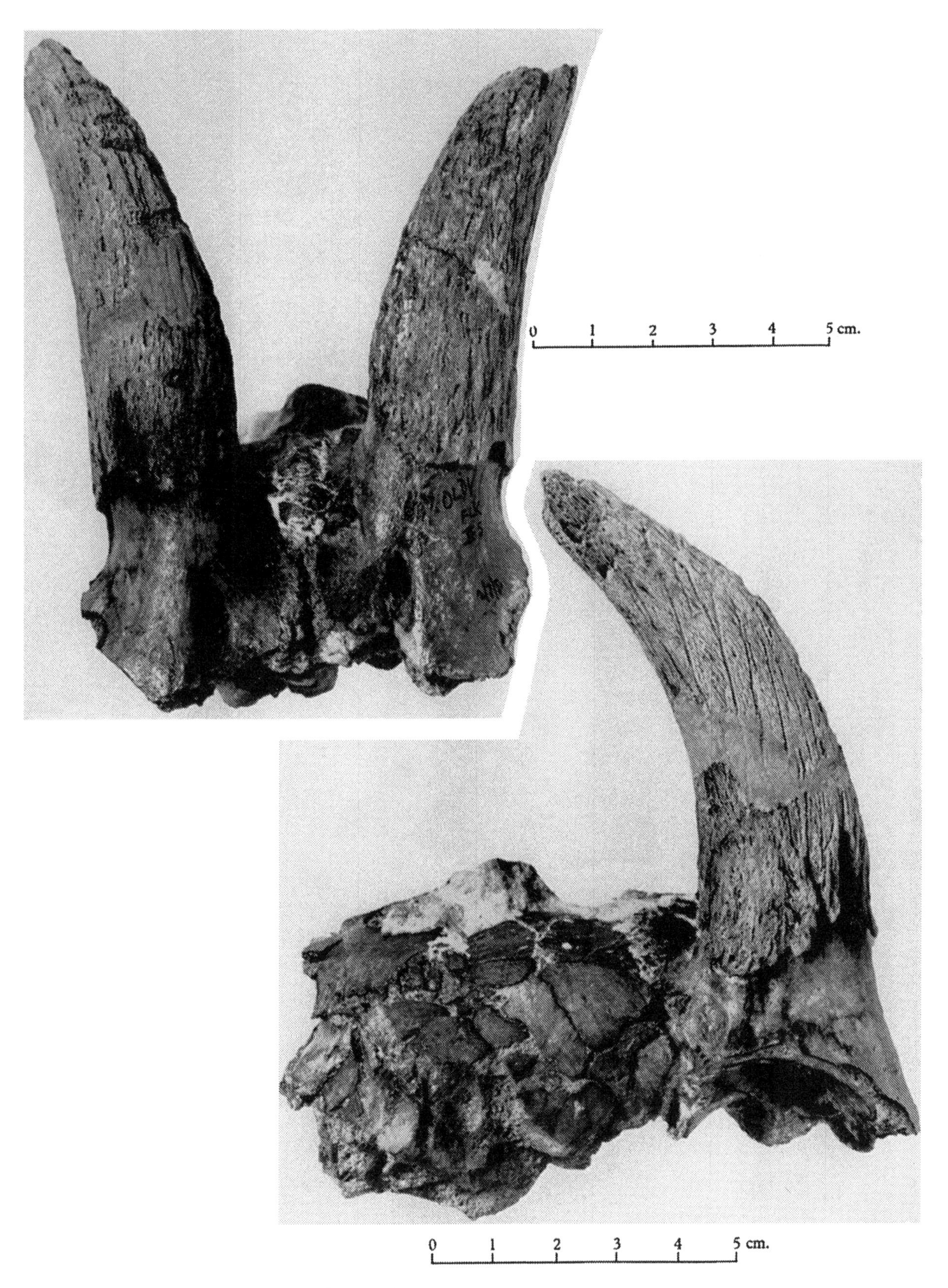

85. *Gazella* sp. (*d*) no. M. 21462. Cranium and horn cores

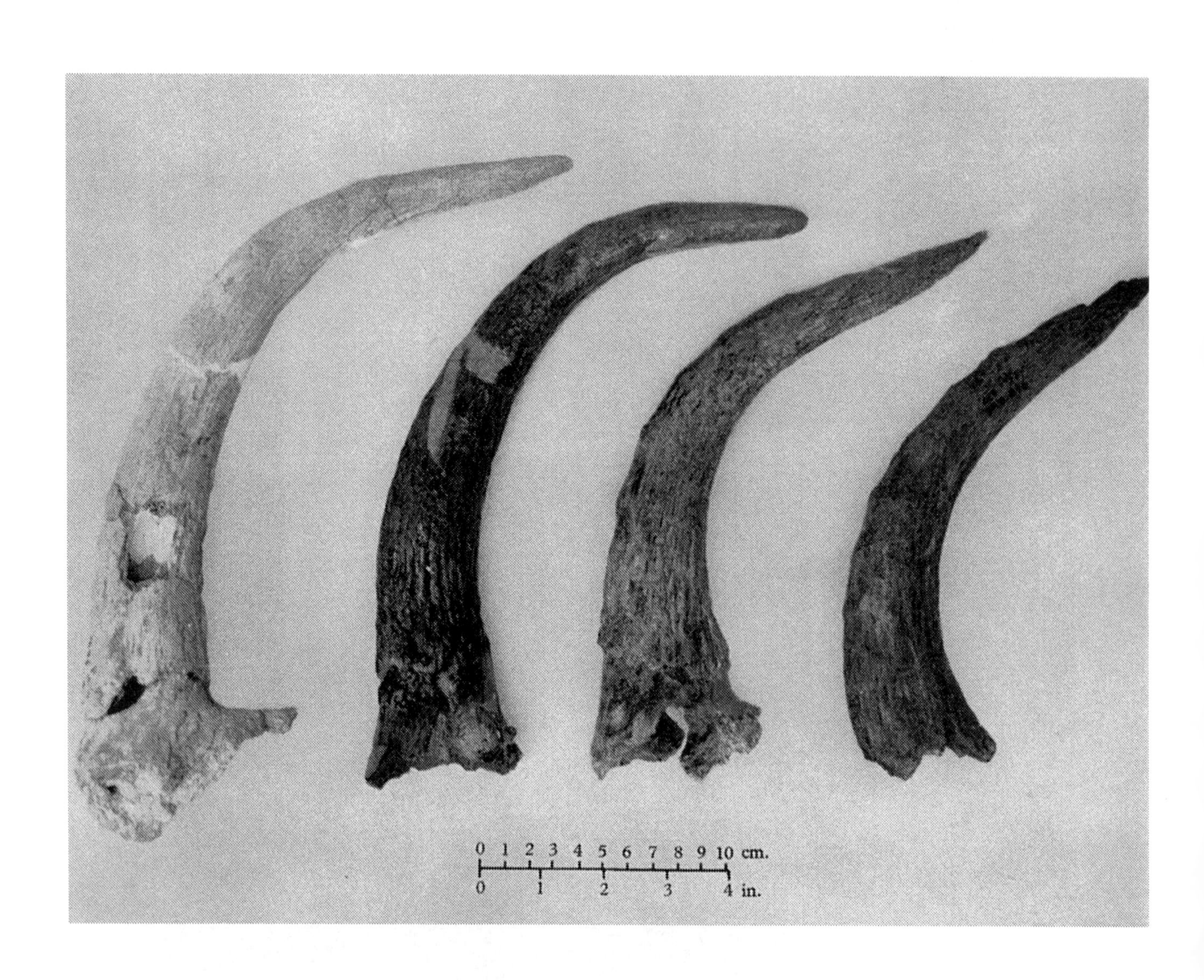

86. *Gazellae* indet. (*j*). Examples of horn cores

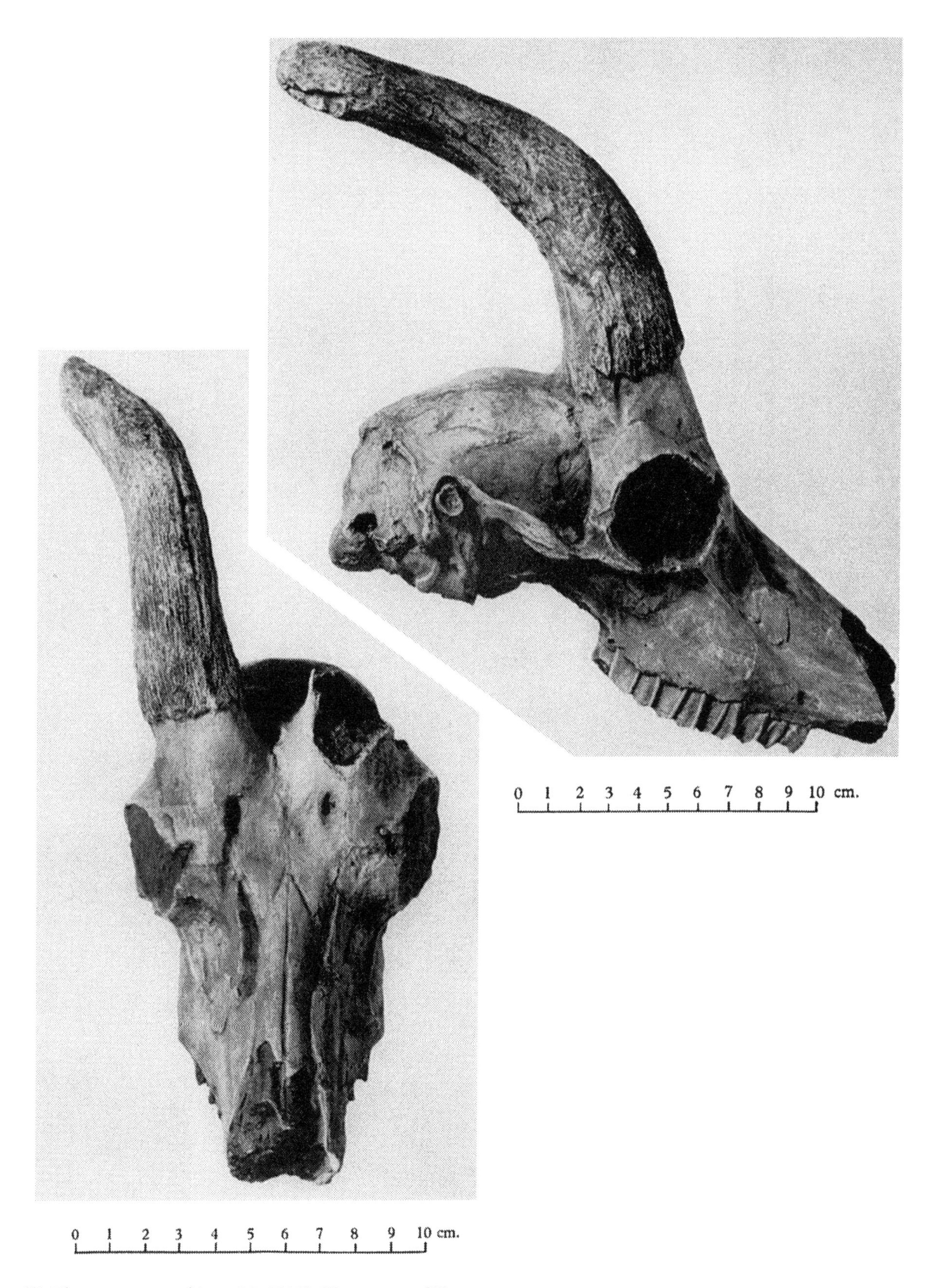

87. *Phenacotragus recki*, no. M. 21460. Plaster cast of Type

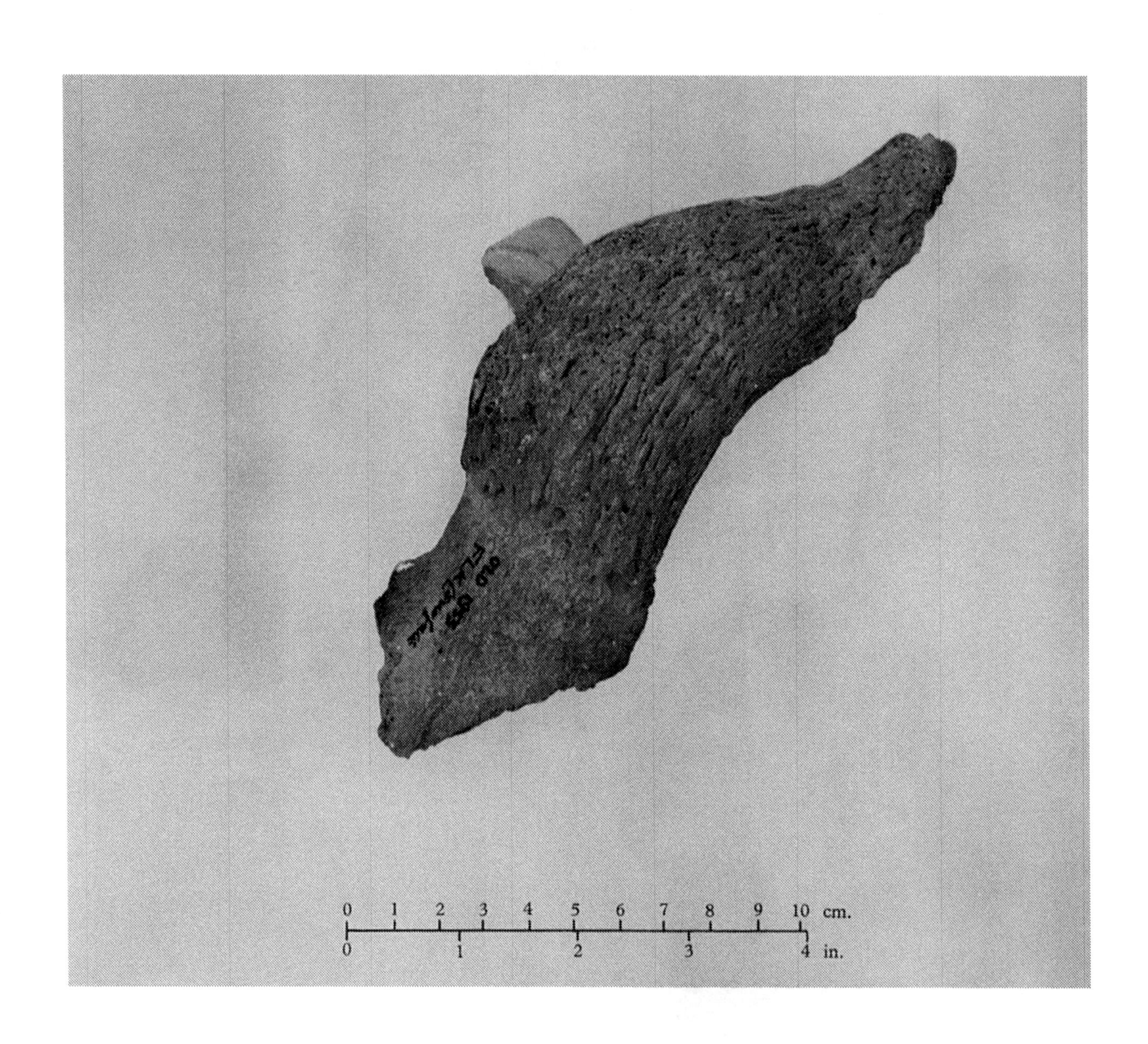

88. *Thaleroceras radiciformis*, no. P.P.F. 4. Side view of left horn core

89. Bovinae, incertae sedis (*a*). Frontlet and horn core

90. Bovinae, incertae sedis (*a*). Side view

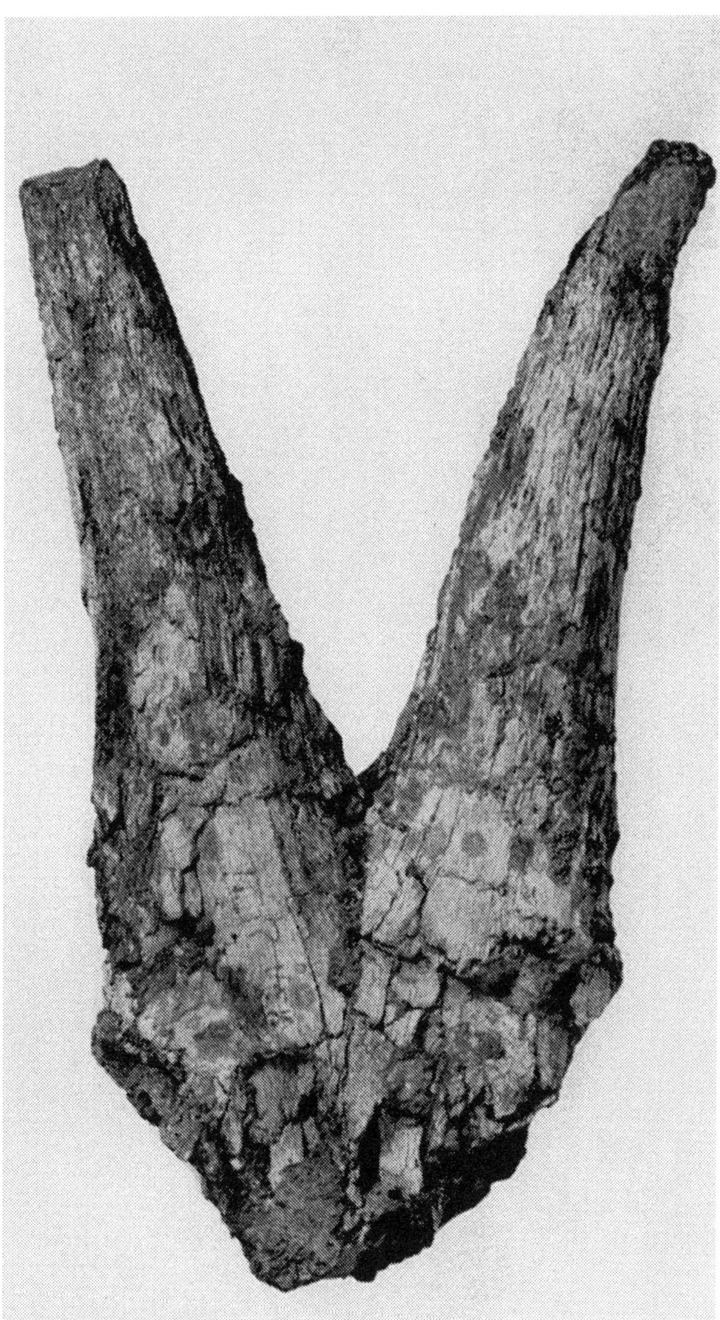

91. Bovinae, incertae sedis (*b*) (cf. Alcelaphini), no. M. 21429. Cranium and horn cores

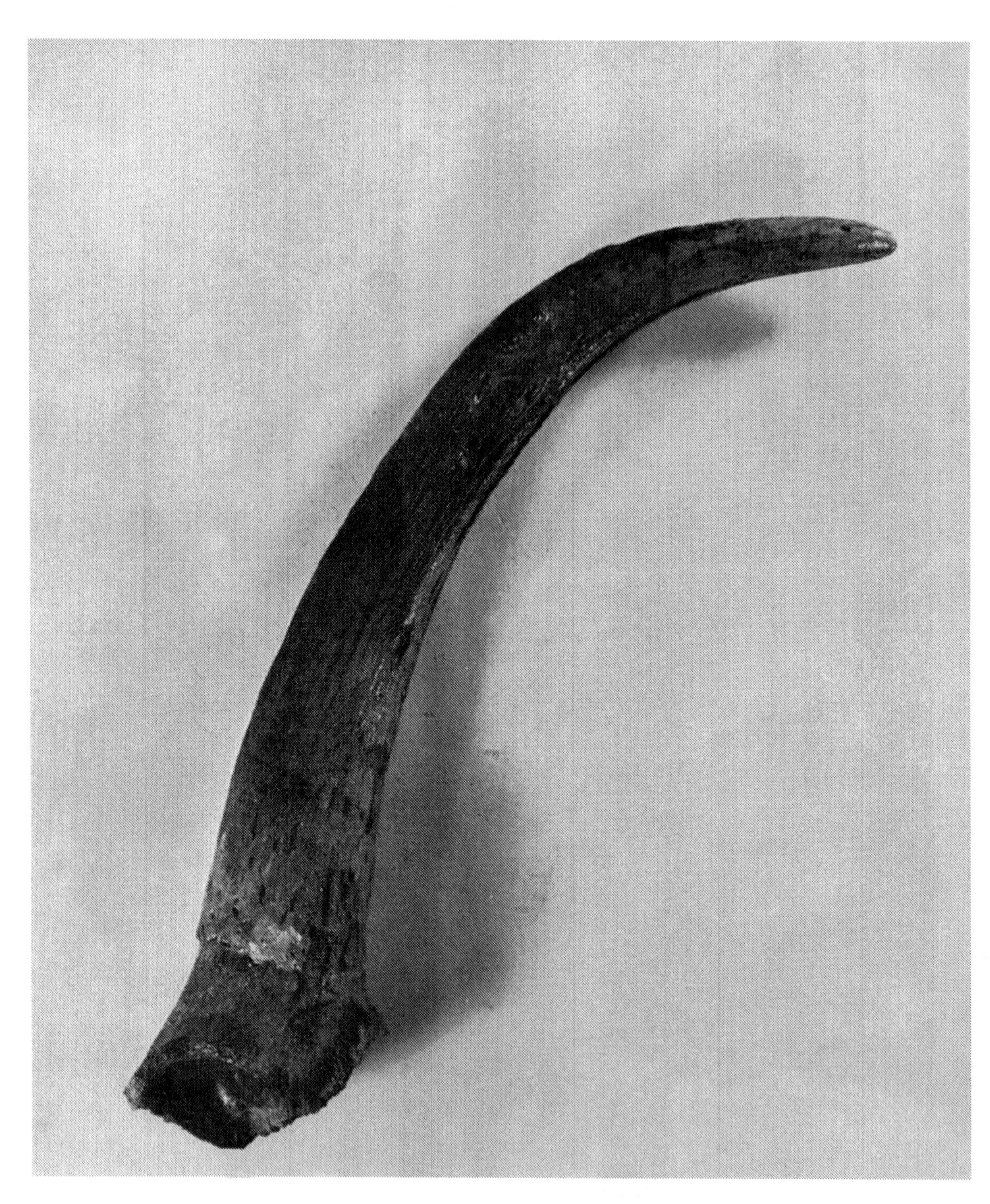

92. Bovinae, incertae sedis (*e*). Alcelaphini indet. Horn core

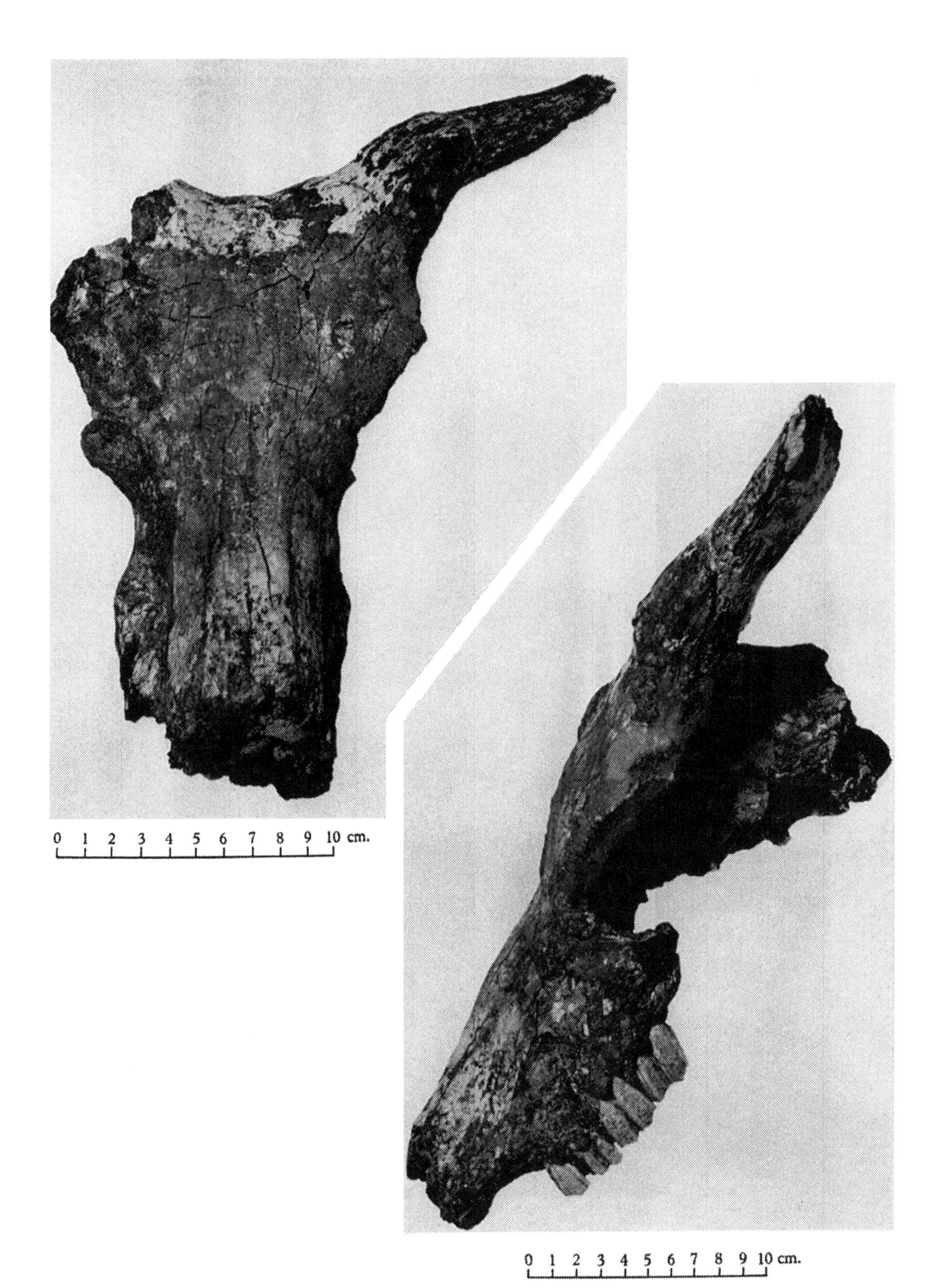

93. *Pultiphagonides africanus*. Type, no. M. 14688. Skull and horn core

94. *Pultiphagonides africanus*. Type, no. M. 14688. Palate

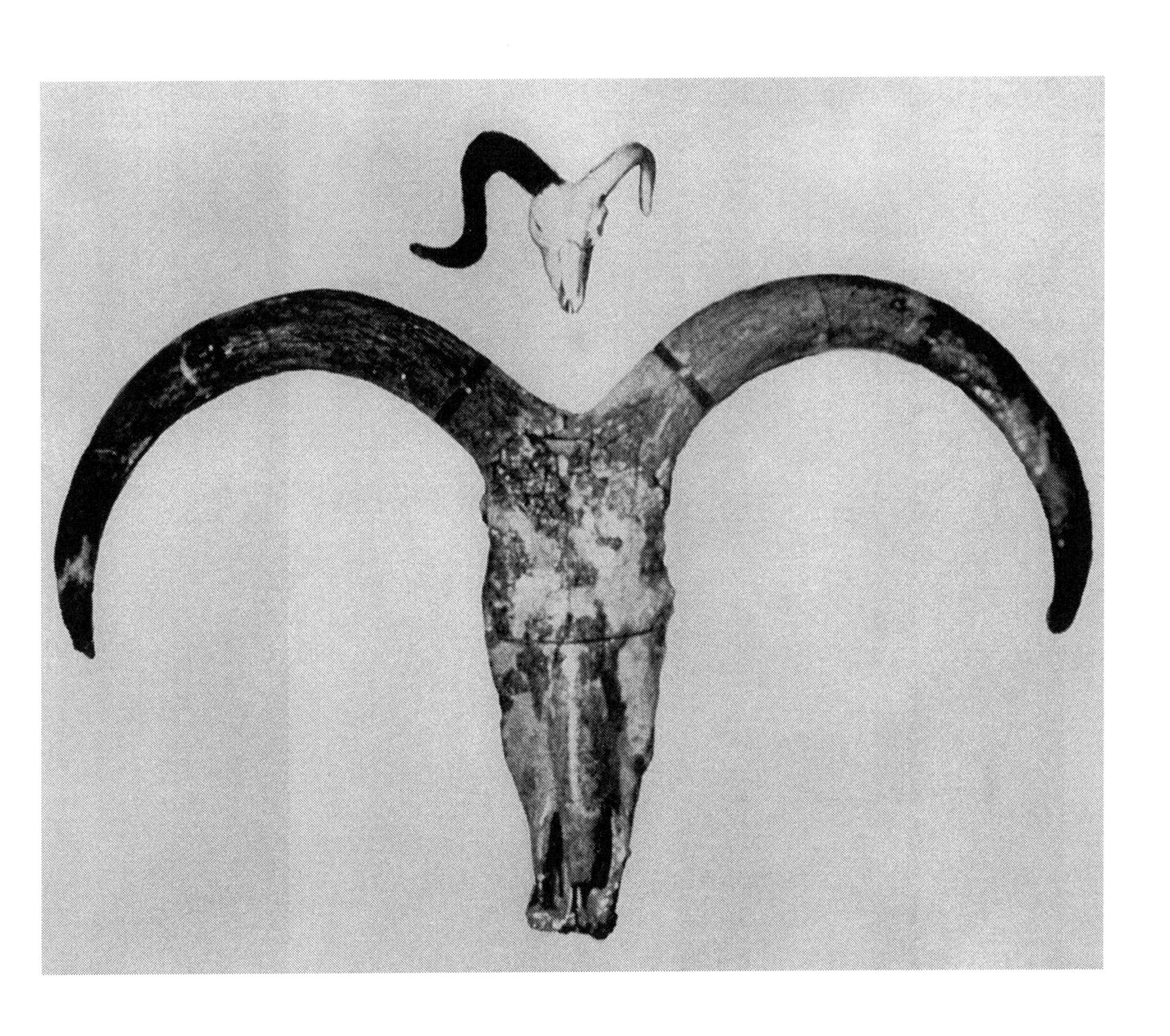

95. *Pelorovis oldowayensis*, no. P.P.F. 8, from site BK II (on view in the Coryndon Memorial Museum), compared with merino ram

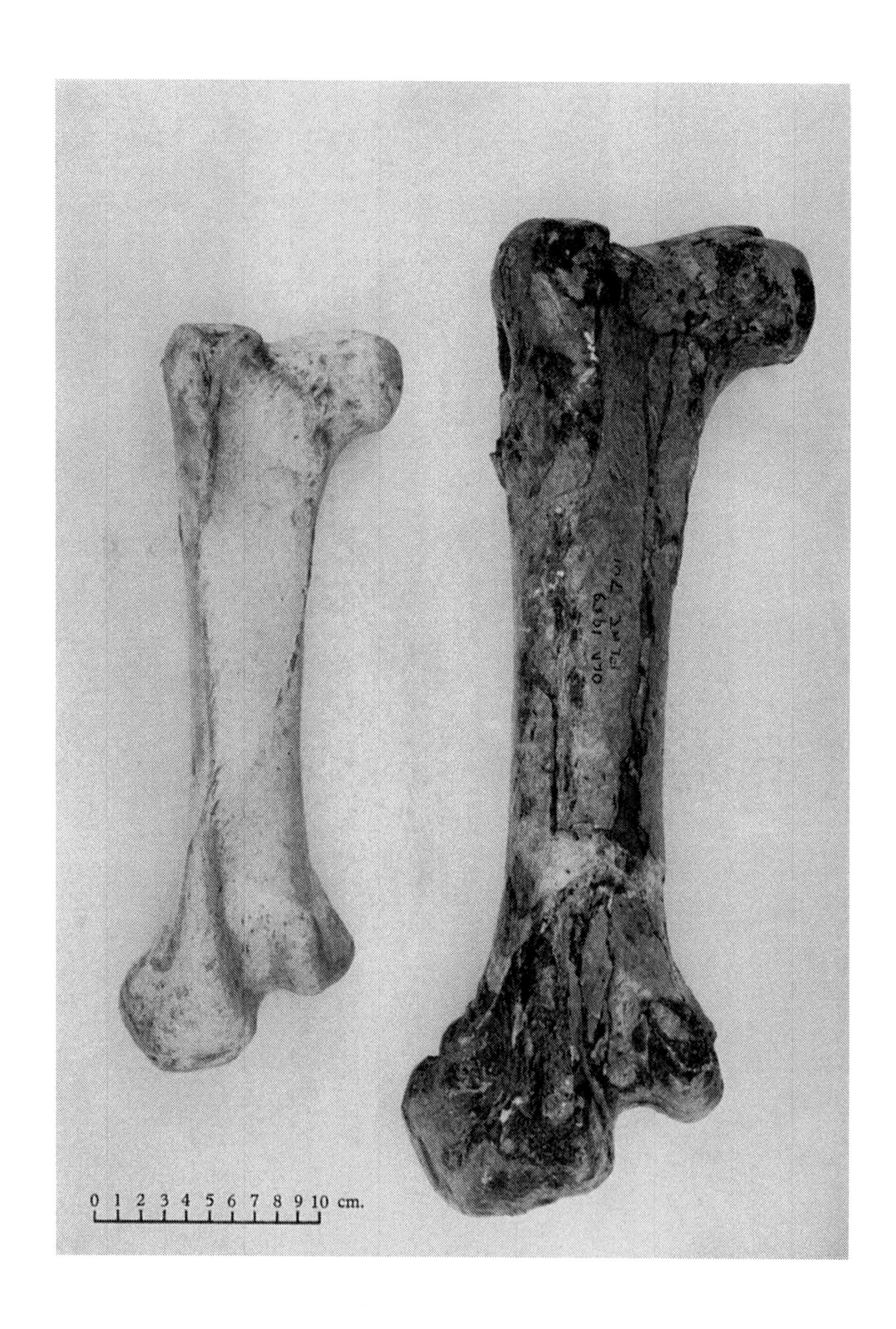

96. Fossil ostrich femur (right), compared with that of modern *Struthio* (left)

97. Fossil ostrich tibio-tarsus and tarso-metatarsus, compared with those of modern *Struthio*. The fossil is on the right in each case

INDEX

INDEX

INDEX

INDEX

INDEX

For EU product safety concerns, contact us at Calle de José Abascal, 56–1°,
28003 Madrid, Spain or eugpsr@cambridge.org.

www.ingramcontent.com/pod-product-compliance
Ingram Content Group UK Ltd.
Pitfield, Milton Keynes, MK11 3LW, UK
UKHW030904150625
459647UK00025B/2885